Principles and Practice
of Soil Science

The Soil as a Natural Resource

R.E. WHITE

Professor of Soil Science
Department of Agriculture and Resource Management
The University of Melbourne, Parkville
Victoria 3052, Australia

THIRD EDITION

b

**Blackwell
Science**

© 1979, 1987, 1997 by
Blackwell Science Ltd
Editorial Offices:
Osney Mead, Oxford OX2 0EL
25 John Street, London WC1N 2BL
23 Ainslie Place, Edinburgh EH3 6AJ
350 Main Street, Malden
  MA 02148 5018, USA
54 University Street, Carlton
  Victoria 3053, Australia

Other Editorial Offices:
Blackwell Wissenschafts-Verlag GmbH
Kurfürstendamm 57
10707 Berlin, Germany

Blackwell Science KK
MG Kodenmacho Building
7–10 Kodenmacho Nihombashi
Chuo-ku, Tokyo 104, Japan

First published 1979
Reprinted 1981, 1983
Second edition 1987
Reprinted 1995
Third edition 1997

Set by Semantic Graphics, Singapore
Printed and bound in Great Britain at
the University Press, Cambridge

The Blackwell Science logo is a
trade mark of Blackwell Science Ltd,
registered at the United Kingdom
Trade Marks Registry

DISTRIBUTORS

Marston Book Services Ltd
PO Box 269
Abingdon
Oxon OX14 4YN
(*Orders*: Tel: 01235 465500
           Fax: 01235 465555)

USA
Blackwell Science, Inc.
Commerce Place
350 Main Street
Malden, MA 02148 5018
(*Orders*: Tel: 800 759 6102
           617 388 8250
           Fax: 617 388 8255)

Canada
Copp Clark Professional
200 Adelaide St West, 3rd Floor
Toronto, Ontario M5H 1W7
(*Orders*: Tel: 416 597-1616
           800 815-9417
           Fax: 416 597-1617)

Australia
Blackwell Science Pty Ltd
54 University Street
Carlton, Victoria 3053
(*Orders*: Tel: 3 9347 0300
           Fax: 3 9349 5001)

A catalogue record for this title
is available from the British Library

ISBN 0-86542-960-X

Library of Congress
Cataloging-in-publication Data

White, R. E. (Robert Edwin), 1937–.
    Principles and practice of soil science :
the soil as a natural resource / R.E. White. — 3rd ed.
        p.   cm.
    Rev. ed. of: Introduction to the principles and practice of soil
science. 2nd ed. 1987.
    Includes bibliographical references and index.
    ISBN 0-86542-960-X.
    1. Soil science.   I. White, R. E. (Robert Edwin), 1937–.
Introduction to the principles and practice of soil science.
II. Title.
S591.W49   1997
631.4—dc21
                                                    97-7175
                                                    CIP

Contents

Colour plate falls between pp. 182 and 183.

Preface to the Third Edition

At the time of preparing the second edition of this book, some 10 years ago, I commented on the 'fundamental contribution that soil science has made to many of the applied sciences associated with management of the environment – either natural or man-made'. In the intervening years, this scientific contribution has continued to be important. I also noted the advent of a new buzzword, *biotechnology*, which was being used in a generic sense to cover applications of molecular biology and genetic engineering techniques to the development of modified organisms, new biologically active chemicals, and improvement in the efficiency of various biological processes. Biotechnology continues to flourish, although expectations of its capacity to deliver new products and processes at commercially realistic costs have become more realistic than they were 10 years ago. In the meantime, a new generic concept has attracted major community and political attention – that of *ecologically sustainable development* (ESD).

There are several ways of attempting to define sustainability (see Chapter 15) and many governments and international agencies have developed strategies to achieve ESD. The conditions which have prompted these initiatives are no different in kind to those pertaining 10 years ago when I wrote in the Preface to the Second Edition: 'There is a growing need for information about soils, their behaviour and their influence on land use. The demand for food and other natural products is rising as the world's population continues to grow; and an increasing proportion of that population has an awakening expectation of an improved quality of life – better food, shelter and recreational areas in which to enjoy their leisure.' The change is one of degree; the previous steady rise in world food production per head has plateaued in the last 10 years, while the area of land per head from which food is produced has continued to decline. So the problem of meeting the legitimate needs of a growing world population, while limiting the impact of development on the quality of the environment, has become more acute.

In this edition of *Principles and Practice of Soil Science*, not only have I updated the scientific content of the text but I have strengthened the treatment of applications of soil science to the solution of practical problems in soil and land management. Thus, in Part 3: Soil Management, new material has been introduced on soil productivity, fertilizers and agro-chemicals, problem soils, soil information systems, and a completely new chapter, 'Soil Quality and Sustainable Land Management'. As before I am indebted to many colleagues who have contributed their knowledge, experience and in some cases visual material, and to several in particular who read parts of this revised edition, namely Tom Addiscott, John Catt, Lindsay Falvey, David Jenkinson, Nick Uren and Dick Webster. My sincere thanks to them all. I am also grateful to Professor Ben Miflin, Director of the Institute of Arable Crops Research at Rothamsted, UK, and Professor David Powlson, Head of the Department of Soil Science at Rothamsted, whose kind provision of facilities during my study leave enabled me to complete this revision. I am also grateful to Ian Sherman and Alice Emmott at Blackwell Science for their enthusiastic support over the past eight months.

Robert E. White

Units of Measurement and Abbreviations used in this Book

SI units

Basic unit	Symbol
metre	m
minute	min
kilogram	kg
second	s
ampere	A
kelvin	K
mole	mol

Derived and SI related units

Unit	Symbol	Value
newton	N	$kg\ m\ s^{-2}$
joule	J	$N\ m$
pascal	Pa	$N\ m^{-2}$
volt	V	$J\ A^{-1}\ s^{-1}$
siemen	S	$A\ V^{-1}$
coulomb	C	$A\ s$
hectare	ha	$10^4\ m^2$
litre	L or dm^3	$10^{-3}\ m^3$
tonne	t	$10^3\ kg$
bar	bar	$10^5\ Pa$
Faraday's constant	F	$96\ 500\ J\ mol^{-1}\ V^{-1}$
Gas constant	R	$8.3143\ J\ K^{-1}\ mol^{-1}$

Non-SI units used in soil science

Physical quantity	Unit	Symbol	Value
length	Ångström	Å	$10^{-10}\ m$
temperature	degree Celsius	°C	$K-273$
concentration	moles per litre	M	$mol\ L^{-1}$
cation exchange capacity	milli-equivalent per 100 gram	meq per 100 g	$cmol$ charge $(+)\ kg^{-1}$

Physical quantity	Unit	Symbol	Value
radioactivity	Curie	Ci	3.7×10^{10} disintegrations s^{-1}
electrical conductivity	millimho per cm	mmho cm^{-1}	$dS\ m^{-1}$

Prefixes to SI and derived SI units

giga	G	10^9
mega-	M	10^6
kilo-	k	10^3
deca-	da	10^1
deci-	d	10^{-1}
centi-	c	10^{-2}
milli-	m	10^{-3}
micro-	μ	10^{-6}
nano-	n	10^{-9}
pico-	p	10^{-12}

Abbreviations used in this book

A	Australian Soil Classification
ACLEP	Australian Collaborative Land Evaluation Program
AEC	Anion exchange capacity
API	Air photo interpretation
AR	Activity ratio
AWC	Available water capacity
BD	Bulk density
BET	Branauer, Emmet and Teller
BIO	Microbial biomass
BP	Before present
c.	about, approximately
CEC	cation exchange capacity

concn.	concentration	MW	Molecular (formula) weight
CPMAS	Cross-polarization, magic angle spinning	NCSS	National Cooperative Soil Survey
CSIRO	Commonwealth Scientific and Industrial Research Organization	NMR	Nuclear magnetic resonance
		NV	Neutralizing value
d	day	o.d.	oven-dry
DDL	Diffuse double layer	*OHP*	Outer Helmholtz plane
DEM	Digital elevation model	OP	Osmotic pressure
DM	Dry matter	PAPR	Partially acidulated phosphate rock
DOC	Dissolved organic carbon	*PBC*	Phosphate buffering capacity
DPM	Decomposable plant material	ppm	parts per million
EC	Electrical conductivity	PR	Phosphate rock
ECEC	Effective cation exchange capacity	*PSR*	Pore space ratio
EI	Soil erodibility index	*PWD*	Potential water deficit
EMR	Electromagnetic radiation	*PWP*	Permanent wilting point
ESD	Ecologically sustainable development	*PZC*	Point of zero charge
ESP	Exchangeable sodium percentage	*Q/I*	Quantity/intensity
ET	Evapotranspiration	*RH*	Relative humidity
FA	Fulvic acid	RHS	Right-hand side
FAO	Food and Agriculture Organization	RPM	Resistant plant material
FC	Field capacity	RPR	Reactive phosphate rock
FESLM	Framework for sustainable land management	*RQ*	Respiratory quotient
		SAR	Sodium adsorption ratio
FTIR	Fourier transform infrared	*SG*	Specific gravity
FYM	Farmyard manure	*SI*	Saturation Index
GIS	Geographical information system	SIR	Substrate-induced respiration
GLC	Gas–liquid chromatography	SLM	Sustainable land management
GRP	Ground rock phosphate	*SMD*	Soil moisture deficit
h	hour	soln	solution
HA	Humic acid	SOM	Soil organic matter
HUM	Humified organic matter	SPAC	Soil–plant–atmosphere continuum
HYV	High-yielding variety	sp., spp.	species (singular; plural)
IAEA	International Atomic Energy Agency	SSEW	Soil Survey of England and Wales
IEP	Isoelectric point	ST	Soil Taxonomy
IHP	Inner Helmholtz plane	TDR	Time domain reflectometer
IOM	Inert organic matter	*TDS*	Total dissolved solids
IR	Infiltration rate	*TEC*	Threshold electrolyte concentration
KE	Kinetic energy	UK	United Kingdom
LF	Leaching fraction	UNEP	United Nations Environment Programme
LHS	Left-hand side	UNESCO	United Nations Education, Scientific and Cultural Organization
LR	Leaching requirement		
LRA	Land resource assessment	USA	United States of America
MAFF	Ministry of Agriculture, Fisheries and Food	USDA	United States Department of Agriculture
MD	Moisture deficit	USLE	Universal soil loss equation

VFA	Volatile fatty acid
VAM	Vesicular-arbuscular mycorrhiza
VD	Vapour density
WCED	World Commission on Environment and Development
XRD	X-ray diffraction
y	year

Some chemical and miscellaneous symbols

2,4-D	2,4-dichlorophenoxyacetic acid
2,4,5-T	2,4,5-trichlorophenoxyacetic acid
ATC	4-amino-1, 2, 4-triazole
ATP	Adenosine triphosphate
DAP	Diammonium phosphate
DCD	Dicyandiamide
DCP	Dicalcium phosphate
DCPD	Dicalcium phosphate dihydrate
DDT	Dichlorodiphenyltrichloroethane
DTPA	Diethylene triamine penta-acetic acid
EDDHA	Ethylene diamine di(*o*-hydroxyphenylacetic acid)
EDTA	Ethylenediamine tetra-acetic acid
GRP	Ground rock phosphate
HA	Hydroxyapatite
IAA	Indoleacetic acid

IBDU	Isobutylidene diurea
MAH	Monocyclic aromatic hydrocarbons
MAP	Monoammonium phosphate
MCP	Monocalcium phosphate
MCPA	(4-chloro-*o*-tolyl) phenoxyacetic acid
OCP	Octacalcium phosphate
PAH	Polycyclic aromatic hydrocarbons
PAM	Polyacrylamide
PCB	Polychlorobiphenyls
PEG	Polyethylene glycol
PVA	Polyvinyl alcohol
PVC	Polyvinyl chloride
PVAc	Polyvinylacetate
SCU	Sulphur-coated urea
SSP	Single superphosphate
TCP	Tricalcium phosphate
TSP	Triple superphosphate
( )	denotes activity
[ ]	denotes concentration
$\simeq$	approximately equal to
$\sim$	of the order of
$<$	less than
$>$	greater than
$\leqslant$	less than or equal to
ln	$\log_e$

Part 1
The Soil Habitat

'Soils are the surface mineral and organic formations, always more or less coloured by humus, which constantly manifest themselves as a result of the combined activity of the following agencies; living and dead organisms (plants and animals) parent material, climate and relief.'

V.V. Dokuchaev (1879), quoted by J.S. Joffe in *Pedology*

'The soil is teeming with life. It is a world of darkness, of caverns, tunnels and crevices, inhabited by a bizarre assortment of living creatures. . .'

J.A. Wallwork (1975) in *The Distribution and Diversity of Soil Fauna*

Redrawn from Reganold J.P., Papendick, R.I. & Parr J.F. (1990) Sustainable Agriculture. *Scientific American* **262(6)**, 112–20.

Chapter 1
Introduction to
the Soil

1.1 Soil in the making

With the exposure of rock to a new environment – following the extrusion and solidification of lava, the uplift of sediments, the recession of water or the retreat of a glacier – a soil begins to form. Decomposition proceeds inexorably towards decreased free energy and increased entropy. The free energy of a closed system, such as a rock fragment, is that portion of its total energy that is available for work, other than work done in expanding its volume. Part of the energy released in a spontaneous reaction, such as rock weathering, appears as entropic energy, and entropy is a measure of the degree of disorder of the system. For example, as the rock weathers, crystalline minerals are broken down into simple molecules and ions, some of which are leached out by water or escape as gases.

Weathering is hastened by the appearance of primitive plants on the rock surfaces. These plants – lichens, mosses and liverworts – can store radiant energy from the sun as chemical energy in the products of photosynthesis. Lichens, which are a symbiotic association of an alga and fungus, are able to 'fix' atmospheric nitrogen and incorporate it into plant protein, and to extract elements from the weathering rock surface. On the death of each generation of these primitive plants, some of the rock elements and a variety of complex organic molecules are returned to the weathering surface where they nourish the succession of organisms gradually colonizing the embryonic soil.

A simple example is that of soil formation under the extensive deciduous forests of the cool humid areas of Europe, Asia and North America, on calcareous deposits exposed by the retreat of the Pleistocene ice cap. The profile development is summarized in Fig. 1.1. The initial state is little more than a thin layer of weathered material stabilized by primitive plants. Within a century or so, as the organo-mineral material accumulates, more advanced species of sedge and grass appear, which are adapted to the harsh habitat. The developing soil is described as a *Lithosol* (Entisol (ST) or Rudosol* (A)). Pioneering micro-organisms and animals feed on the dead plant remains and gradually increase in abundance and variety. The litter deposited on the surface is mixed into the soil by burrowing animals and insects, where its decomposition is hastened. The eventual appearance of larger plants – shrubs and trees – with their deeper rooting habit, pushes the zone of rock weathering farther below the soil surface. After a few hundred more years, a *brown forest soil* (Inceptisol (ST) or Tenosol (A)) emerges. We return to the topic of soil formation, and the wide range of soils that occur in the landscape, in Chapters 5 and 9.

1.2 Concepts of soil

The soil is at the interface between the *atmosphere* and *lithosphere* (the mantle of rocks making up the Earth's crust). It also has an interface with bodies of fresh and salt water (collectively called the *hydrosphere*). The soil sustains the growth of many plants and animals, and so forms part of the *biosphere*.

There is little merit in attempting to give a rigorous definition of soil because of the complexity of its make-up and of the physical, chemical and biological forces to which it is exposed. Nor is it necessary to do

* See Box 1.1 for a discussion of soil names.

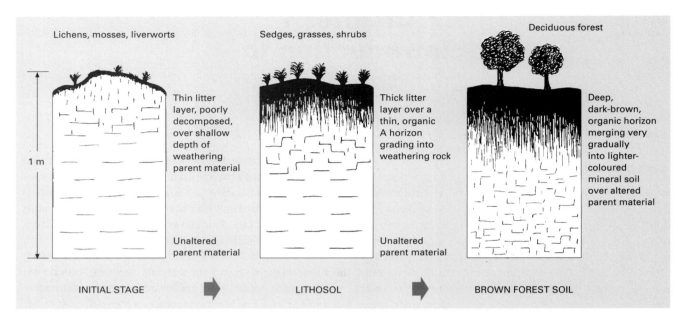

Fig. 1.1 Stages in soil formation on a calcareous parent material in a humid temperature climate.

so, for soil means different things to different users. For example, to the geologist and engineer, the soil is no more than finely divided rock material. The hydrologist may see the soil as a storage reservoir affecting the water balance of a catchment, while the ecologist may be interested in only those soil properties that influence the growth and distribution of different plant species. The farmer is naturally concerned about the many ways in which the soil can influence the growth of crops and the health of livestock, although frequently their interest does not extend below the depth of soil disturbed by a plough (20–25 cm).

In view of this wide spectrum of potential user need, it is appropriate when introducing the topic of soil to readers, perhaps for the first time, to review briefly the evolution of man's relationship with the soil and identify some of the past and present *concepts* of soil.

Soil as a medium for plant growth

Man's exploitation of the soil for food production be-

gan two or three thousand years after the close of the last Pleistocene ice age, which occurred about 10 000 years BP (before present). Neolithic man and primitive agriculture spread outwards from settlements in the fertile crescent embracing the ancient lands of Mesopotamia and Canaan (Fig. 1.2) and reached as far as China and the Americas within a few thousand years. In China, for example, the earliest records of soil survey (4000 BP) show how the fertility of the soil was used as a basis for levying taxes on the landholders. The study of the soil was a practical exercise of everyday life, and the knowledge of soil husbandry that had been acquired by Roman times was faithfully passed on by peasants and landlords, with little innovation, until the early 18th century.

From that time onwards, however, the rise in demand for agricultural products in Europe was dramatic. Conditions of comparative peace, and rising living standards as a result of the Industrial Revolution, further stimulated this demand throughout the 19th century. The period was also one of great discoveries in physics and chemistry, the implications of which sometimes burst with shattering effect on the conservative world of agriculture. In 1840, von Liebig established that plants

Box 1.1 Soil variability, description and classification.

There is a remarkable range of soil types in the landscape. This results from the almost infinite variation in parent material, climate, vegetation, other organisms and topography, and the time for which these factors have combined to influence soil formation (Chapter 5). To bring order to such disorder, and to facilitate the acquisition and dissemination of knowledge about soils, soil scientists have developed ways of *classifying* soils. Individual soils are described in terms of their properties, and possibly their mode of formation, and similar soils are grouped into classes that are given distinctive names. However, unlike the plant and animal kingdoms, there are no soil 'individuals' – the boundaries between soils in the landscape are not sharp. Partly because of the difficulty in setting class limits, and because of the evolving nature of this young science, no universally accepted system of classifying (and naming) soils exists. For many years, Great Soil Group names, based on the USDA Classification of Baldwin *et al.* (1938) (Section 5.3), held sway. But in the last 25 years, new classifications and a galaxy of new soil names have evolved (Chapter 14). Some of these classifications (e.g. Soil Taxonomy (Soil Survey Staff, 1975) and the World Soil Classification (FAO–Unesco, 1974, 1988)) purport to be international. Others such as the Australian Soil Classification (Isbell, 1996) and the Soil Classification for England and Wales (Avery, 1980) are national in focus. This diversity of classifications creates problems for non-specialists in naming soils and understanding what a particular soil name connotes. In this book, the more descriptive and (to most) more familiar Great Soil Group names will be used. Where possible, the approximate equivalent at the Order or Suborder level in Soil Taxonomy (ST) and the Australian Soil Classification (A) will be given, in that order, in parentheses.

absorbed nutrients as inorganic compounds from the soil, although he insisted that plants obtained their nitrogen (N) from the atmosphere: Laws and Gilbert at Rothamsted subsequently demonstrated that plants (except legumes) absorbed inorganic N from the soil. In the 1850s Way discovered the process of cation exchange in soils. During the years from 1860 to 1890 eminent bacteriologists including Pasteur, Warington and Winogradsky elucidated the role of microorganisms in the decomposition of plant residues and the conversion of ammonia to nitrate.

Over the same period, botanists such as von Sachs and Knop, by careful experiments in water culture and analysis of plant ashes, identified the major elements which were essential for healthy plant growth. Agricultural chemists drew up balance sheets of the quantities of these elements taken up by crops and, by inference, the quantities which should be returned to the soil in fertilizers or animal manure to sustain growth. This approach, whereby the soil is regarded as a relatively inert medium providing water, mineral* ions and phys-

* The term 'mineral' is used in two contexts: first, as an adjective referring to the inorganic constituents of the soil (ions, salts and particulate matter); second, as a noun referring to specific inorganic compounds found in rocks and soil, such as quartz and feldspars (Chapter 2).

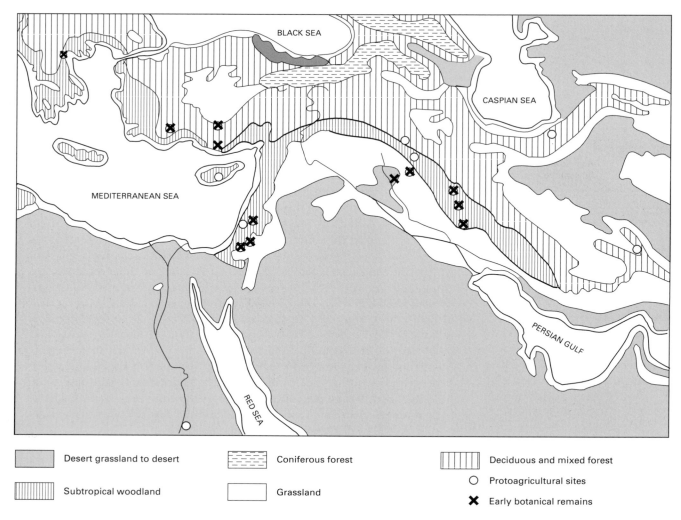

BLACK SEA

CASPIAN SEA

MEDITERRANEAN SEA

PERSIAN GULF

RED SEA

	Desert grassland to desert		Coniferous forest		Deciduous and mixed forest
	Subtropical woodland		Grassland	○	Protoagricultural sites
				✗	Early botanical remains

Fig. 1.2 Sites of primitive settlements in the Middle East (after Gates, 1976).

ical support for plants, has been called the '*nutrient bin*' concept.

Soil and the influence of geology

The pioneering chemists who investigated a soil's ability to supply nutrients to plants tended to see the soil as a medium in which chemical and biochemical reactions occurred. They little appreciated soil as part of the landscape, moulded by natural forces acting on the land surface. Great contributions were made to our knowledge of soil by geologists in the late 19th century. Geologists defined the mantle of loose, weathered material on the Earth's surface as the *regolith*, of which only the upper few decimetres could be called soil, since this was the maximum depth to which organic material was incorporated (Fig. 1.3). Below the soil was the *subsoil* which was largely devoid of organic matter. However, the mineral matter of both soil and subsoil was recog-

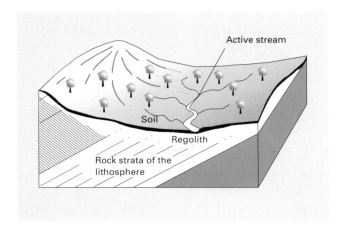

Fig. 1.3 Soil development in relation to the landscape and underlying regolith.

nized as being derived from the weathering of underlying rock. This led to intensive investigations of the relationship between soils and the geology of the underlying rocks, with the result that soils were loosely classified in geological terms, such as loessial, glacial, alluvial, granitic or marly (derived from limestone).

The influence of Russian soil science

Russia at the end of the 19th century was a vast country covering many climatic zones in which crop production was limited not so much by the fertility of the soils, but by the primitive methods of agriculture. Early Russian soil science was therefore concerned not with soil fertility, but with observing soils in the field and studying relationships between the properties of a soil and the environment in which it had formed. From 1870 onwards, Dokuchaev and his school emphasized the distinctive features of a soil that developed gradually and distinguished it from the undifferentiated weathering rock or *parent material* below. This was the beginning of the science of *pedology*.*

Following the Russian lead, scientists in other countries began to appreciate that factors such as climate,

* From the Greek word for ground or earth.

parent material, vegetation, relief and time, interacted in many ways to produce an almost infinite variety of soil types. For any particular combination of these *soil-forming factors* (Chapter 5), a unique physicochemical and biological environment was established that led to the development of a distinctive soil body – the process of *pedogenesis*. A set of new terms was developed to describe fundamental soil features, such as:

• the *soil profile*, constituting the vertical dimension exposed by excavation from the surface to the parent material, and

• *soil horizons* – layers in the soil distinguished on the basis of colour, hardness, texture, the occurrence of included structures and other visible or tangible properties. The upper layer, from which materials are generally washed downwards, is described as *eluvial*; lower layers in which these materials accumulate are called *illuvial*.

In 1932, an international gathering of soil scientists adopted the notation of A and B for the eluvial and illuvial horizons, and C horizon for the parent material (Fig. 1.4). The A and B horizons comprise the *solum*. If unweathered rock exists below the parent material it is called *bedrock* R. Organic litter on the surface, not incorporated in the soil, is designated the L layer.

Soil genesis is now known to be much more complex than this early work suggested. For example, many soils are *polygenetic* in origin; that is, they have undergone successive phases of development due to changes in climate and other environmental factors over time. In other cases, two or more *layers* of different parent material are found in the one soil profile. Nevertheless, the Russian approach was a considerable advance on traditional thinking, and the recognition of the intimate relationship between soil and the environment encouraged soil scientists to survey and map the distribution of soils. The wide range of soil morphology that was revealed in turn stimulated studies of pedogenesis, an understanding of which, it was believed, would enable the copious field data on soils to be collated in a systematic manner. Thus, Russian soil science provided the inspiration for many of the early attempts at *soil classification* (Section 14.4).

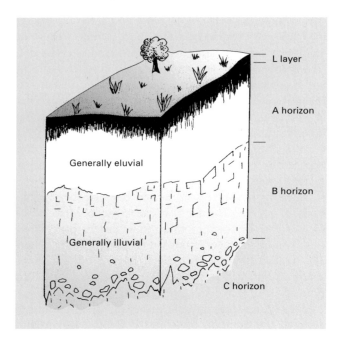

L layer

A horizon

Generally eluvial

B horizon

Generally illuvial

C horizon

Fig. 1.4 Diagram of soil profile.

A contemporary view of soil

Between the two World Wars of the 20th century, the philosophy of the soil as a 'nutrient bin' was much in evidence, particularly in the western world. More and more land was brought into cultivation, much of which was marginal for crop production because of limitations of climate, soil and topography. With the balance between crop success and failure made even more precarious than in favoured areas, the age-old problems of wind and water erosion, encroachment by weeds and the accumulation of salts in irrigated lands became more serious. Since 1945, demand for food, fibre and forest products from an escalating world population (now > 6 billion) has led to increased use of fertilizers to improve crop yields, and pesticides to control pests and diseases (Chapter 12). Such practices have resulted in some accumulation of undesirable fertilizer and pesticide residues in soil, and in increased losses of soluble constituents such as nitrate and phosphates to surface

waters and groundwater. There has also been widespread dispersal of the very stable pesticides throughout the biosphere and their accumulation to concentrations that are potentially toxic to some species of birds and fish.

More recently, however, soil scientists, producers and planners have acknowledged the need to compromise between maximizing crop production and conserving a valuable natural resource. The trend now is towards maintaining the soil's natural state by causing minimum disturbance when crops are grown, matching fertilizer additions more closely to crop demand in order to reduce losses, introducing legumes that can fix nitrogen from the air, and returning plant residues and waste materials to the soil to supply some of a crop's nutrient requirements. In short, more emphasis is being placed on the soil as a natural body (Box 1.2) and on the concept of *sustainable land management* (Chapter 15).

1.3 Components of the soil

As we have seen, a combination of physical, chemical and biotic forces acts on organic materials and weathered rock to produce a soil with a porous fabric that retains water and gases. The water contains dissolved organic and inorganic solutes and is called the *soil solution*. While the soil air consists primarily of nitrogen and oxygen, it usually contains higher concentrations of carbon dioxide than the atmosphere, and traces of other gases that are by-products of microbial metabolism. The relative proportions of the four major components – *mineral matter*, *organic matter*, *water* and *air*–may vary widely, but generally lie within the ranges indicated in Fig. 1.5. These components are discussed in more detail in the subsequent chapters of Part 1.

1.4 Summary

Soil forms at the interface between the atmosphere and the consolidated or loose deposits of the Earth's crust. Physical and chemical weathering, denudation and redeposition combined with the activities of a succession of colonizing plants and animals moulds a distinctive

Box 1.2 Soil as a natural body.

A soil is clearly distinguished from inert rock material by:
- the presence of plant and animal life;
- a structural organization that reflects the action of pedogenic processes, and;
- a capacity to respond to environmental change which can alter the balance between gains and losses within the profile, and predetermine the formation of a different soil in equilibrium with the new set of environmental conditions.

The last point indicates that soil has no fixed inheritance, because it depends on the conditions prevailing during its formation. Nor is it possible to define unambiguously the boundaries of the soil body. The soil atmosphere is continuous with the air above ground, many soil organisms can live as well on the surface as within the soil, the litter layer usually merges gradually with decomposed organic matter in the soil, and likewise the boundary between soil and parent material is difficult to demarcate. It is common, therefore, to speak of the soil as a *three-dimensional body that is continuously variable in time and space.*

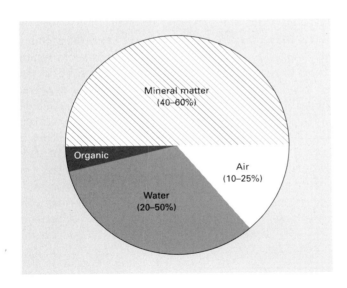

Fig. 1.5 Proportions of the main soil components by volume.

of soil material into a series of *horizons* that constitute the *soil profile*. The horizons are distinguished in the field by their visible and tangible properties such as colour, hardness, texture and structural organization. The intimate mixing of mineral and organic materials to form a porous fabric, permeated by water and air, creates a favourable habitat for a variety of plant and animal life. Soil is a fragile component of the environment. Its use for food and fibre production, and waste disposal, must be managed in a way that minimizes the off-site effects of these activities and preserves the soil for future generations.

References

Avery B. W. (1980) *Soil Classification for England and Wales*. Soil Survey Technical Monograph No. 14. Rothamsted Experimental Station, Harpenden.

Baldwin H., Kellogg C. W. & Thorpe J. (1938) Soil classification, in *Soils and Man*. US Government Printing Office, Washington DC.

FAO–Unesco (1974–78) *Soil Map of the World 1–5,000,000* Volume I–X. Unesco, Paris.

FAO–Unesco (1988) *Soil Map of the World. Revised Legend*. World Resources Report 60. FAO, Rome.

soil body from the milieu of rock minerals in the parent material. The process of soil formation, called *pedogenesis*, culminates in a remarkably variable differentiation

Gates C. T. (1976) China in a world setting: agricultural response to climatic change. *Journal of the Australian Institute of Agricultural Science* **42**, 75–93.

Isbell R. F. (1996) *The Australian Soil Classification*. Australian Soil and Land Survey Handbook. CSIRO Publishing, Melbourne.

Soil Survey Staff (1975) *Soil Taxonomy. A basic system of soil classification for making and interpreting soil surveys*. Agriculture Handbook No. 436. United States Department of Agriculture, Washington DC.

Further reading

Jenny H. (1980) *The Soil Resource – Origin and Behaviour*. Springer, New York.

Marschner H. (1995) *Mineral Nutrition of Higher Plants*, 2nd edn. Academic Press, London.

Simonson R. W. (1968) Concepts of Soil. *Advances in Agronomy* **20**, 1–47.

Wild A. (1988) (Ed.) *Russell's Soil Conditions and Plant Growth*, 11th edn. Longman Scientific & Technical, Harlow.

Chapter 2
The Mineral Component of the Soil

2.1 The size range

Mineral fragments in soil vary enormously in size from boulders and stones down to sand grains and minute particles that are beyond the resolving power of an optical microscope ($< 0.2\,\mu$m in length or diameter). Particles smaller than *c.* 1 μm are classed as *colloidal*. If they do not settle quickly when mixed with water, they are said to form a colloidal solution or *sol*; if they settle within a few hours they form a suspension. Colloidal solutions are distinguished from true solutions (dispersions of ions and molecules) by the *Tyndall effect*. This occurs when the path of a beam of light passing through the solution can be seen from either side at right angles to the beam, indicating a scattering of the light rays.

An arbitrary division is made by size-grading soil into material:

- that passes through a sieve with 2-mm diameter holes – the *fine earth*;
- that is retained on the sieve (> 2 mm) – *the stones or gravel*, and;
- fragments > 600 mm which are called *boulders*.

The separation by sieving is carried out on air-dry soil which has been gently ground by mortar and pestle, or crushed between wooden rollers, to break up the aggregates of smaller particles. Air-dry soil is soil allowed to dry in air at ambient temperatures.

Particle-size distribution of the fine earth

The distribution of particle sizes determines the soil texture, which may be assessed subjectively in the field or more rigorously by particle-size analysis in the laboratory.

Size classes

All soils show a continuous spectrum of particle sizes, called a *frequency distribution*, which is obtained by plotting the number (or mass) of particles of a given size against the actual size. When the number or mass of particles in each size class is summed sequentially we obtain a cumulative frequency distribution of soil particle sizes, some examples of which are given in Fig. 2.1. In practice, it is convenient to subdivide the continuous distribution into several class intervals that define the size limits of the *sand*, *silt* and *clay* fractions. The extent of this subdivision, and the class limits chosen, vary

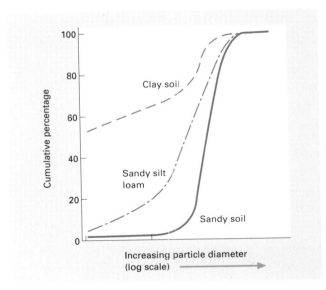

Fig. 2.1 Cumulative frequency distributions of soil particle sizes in a typical clay, sandy silt loam and sandy soil.

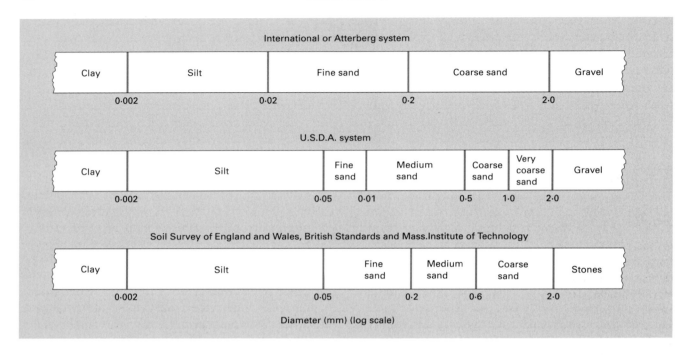

Fig. 2.2 Particle-size classes most widely adopted internationally.

from country to country and even between institutions within countries. The major systems in use are those adopted by the Soil Survey Staff of the USDA, the British Standards Institution and the International Society of Soil Science. These are illustrated in Fig. 2.2. All three systems set the upper limit for clay at 2 μm, but differ in the upper limit chosen for silt and the way in which the sand fraction is subdivided.

Field texture

A soil surveyor assesses the texture of a soil by moistening a sample until it glistens. It is then kneaded between fingers and thumb until the aggregates are broken down and the soil grains thoroughly wetted. The proportions of sand, silt and clay are estimated according to the following qualitative criteria.

• *Coarse sand* grains are large enough to grate against each other and can be detected individually by sight and feel.

• *Fine sand* grains are much less obvious, but when they comprise more than about 10% of the sample they can be detected by biting the sample between the teeth.

• *Silt* grains cannot be detected by feel, but their presence makes the soil feel smooth and silky and only slightly sticky.

• *Clay* is characteristically sticky, although some dry clays, especially of the expanding lattice type (Section 2.4), require much moistening and kneading before they develop their maximum stickiness.

High organic matter contents tend to reduce the stickiness of clay soils and to make sandy soils feel more silty. Finely divided calcium carbonate also gives a silt-like feeling to the soil.

Depending on the estimated proportions of sands, silt and clay, the soil can be assigned to a *textural class* according to a triangular diagram as in Fig. 2.3. The triangle in Fig. 2.3a is used by the Soil Survey of England and Wales and is based on the British Standards system of particle-size grading (Fig. 2.2); the one in Fig. 2.3b is used in Australia and is based on the International system (Fig. 2.2). The American system is very similar to the British Standards system. Note that in these systems there are 11 textural classes, but the Australian system has broader classes for the silty clays, silty clay loams and silty loams than the British system.

Soil surveyors become expert at field texturing after years of experience, which includes checking their field assessments of texture against a laboratory analysis of a soil's particle-size distribution.

Particle-size analysis in the laboratory

The success of the method relies upon the complete disruption of soil aggregates and the addition of chemicals that ensure the dispersion of the soil colloids in water. Full details of the methods employed are given in standard texts, for example Klute (1986) and Rayment and Higginson (1992). The sand particles are separated by sieving; silt and clay are separated using the differences in their settling velocities in the suspension. The principle of the latter technique is outlined in Box 2.1 (p. 16).

The result of the particle-size analysis is expressed as the mass of the individual fractions per 100 g of oven-dry soil (fine earth only). Oven-dry soil is soil dried to a constant weight at 105°C. By combining the coarse and fine sand fractions, the soil may be represented by one point on the triangular diagrams of Fig. 2.3(a,b). Alternatively, by stepwise addition of particle-size percentages, graphs of cumulative percentage against particle diameter of the kind shown in Fig. 2.1 are obtained.

2.2 The importance of soil texture

Soil scientists are primarily interested in the texture of the fine earth of the soil. Nevertheless, in some soils the size and abundance of the stones cannot be ignored because they can have a marked influence on the soil's suitability for agriculture. As the stone content increases, the soil will hold less water than a stoneless soil of the same fine-earth texture, so that crops become more susceptible to drought. Conversely, the drainage of such soils may be better under wet conditions. Stoniness also determines the ease, and to some extent the cost, of cultivation as well as the abrasive effect of the soil on tillage implements.

Texture is one of the most stable of soil properties and is a useful index of several other properties that determine a soil's agricultural potential. Fine- and medium-textured soils, such as the clays, clay loams, silty clays and silty clay loams, are generally more desirable than coarse textured soils because of their superior retention of nutrients and water. Conversely, where high rates of infiltration and good drainage are required, as for irrigation or liquid waste disposal, sandy or coarse-textured soils are preferred. In farming terms, clay soils are described as 'heavy' and sandy soils as 'light'. This does not refer to their mass per unit volume, which may be greater for a sandy soil than a clay, but to the power required to draw a plough or other implements through the soil. Because it is easy to estimate, and is routinely measured for soil surveys, texture (and more specifically clay content) has been used as a 'surrogate' variable for other soil properties that are less easily measured; for example, the cation exchange capacity (Section 2.5).

Texture has a pronounced effect on *soil temperature*. Clays hold more water than sandy soils, and the presence of water considerably modifies the heat required to change a soil's temperature because:
• its specific heat capacity is three to four times that of the soil solids, and
• considerable latent heat is either absorbed or evolved during a change in the physical state of water, for example, from ice to liquid or vice versa. Thus, the temperature of wet clay soils responds more slowly than that of sandy soils to changes in air temperature in spring and autumn (Section 6.6).

Texture should not be confused with *tilth*, of which it has been said that a good farmer can recognize it with his boot, but no soil scientist can describe it (Russell, 1971). Tilth refers to the condition of the surface of ploughed soil prepared for seed sowing: how sticky it is when wet and how hard it sets when dry. Tilth therefore

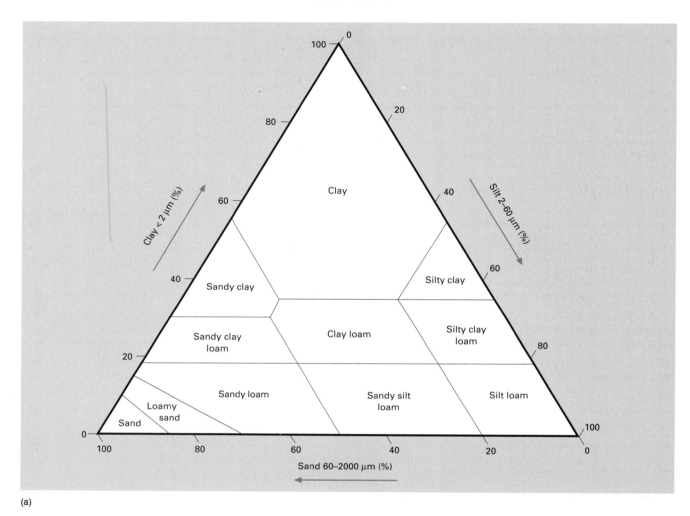

(a)

Fig. 2.3 (a) Triangular diagram of soil textural classes adopted in England and Wales (after Hodgson, 1974). *Continued on p. 15.*

depends on the soil's consistence (Box 4.2) which reflects both texture and structure. The action of frost in cold climates can break down the massive clods left on the surface of a heavy clay soil after autumn ploughing, producing a mellow 'frost tilth' of numerous small granules (Section 4.2).

2.3 Minerology of the sand and silt fractions

Simple crystalline structures

Sand and silt consist almost entirely of the resistant residues of *primary* rock minerals, although small amounts of *secondary* minerals (salts, oxides and hydroxides) formed by weathering also occur. The primary rock minerals are predominantly *silicates*, which have a crystalline structure based upon a simple unit – the *silicon tetrahedron*, SiO_4^{4-} (Fig. 2.5). An electrically

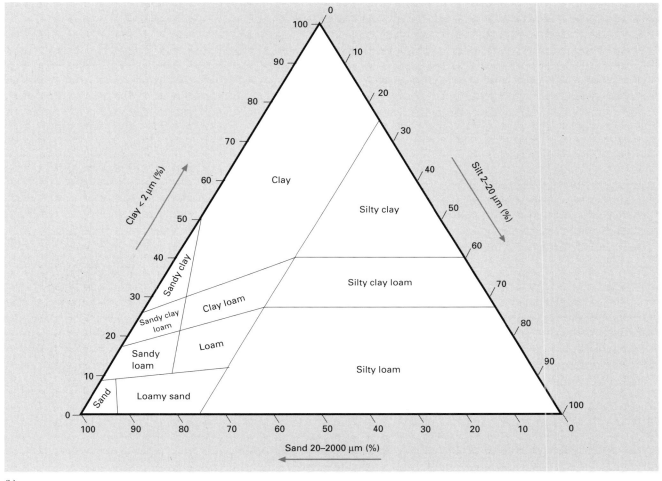

Fig.2.3 (b) Triangular diagram of soil textural classes adopted in Australia (after McDonald *et al.* 1990).

neutral crystal is formed when cations, such as Al^{3+}, Fe^{3+}, Fe^{2+}, Ca^{2+}, Mg^{2+}, K^+ and Na^+, become co-valently bonded to the oxygen atoms in the tetrahedron and the surplus valencies of the oxygen anions in the SiO_4^{4-} group are satisfied.

Coordination number

The packing of the oxygen atoms, which are by far the largest of the more abundant elements in the silicates, determines the crystalline dimensions. In quartz, for example, oxygen occupies 98.7% and silicon only 1.3% of the mineral volume. The size ratio of Si to O is such that four oxygens can be packed around one silicon. Larger cations can accommodate more oxygen atoms, the ratio

cation radius/oxygen radius

determining the *coordination number* of the cation (Table 2.1). Many of the common metal cations have radius ratios between 0.4 and 0.7 which means that an

A rigid particle falling freely through a liquid of a lower density will attain a constant velocity when the force opposing movement is equal and opposite to the force of gravity acting on the particle. The frictional force acting vertically upward on a spherical particle is calculated from Stokes' law. The net gravitational force acting downwards is equal to the weight of the submerged particle. At equilibrium, these expressions can be combined to give an equation for the terminal settling velocity v, as

$$v = \frac{2}{9}\frac{g}{\eta}(\rho_p - \rho)\, r^2 \qquad\qquad\qquad \text{(B2.1.1)}$$

where g is acceleration due to gravity, η is the coefficient of viscocity of the liquid (water) which varies with temperature, ρ_p is the density of the particle, ρ is density of water, and r is the radius of the particle.

When all the constants in this equation are collected into one term A, we derive the simple relationship

$$v = Ar^2 = \frac{h}{t} \qquad\qquad\qquad \text{(B2.1.2)}$$

where h is the depth, measured from the liquid surface, below which all particles of radius r will have fallen in time t. To illustrate the use of Equation B2.1.2, we can calculate that all particles $> 2\,\mu m$ in diameter settling in a suspension at 20°C will fall below a depth of 10 cm in 8 h. Thus, by sampling the mass per unit volume of the suspension at this depth after 8 h, the amount of clay present can be calculated. The suspension density (mass per unit volume) at a particular depth can be measured in one of several ways:
• by withdrawing a sample volume of the suspension, evaporating to dryness, and weighing the mass of sediment – the *pipette method*,
• by using a Bouyoucos hydrometer in the suspension, or
• by calculating the loss in weight of a bulb of known volume when immersed in the suspension – the *plummet balance method*, illustrated in Fig. 2.4.
Note that constant temperature should be maintained (because of the effect of temperature change on the viscosity of water), and also that simplifying assumptions are made in the calculation of settling velocity by Equation B2.1.2. In particular, note that:
• clay and silt particles are not smooth spheres, they have irregular plate-like shapes, and
• the density of the particle varies with its mineralogy (Section 2.3).
In practice, we take an average value for ρ_p of 2.65 g cm^{-3}, and speak of the 'equivalent spherical diameter' of the particle size being measured.

Box 2.1 Measurement of silt and clay by sedimentation.

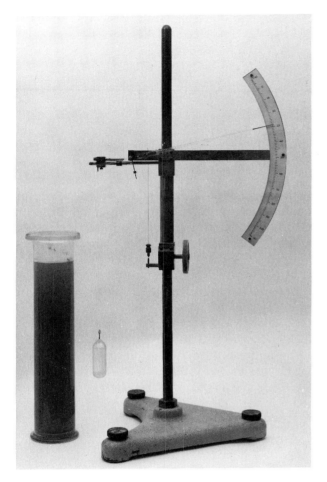

Fig. 2.4 A settling soil suspension and plummet balance (courtesy of J. Loveday).

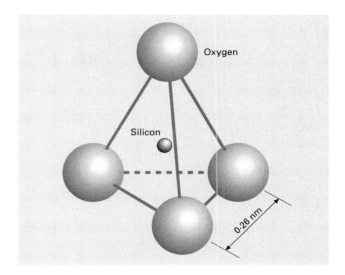

Fig. 2.5 Diagram of a silicon tetrahedron (interatomic distances not to scale).

Table 2.1 Cation/oxygen radius ratios and coordination numbers for common elements in the silicate minerals. After Marshall, 1964.

Radius of cation/radius of oxygen							
Coordination No. 3		Coordination No. 4		Coordination No. 6		Coordination No. 8 and greater	
B	0.15	Be	0.26	Al	0.43	Na	0.74
				Ti	0.48	Ca	0.80
		Si	0.30	Fe^{3+}	0.51		
		Mn^{4+}	0.39	Mg	0.59	Sr	0.96
				Fe^{2+}	0.63		
				Zr^{4+}	0.69	K	1.01
				Mn^{2+}	0.69	Ba	1.08

octahedral arrangement of six oxygen atoms around the cation (coordination number 6) is preferred. The larger alkali and alkaline earth cations have radius ratios > 0.7 and therefore form complexes of coordination number 8 or greater. Aluminium, which has a cation/oxygen radius ratio close to the maximum for coordination number 4 and the minimum for coordination number 6, can exist in either fourfold (IV) or sixfold (VI) coordination.

Isomorphous substitution

Elements of the same valency and coordination number frequently substitute for one another in a silicate structure – a process called *isomorphous substitution*. The structure remains electrically neutral. However, when elements of the same coordination number but different valency are exchanged, there is an imbalance of charge.

The most common substitutions are Mg^{2+}, Fe^{2+} or Fe^{3+} for Al^{3+} in octahedral coordination, and Al^{3+} for Si^{4+} in tetrahedral coordination. The excess negative charge is neutralized by the incorporation of additional cations, such as K^+, Na^+, Mg^{2+} or Ca^{2+} into the crystal lattice, or by structural arrangements that allow an internal compensation of charge (see chlorites and smectites).

More complex crystalline structures

Chain structures

These are represented by the *pyroxene* and *amphibole* groups of minerals, which collectively make-up the ferromagnesian minerals. In the pyroxenes, each silicon tetrahedron is linked to adjacent tetrahedra by the sharing of two out of three basal oxygen atoms to form a single extended chain (Fig. 2.6). In the amphiboles, two parallel pyroxene chains are linked by the sharing of an oxygen in every alternate tetrahedron (Fig. 2.7). Cations such as Mg^{2+}, Ca^{2+}, Al^{3+} and Fe^{2+} are ionic–covalently bonded to the oxygen ions to neutralize the surplus negative charge.

Sheet structures

It is easy to visualize an essentially one-dimensional chain structure being extended in two dimensions to

form a sheet of silicon tetrahedra linked by the sharing of all the basal oxygen atoms. When viewed from above the sheet, the bases of the linked tetrahedra form a network of hexagonal holes (Fig. 2.8). The apical O atoms (superimposed on the Si atoms in Fig. 2.8) form ionic–covalent bonds with other metal cations by, for example, displacing OH groups from their coordination positions around a trivalent aluminium ion. (Because the unhydrated proton is so small, OH occupies virtually the same space as O.) When aluminium octahedral units (shown diagrammatically in Fig. 2.9) are linked by the sharing of edges, they form an *alumina* sheet. Aluminium atoms normally occupy only two-thirds of the available octahedral positions – this is a *dioctahedral* structure, characteristic of the mineral *gibbsite*, $[Al_2OH_6]_n$. If Mg is present instead of Al, however, all the available octahedral positions are filled – this is a *trioctahedral* structure, characteristic of the mineral *brucite* $[Mg_3(OH)_6]_n$.

The bonding together of silica and alumina sheets through the apical O atoms of the silicon tetrahedra causes the bases of the tetrahedra to twist slightly so that the cavities in the sheet become trigonal rather than hexagonal in shape. When the two silica sheets sandwich one alumina sheet, the result is a 2 : 1 layer

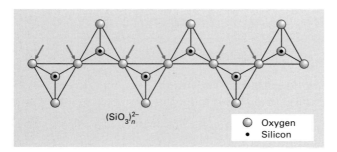

$(SiO_3)_n^{2-}$

○ Oxygen
• Silicon

Fig. 2.6 Single chain structure of silicon tetrahedra as in a pyroxene. The arrows indicate oxygen atoms shared between adjacent tetrahedra (after Bennett and Hulbert, 1986).

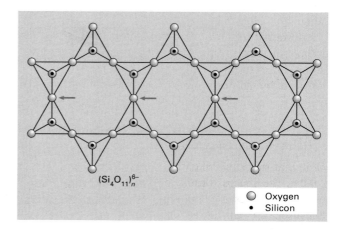

$(Si_4O_{11})_n^{6-}$

○ Oxygen
• Silicon

Fig. 2.7 Double chain structure of silicon tetrahedra as in an amphibole. The arrows indicate oxygen atoms shared between chains (after Bennett and Hulbert, 1986).

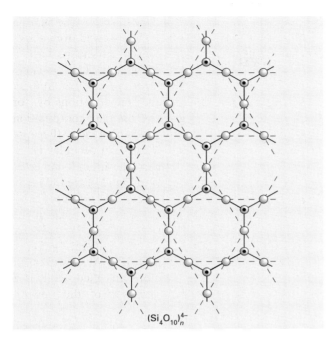

Fig. 2.8 A silica sheet in plan view showing the pattern of hexagonal holes (after Fitzpatrick, 1971).

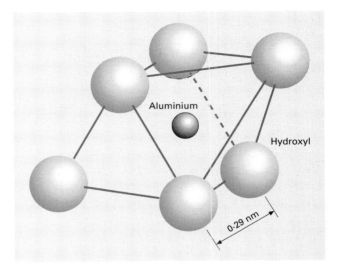

Fig. 2.9 Diagram of an aluminium octahedron (interatomic distances not to scale).

lattice (or *phyllosilicate*) structure characteristic of the micas, chlorites and many soil clay minerals (Section 2.4). Some characteristics of phyllosilicate structures are given in Box 2.2.

Additional structural complexity is introduced by isomorphous substitution. In the dioctahedral mica *muscovite*, for example, one-quarter of the tetrahedral Si is replaced by Al resulting in a net 2 mol of negative charge per unit cell (Box 2.3). Muscovite has the structural composition

$$[(OH)_4(Al_2Si_6)^{IV} Al_4^{VI}O_{20}]^{2-} 2K^+.$$

Note that the negative charge is neutralized by K^+ ions held in the spaces formed by the juxtaposition of the trigonal cavities of adjacent silica sheets (Fig. 2.10). Because the isomorphous substitution in muscovite occurs in the tetrahedral sheet, the negative charge is distributed over only three surface O atoms. The magnitude of the layer charge and its localization are suffi-

cient to cause cations of relatively low ionic potential*, such as K^+, to lose their water of hydration. The unhydrated K^+ ions have a diameter comparable to that of the ditrigonal cavity formed between opposing siloxane surfaces and their presence provides very strong bonding between the layers. Such complexes, where an unhydrated ion forms an ionic–covalent bond with atoms of the mineral surface, are called *inner-sphere* complexes.

Biotite is a trioctahedral mica which has Al substituted for Si in the tetrahedral sheet and Fe^{2+} for Mg^{2+} in the octahedral sheet. Again, because of the localization of charge in the tetrahedral sheet, unhydrated K^+ ions are retained in the interlayer spaced to give a unit cell formula of

$$[(OH)_4(Al_2Si_6)^{IV} (Mg,Fe)_6^{VI}O_{20}]^{2-} 2K^+.$$

The most complex of the two-dimensional structures belongs to the *chlorites*, which have a brucite layer and sandwiched between two mica layers. In the type-mineral *chlorite*, the negative charge of the two biotite layers is neutralized by a positive charge in the brucite, developed due to the replacement of two-thirds of the

*Determined by the charge to radius ratio.

Box 2.2 Model layer-lattice structures.

Phyllosilicates are silicate minerals essentially composed of two-dimensional tetrahedral and octahedral sheets stacked in regular arrays in the c direction (for crystal axes, see Fig. B2.2.1 which shows the generalized structure of a phyllosilicate). The following terminology is used:

• a single *plane* of atoms (such as O and OH),

• a *sheet* is a combination of planes of atoms (such as Si tetrahedral sheet),

• a *layer* is a combination of sheets (such as two silica sheets combined with one alumina sheet in mica),

• a *crystal* is made up of one or more layers,

• planes of atoms are repeated at regular distances in multilayer crystals, which gives rise to a d spacing, or *basal spacing* in the phyllosilicates,

• between the layers is *interlayer* space which may be occupied by water or solute molecules, and

• phyllosilicates generally have large *planar surface* and small *edge faces*.

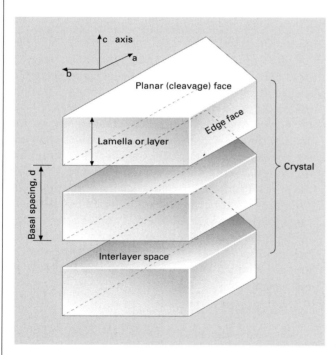

Fig. B2.2.1 General structure of a phyllosilicate crystal.

Box 2.3 Moles of charge and equivalents.

Before the introduction of SI units, charges on soil minerals were expressed in terms of an *equivalent weight*, which is defined as the mass (or weight) of an element which carries the same magnitude of charge (+) or (–) as 1 g of H^+ ions. The equivalent weight of an element is its atomic mass (g) divided by its valency. For H in the form of H^+, the equivalent weight is obviously 1; for K as K^+ the equivalent weight is 39/1, and for Ca as Ca^{2+}, the equivalent weight is 40/2 = 20. The cation exchange capacity of a mineral (Section 2.5) was expressed in milli-equivalents (meq) per 100 g. With SI units, the recommended unit of charge for cations, anions and surfaces is the mole (abbreviated mol) of charge, which is the same as an equivalent weight. Thus, the H^+ ion comprises 1 mol of charge, and the moles of charge for other cations and anions, are given by their molar mass divided by the valency. For soils, the most appropriate SI unit is the *centimol (cmol) of charge (+) or (–) kg⁻¹*, which is numerically equal to meq/100 g.

Mg atoms by Al (Fig. 2.11). Chlorite is an example of a regular mixed layer mineral. Predictably, the bonding between layers is strong, but the high content of Mg renders this mineral susceptible to dissolution in acidic solutions. For the same reason, and also because of its ferrous iron content, biotite is much less stable than muscovite.

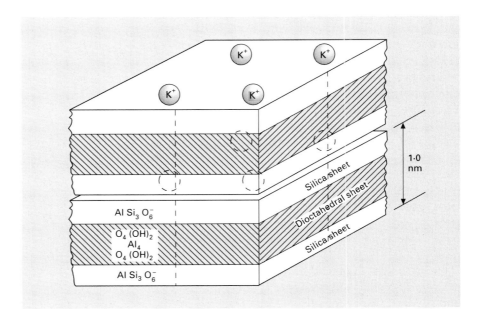

Fig. 2.10 Structure of muscovite mica–a 2 : 1 non-expanding, layer-lattice mineral.

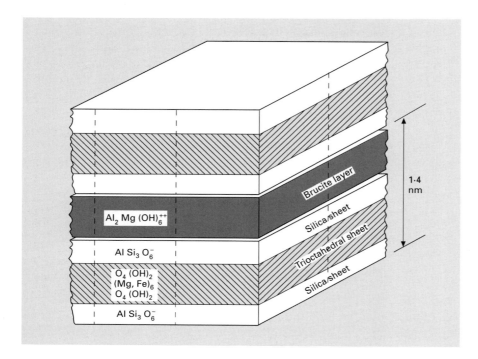

$$Al_2 Mg (OH)_6^{++}$$

$$Al Si_3 O_6^-$$

$$O_4 (OH)_2$$
$$(Mg, Fe)_6$$
$$O_4 (OH)_2$$

$$Al Si_3 O_6^-$$

Brucite layer

Silica sheet

Trioctahedral sheet

Silica sheet

1·4 nm

Fig. 2.11 Structure of chlorite–2 : 2 or mixed layer mineral.

Three-dimensional structures

The most important silicates in this group are *silica* and the *feldspars*. Silica minerals consist entirely of polymerized silicon tetrahedra of general composition $(SiO_2)_n$. Silica occurs as the residual mineral *quartz*, which is very inert, and as a secondary mineral precipitated after the hydrolysis of more complex silicates. Secondary silica initially exists as amphorous opal which dehydrates over time to form microcrystalline quartz, known as *flint* or *chert*. Silica also occurs as amorphous or microcrystalline silica of biological origin. For example, the *diatom*, a minute aquatic organism, has a skeleton of almost pure silica. Silica absorbed from the soil by terrestrial plants (grasses and hardwood trees in particular) forms opaline structures called *phytoliths*, which are returned to the soil when the plant dies.

Quartz or flint fragments of greater than colloidal size are very insoluble, and hence are abundant in the sand and silt fractions of many soils. The feldspars, on the other hand, are chemically more reactive and rarely comprise more than *c.* 10% of the sand fraction of mature soils. Of these, the potash feldspars are more resistant to weathering than the Ca and Na feldspars.

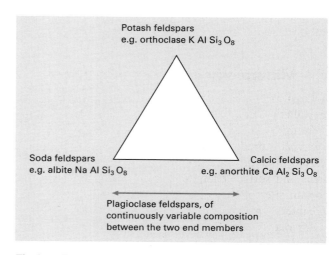

Potash feldspars
e.g. orthoclase $K Al Si_3 O_8$

Soda feldspars
e.g. albite $Na Al Si_3 O_8$

Calcic feldspars
e.g. anorthite $Ca Al_2 Si_3 O_8$

Plagioclase feldspars, of continuously variable composition between the two end members

Fig. 2.12 Type minerals of the feldspar group.

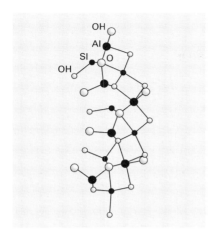

Fig. 2.13 Projection along the imogolite *c* axis showing the curvature produced as the Al octahedral sheet distorts to accommodate the Si tetrahedra (after Wada, 1980).

Their structure consists of a three-dimensional framework of polymerized silicon tetrahedra in which some Si is replaced by Al. The cations balancing the excess negative charge are all of high coordination number, such as K^+, Na^+ and Ca^{2+}, and less commonly, Ba^{2+} and Sr^{2+}. The range of composition encountered is shown in Fig. 2.12. Details of the structure, composition and chemical stability of the feldspars are given in specialist texts by Loughnan (1969) and Nahon (1991).

2.4 Mineralogy of the clay fraction

A large assortment of minerals of varying degrees of crystallinity occurs in the clay fraction of soils. Broadly, these minerals may be divided into the *crystalline clay minerals* – predominantly phyllosilicates – and the *accessory minerals*. Because of their large specific surface areas and surface charges, these minerals are very important sites of chemical and physical reactions in soil (Chapters 6 and 7).

For many years the small size of clay particles prevented scientists from elucidating their mineral structure. It was thought that the clay fraction consisted of inert mineral fragments enveloped in an amorphous gel of hydrated sesquioxides ($Fe_2O_3.nH_2O$ and $Al_2O_3.nH_2O$) and silicic acid ($Si(OH)_4$). The surface gel was amphoteric, the balance between acidity and basicity being dependent on the pH of the soil. Between pH 5 and 8, the surface was usually negatively charged (proton-deficient), which could account for the observed cation exchange properties of soil. During the 1930s, however, the crystalline nature of the clay minerals was established unequivocally by X-ray diffraction (XRD) (Box 2.4). Most of the minerals were found to have a plate-like structure similar to that of the micas and chlorite. The various mineral groups were identified from their characteristic *basal spacings* (Fig. B2.2.1), as measured by XRD, and their unit cell compositions deduced from elemental analyses. Subsequent studies using the scanning and transmission electron microscopes have fully confirmed the conclusions of the early work.

In addition, within the accessory minerals there are the weathered residues of resistant primary minerals that have been comminuted to colloidal size, and soil minerals synthesized during pedogenesis. The latter mainly comprise Al and Fe hydroxides and oxyhydroxides which occur as discrete particles or as thin coatings on the clay minerals. The crystallinity of these minerals varies markedly depending on their mode of formation, the presence of other elements as inclusions, and their age. Some, such as the iron hydroxide *ferrihydrite*, were previously thought to be amorphous, but are now known to form extremely small crystals and to possess *short-range order*; that is, their structure is regular over distances of a few nanometres, but disordered at a larger scale (tens of nanometres).

The crystalline clay minerals

Most clay minerals have a layer-lattice structure but a small group – the *sepiolite–palygorskite* series – has a chain structures and another group – the *allophanes* – form hollow spherical crystals. Palygorskite and sepiolite are unusual in having very high Si/Al ratios, with Mg occupying most of the octahedral positions. Sepio-

The technique of X-ray diffraction (XRD) involves directing a beam of X-rays (electromagnetic radiation of wavelength from 0.1 to 10 nm) at a clay fraction sample. Monochromatic X-rays whose wave length is of the same order as the spacings of atomic planes in the crystals (0.1–0.2 nm) are most useful. The clay sample can be a powder sample or a suspension dried on to a glass slide, which gives a preferred orientation of the plate-like crystals. As the X-rays penetrate a crystal, a small amount of their energy is absorbed by the atoms which become 'excited' and emit radiation in all directions. Radiation from atomic planes that is in phase will form a coherent reflected beam that can be detected by X-ray sensitive film. For a beam of parallel X-rays of wavelength λ, striking a crystal surface at an angle θ, the necessary condition for the reflected radiation from atomic planes to be in phase is

$$n\lambda = 2\,d\sin\theta \tag{B2.4.1}$$

where n is an integer and d is the characteristic atomic spacing. Equation (B2.4.1) is a mathematical statement of Bragg's law.

Box 2.4 Identification of minerals in the clay fraction.

lite is very rare in soil and palygorskite only survives in soils of semi-arid and arid regions. They are not discussed further.

Under mild (generally physical) weathering conditions, clay minerals may be *inherited* as colloidal fragments of primary layer silicates, such as the micas and chlorites. Under more intense weathering, the primary minerals may be *transformed* to secondary clay minerals, as when soil illites and vermiculites are formed by the leaching of interlayer K from primary micas. *Neoformation* of clay minerals is a feature of intense weathering, or the diagenesis of sedimentary deposits (Section 5.2), when minerals completely different from the original primary minerals are formed. When the soluble silica concentration in the weathering environment is high, 2 : 1 layer minerals like the smectites are likely to form. Leaching and removal of silica, however, can produce kaolinite and aluminium hydroxide. Increased lattice negative charge due to isomorphous substitution of Al for Si in the smectites leads to K^+ being the favoured interlayer cation with the resultant formation of illite.

Minerals with a Si/Al mole ratio $\leqslant$ 1

Three groups of clay minerals – *imogolite*, the *allophanes* and the *kaolinites* – have Si/Al ratios $\leqslant$ 1.

Imogolite and allophane are most commonly found in relatively young soils (< 10 000 years) formed on volcanic ash and pumice (order Andisol (ST)). They have also been identified in the B horizon of podzols (order Spodosol (ST) or Podosol (A)). Both minerals appear amorphous by XRD, but high-resolution electron microscopy has revealed their crystalline nature, so they are properly called short-range order minerals. Because of their hollow crystal structures, they have very high specific surfaces (Table 2.2) and are highly reactive, especially towards organic anions and phosphate.

Imogolite has the structural formula:

$6OH\ 4Al^{VI}\ 6O\ 2Si\ 2OH$

and occurs in 'threads' 10–30 nm in diameter and several microns long. Each thread consists of several tubular crystals with inner and outer diameters c. 1 and 2 nm, respectively. They are curved to permit a reduced

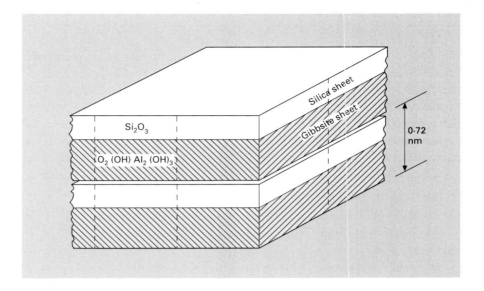

Fig. 2.14 Structure of kaolinite–a 1 : 1 layer-lattice mineral with hydrogen-bonding between layers.

number of silicon tetrahedra to bond to the Al octahedral sheet, as shown in Fig. 2.13.

Allophane exists as hollow, nearly spherical particles of diameter 3.5–5 nm. One kind of allophane, which has a Si/Al ratio of 0.5 and the structural formula

$$2H_2O, 4OH\ 3Al^{VI}\ 2O\ 4OH(2Si, Al)^{IV}\ 3O, 2OH, H_2O$$

contains most of its Al in sixfold coordination and has charge properties very similar to imogolite; that is, very little permanent negative charge due to isomorphous substitution, but variable positive and negative charge due to proton association or dissociation at surface OH groups (Section 7.1). At the other extreme, allophane with a Si/Al ratio of 1 and the structural formula

$$H_2O, 2OH\ Al^{VI}\ O, 2OH\ H_2O(2Si, Al)^{IV}\ 3O, 2OH, H_2O$$

contains half its Al in the tetrahedral sheet and half in the octahedral sheet. Although a large layer charge arises because of the substitution of Al for Si, it is neutralized to a variable extent, depending on the ambient pH, by the association of protons at surface OH groups. Thus, the allophanes and imogolite have pH-dependent surface charges at pH > 4.5.

Minerals of the *kaolinite* group have a well-defined 1 : 1 layer-lattice structure formed by the sharing of O atoms between one silicon tetrahedral sheet and one dioctahedral gibbsite sheet. The unit cell composition is $Si_2^{IV}\ O_5\ Al_2^{VI}\ OH_4$. The common mineral of the group is *kaolinite*, the dominant clay mineral in many weathered tropical soils. *Halloysite* is found in weathered soils formed on volcanic ash parent material.

Kaolinite has a basal spacing fixed at 0.72 nm due to hydrogen-bonding between the hydrogen and oxygen atoms of adjacent lamellae (Fig. 2.14). The layers are stacked fairly regularly in the *c* direction to form crystals from 0.05 to 2 μm thick, the larger crystals occurring in relatively pure deposits of China Clay which is used for pottery. The crystals are hexagonal in plan view and usually larger than 0.2 μm in diameter, as shown in the electron-micrograph of Fig. 2.15. Halloysite has the same structure as kaolinite, with the addition of two layers of water molecules between the crystal layers which increases the basal spacing of the mineral to 1 nm. The presence of this hydrogen-bonded water so alters the distribution of stresses within the crystal that the layers curve to form a tubular structure.

There is some isomorphous substitution of Al for Si in kaolinite which produces < 0.005 mol of negative

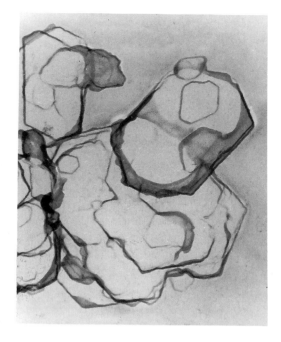

1 μm

Fig. 2.15 Electron micrograph of kaolinite crystals.

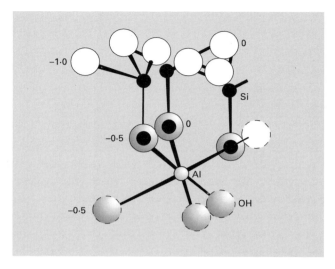

Fig. 2.16 Charges on the edge face of kaolinite at high pH (after Hendricks, 1945).

charge per unit cell. In addition, a variable charge can develop at the crystal edge faces (Fig. 2.16) due to the association or dissociation of protons at exposed O and OH groups. At pH > 9, this can contribute as much as 8×10^{-4} cmol of charge (–) per m^2 of edge area (*cf.* Table 2.3).

Minerals with a Si/Al mole ratio of 2

This category comprises the *mica*, *vermiculite* and *smectite* groups of clay minerals, which all have 2 : 1 layer-lattice structures. The minerals differ mainly in the extent and location of isomorphous substitution, and hence in the type of interlayer cation that predominates.

Within the mica group, the dioctahedral mineral *illite* has substitution of Al for Si in the tetrahedral sheets and some substitution of Mg and Fe^{2+} for Al in the octahedral sheet. An average of 1.6 mol of negative charge per unit cell of formula

$$[(OH)_4(Si_{7.1}Al_{0.9})^{IV} (Al_{3.3}Mg_{0.7})^{VI}O_{20}]^{-1.6} \, 1.6K^+$$

and its location partly in the tetrahedral sheets, are sufficient to favour the formation of inner-sphere complexes with cations, primarily K^+, which fits into the ditrigonal cavities between opposing siloxane surfaces. The basal spacing is characteristically 0.9–1.01 nm. As illite weathers and the K^+ is gradually replaced by cations of higher ionic potential such as Ca^{2+} and Mg^{2+}, which remain partially hydrated in the interlayer regions, the mineral expands to a basal spacing of 1.4–1.5 nm. The additional spacing is equivalent to a bimolecular layer of water molecules between the mineral layers. Consequently, the interlayer bonding is weaker than in the primary micas and the stacking of the layers to form crystals is much less regular. These minerals are sometimes called *hydrous micas*.

Vermiculites are trioctahedral minerals with both di- and trivalent cations occupying all the available sites in the octahedral sheet. This results in a net positive charge which in part neutralizes the negative charge developed through substitution of Al for Si in the tetra-

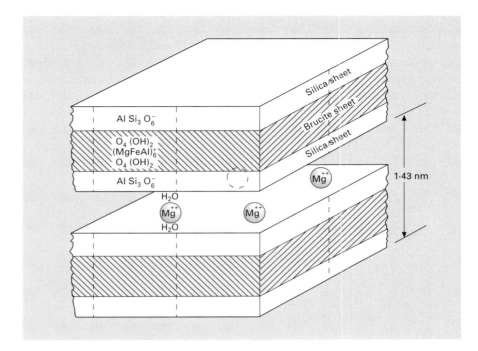

Fig. 2.17 Structure of vermiculite–basal spacing fixed in the presence of partially hydrated Mg^{2+} ions.

hedral sheet. The net layer charge of the vermiculites is therefore slightly lower than that of the soil micas at 1.2–1.9 mol of negative charge per unit cell. As a result, ions of lower ionic potential are not likely to form *inner-sphere* complexes and partially hydrated Ca^{2+} and Mg^{2+} are the dominant interlayer cations. The basal spacing of a Mg-vermiculite, shown in Fig. 2.17, is typically 1.43 nm, but collapses to 1 nm on heating. Thus, the vermiculites show limited reversible swelling. In a moderately acid environment, hydrated Al^{3+} ions replace the Ca^{2+} and Mg^{2+} in the interlayers and 'islands' of poorly ordered $Al(OH)_3$ may form. This gives rise to an aluminous chlorite type of clay mineral.

Smectite minerals exhibit a wide range of composition with isomorphous substitution occurring in both the tetrahedral and octahedral sheets. The layer charge is lower (0.4–1.2 mol of negative charge per unit cell) than in either the soil micas or vermiculites. Two smectite minerals – *nontronite* and *beidellite* – have isomorphous substitution in the tetrahedral sheet, but the commonest smectite in soil, *montmorillonite*, is a dioc-

tahedral mineral with substitution in the octahedral sheet only. It has the unit cell formula

$$[(OH)_4 Si_8{}^{IV} (Al_{3.3} Mg_{0.7})^{VI} O_{20}]^{-0.7} 0.35\, Ca^{2+}.$$

When isomorphous substitution occurs in the octahedral sheet, the excess negative charge is distributed over 10 surface O atoms and the field strength is weak. Consequently, there is no tendency for cations of low ionic potential to dehydrate and to draw adjacent mineral layers close together, as occurs in the micas. Rather, the cations neutralizing the layer charge remain hydrated and may form *outer-sphere* complexes with the surface, as in the case of Ca-montmorillonite. The interlayer cations are freely exchangeable with other cations in solution, and depending on the nature of the exchangeable cations, the basal spacing may vary from *c*. 1.5 to 4 nm, as illustrated in Fig. 2.18. Ca-montmorillonite, a form common in soil, has a basal spacing of 1.9 nm at full hydration (relative humidity > 98%), when three molecular layers of water exist between the mineral layers.

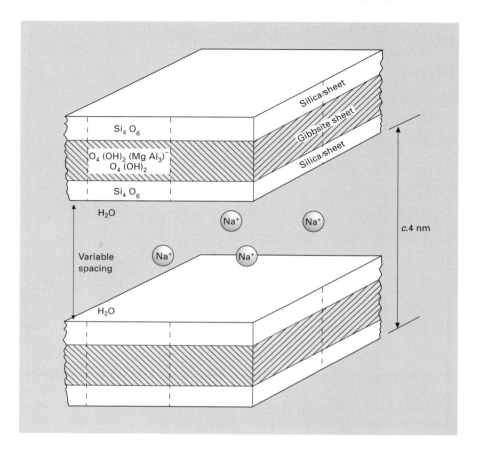

Si$_4$ O$_6$

O$_4$ (OH)$_2$ (Mg Al$_3$)$^-$
O$_4$ (OH)$_2$

Si$_4$ O$_6$

H$_2$O

Variable
spacing

H$_2$O

Na$^+$ Na$^+$

Na$^+$ Na$^+$

Silica sheet

Gibbsite sheet

Silica sheet

c.4 nm

Fig. 2.18 Structure of montmorillonite–
basal spacing variable depending on the
dominant exchangeable cation.

Because of the relatively weak interlayer bonding and
the free movement of water and cations into and out of
this region, the smectites have been called expanding-
lattice clays. The stacking of the layers is very irregular
and the average crystal size is therefore much smaller
than in the micas and kaolinites. The planar interlayer
surfaces provide a large internal area which augments
the external cleavage face area so that the total surface
area of a smectite clay is very large indeed (Table 2.2).

Mixed layer or interstratified minerals

Regular and irregular interstratification of clay minerals
produces mixed layer minerals. The chlorites, for example, are formed by the *regular* interstratification of brucite or gibbsite layers and biotite mica. The basal spacing is a consistent 1.4 nm. With the improvement of
XRD techniques, *irregular* interstratification has been
found to occur much more frequently in soil clay minerals than was previously believed. Mixed layers of the
2 : 1 minerals are most common, especially intergrades
of vermiculite and illite, and smectite and illite. However, interstratification of montmorillonite (2 : 1) and
kaolinite (1 : 1) occurs in acid environments, and
anauxite is an example of a 1 : 1 mixed layer mineral in
which a double silica sheet is inserted at random between kaolinite layers.

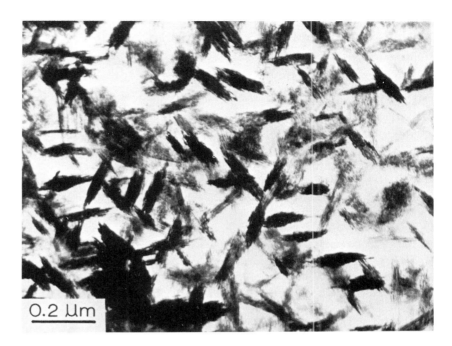

Fig. 2.19 A transmission electron micrograph showing clusters of many small acicular goethite crystals (courtesy of A. Suddhiprakarn and R.J. Gilkes).

Accessory minerals

Accessory minerals of the clay fraction are predominantly *free oxides* and *hydroxides* – compounds in which a single cation species is coordinated with O and/or OH, which include various forms of silica, iron oxides, aluminium oxides, manganese oxides and oxides of titanium. *Calcite* ($CaCO_3$) and dolomite ($Ca,MgCO_3$) can occur in the clay fraction of soils formed on chalk and limestone, and in high pH soils of arid regions (when gypsum $CaSO_4.2H_2O$ may also occur). Accumulations of these carbonate minerals as hard layers within the soil profile are called *calcrete* or *dolocrete*.

Silica

The various forms of silica are described in Section 2.3. Silicic acid in solution has a marked tendency to polymerize as the pH rises. Thus, colloidal silica gel can occur as a cementing agent in the lower B and C horizons of leached soils (Section 4.3), where it may accumulate to such an extent as to form *silcrete*. Amorphous silica also occurs in soils formed on recent deposits of volcanic ash, as has been observed in Japan, New Zealand and Hawaii.

Manganese oxides

There is a variety of oxides and hydroxides of manganese in soil because Mn can occur in oxidation states of 2, 3 and 4. A continuous range in composition from manganous oxide MnO, to manganese dioxide, MnO_2, is possible, with several stable and metastable minerals having been identified. For example, the most stable form of MnO_2 is *pyrolusite*, but the commonest Mn oxides found in soil are the birnessite group in which Mn^{2+}, Mn^{3+} and Mn^{4+} ions are bonded to O^{2-} and OH^-. Small black deposits (1–2 mm diameter) of manganese oxides are common in soils that experience alternating aerobic and anaerobic conditions (Section 8.4). The surface of these minerals has a high affinity for heavy metals, especially Co and Pb, and the availability of Co to plants is often controlled by the solubility of these Mn oxides.

Iron oxides and oxyhydroxides

Iron oxides tend to accumulate in soils that are highly weathered, especially those derived from the more basic rocks (Section 5.2). They have strong colours ranging from yellow to reddish-brown to black, even when disseminated through the soil profile. The most common mineral is the oxyhydroxide *goethite* (α-FeOOH), which can precipitate directly from the soil solution as small acicular crystals, generally in clusters (Fig. 2.19). Because it is the most thermodynamically stable oxide of iron, it also forms by the slow transformation of ferrihydrite and lepidocrocite (see below). *Ferrihydrite* ($Fe_2O_3.2FeOOH.nH_2O$), which has a rusty red colour, readily precipitates in drainage ditches and in the B horizon of podzols (order Spodosol (ST) or Podosol (A)): it is a necessary precursor of *hematite* (α-Fe_2O_3) formation by precipitation from solution. Because ferrihydrite formation is inhibited by the presence of organic ligands in solution, hematite (and hence red-coloured soils) are more common in regions where organic matter is rapidly oxidized than where it is not.

In poorly drained soils, where Fe^{2+} and Fe^{3+} ions can coexist in solution (Section 8.4), rapid oxidation of Fe^{2+} leads to the precipitation of *lepidocrocite* (γ-FeOOH), which through crystal rearrangement slowly reverts to goethite. If the oxidation is slow, however, partial dehydration of the oxide can occur and the ferrimagnetic mineral *maghemite* (γ-Fe_2O_3) forms. The latter is therefore more common in soils of the tropics and subtropics.

With the exception of maghemite and hematite, which also form during weathering of the primary mineral *magnetite* (Fe_3O_4), all the iron oxides form by precipitation from solution. Inclusions of other elements can therefore occur, especially Al which has been found to replace as much as 30% of the Fe in goethite on a molar basis. Al is also found in the crystal structure of hematite and maghemite. Elements such as Si, Ti, V, Zn, Cu, Mo and P also occur as inclusions in iron oxides.

Aluminium oxides and oxyhydroxides

The aluminium oxides have a non-distinctive, greyish-white colour which is easily masked in soils except where large concentrations occur as in bauxite ores. In acid soils, deposits of short-range order aluminium hydroxide form in the interlayers of expanding lattice clays such as the vermiculites and smectites, and as surface coatings on clay minerals generally. This poorly ordered material slowly crystallizes to *gibbsite* (γ-$Al(OH)_3$), the principle aluminium hydroxide mineral in soil. At temperatures $> 150°C$, gibbsite undergoes dehydration to form the oxyhydroxide *boehmite* (α-AlOOH), which is common in bauxite deposits but of uncertain occurrence in soil.

The presence of abundant iron and aluminium oxides (the *sesquioxides*) has a profound effect on the physical and chemical properties of soils formed under the intense weathering conditions of the tropics (Section 9.4). Consequently, the oxide content has been used in soil classification at a high level – an *oxic* horizon is characteristic of the order Oxisol (ST), and an iron oxide content $> 5\%$ (of the fine earth) in a soil's B2 horizon is an essential criterion for the order Ferrosol (A) (Box 1.1). A mixture of sodium citrate and sodium dithionite (sometimes with sodium bicarbonate as well) is widely used to extract the free iron oxides from soil. Acid ammonium oxalate extracts the poorly crystalline Fe oxides as well as some of the short-range order Al oxides, imogolite and allophane. Details of the methods are given in Rayment and Higginson (1992). The ratio of oxalate-extractable Fe to citrate–dithionite extractable Fe, which quantifies the proportion of reactive to less reactive Fe, has been used as an index of a soil's P sorption capacity (Section 7.3).

Titanium oxides

The main titanium oxides found in soil are *rutile* and *anatase* (both TiO_2) and *ilmenite* ($FeTiO_3$). Titanium also occurs in some zirconium oxides. On average, these oxides comprise 1–3% of the clay fraction but may be more abundant in weathered tropical soils. They exist mainly as residual minerals, although there is some evidence for their synthesis during soil formation. Because of their resistance to weathering these minerals have been used as 'markers' in studies of rates of soil formation, although in such studies it is important to identify a specific Ti mineral species and measure its concentration in both the soil and its parent material.

2.5 Surface area and surface charge

Specific surface area and adsorption

We can demonstrate that the smaller the size of an object, imagined as an ideal sphere, the greater the ratio of its surface to volume. Thus, when the radius $r = 1$ mm

$$\frac{\text{surface area}}{\text{volume}} = \frac{4\pi \, r^2}{4/3\pi \, r^3} = 3 \text{ mm}^{-1} \tag{2.1}$$

and when $r = 0.001$ mm (or 1 μm)

$$\frac{\text{surface area}}{\text{volume}} = \frac{4\pi \times 10^{-6}}{4/3\pi \times 10^{-9}} = 3 \times 10^3 \text{ mm}^{-1} \tag{2.2}$$

The ratio of surface area to volume defines the specific surface: in practice, the *specific surface area* of a soil particle is measured as the surface area per unit mass, which implies a constant particle density. For reasons outlined in Box 2.5, a large specific surface increases the adsorption of molecules on the surface. The phenomenon has been exploited to develop methods for measuring the specific surface areas of soil minerals, as described in Box 2.5. Representative values for the specific surfaces areas of sand, silt and clay-size minerals are given in Table 2.2. Note the large range in specific surface area, even within the clay fraction, from as little as 5 m^2 g^{-1} for kaolinite to 750 m^2 g^{-1} for Na-montmorillonite. The latter value is about the maximum for completely dispersed clay of 1-nm layer thickness.

Table 2.2 Specific surface areas according to mineral type and particle size. After Fripiat, 1965.

Mineral or size class	Specific surface (m^2 g^{-1})	Method of measurement
Coarse sand	0.01	Low temperature N$_2$ adsorption (BET*)
Fine sand	0.1	
Silt	1.0	
Kaolinites	5–100	
Hydrous micas (illites)	100–200	BET
Vermiculites and mixed layer minerals	300–500	Ethylene glycol adsorption
Montmorillonite (Na-saturated)	750	Ethylene glycol adsorption
Iron and aluminium oxides and oxyhydroxides	100–300	BET
Allophanes	1000	Includes internal and external surfaces

*BET, Branauer, Emmet and Teller.

Surface charges

In addition to the adsorption of neutral molecules, cations and anions are adsorbed at clay and oxide surfaces as a result of:
• the permanent negative charges of the clays due to isomorphous substitution, and
• the amphoteric properties of the edge faces of kaolinite, the surfaces of imogolite, and allophane and the oxides and oxyhydroxides of Fe and Al.

The permanent charge due to isomorphous substitution can be expressed as a surface density of charge by dividing the layer charge per unit cell by the surface area per unit cell. In practice, the charge on the mineral surface is measured as the difference between the mols of charge contributed per unit mass of mineral by the cations and anions adsorbed from an electrolyte solution of known pH. The cation and anion charge adsorbed gives rise respectively to the *cation exchange capacity* (*CEC*) and *anion exchange capacity* (*AEC*) of the mineral in milli-equivalents (meq) per 100 g, or

When a boulder is reduced to small rock fragments by weathering, part of the energy put in is conserved as free energy associated with the new particle surfaces. The higher the specific surface area of a substance, therefore, the greater is its surface free energy. Because free energy tends to a minimum (second law of thermodynamics), the surface energy is reduced by the work done in attracting molecules to the surface. This results in the phenomenon of *adsorption*. The adsorption of nitrogen gas at very low temperatures on thoroughly dry clay is used to measure the specific surface of the clay (the Branauer, Emmet and Teller or BET method) since the adsorbed N_2 forms a monomolecular layer with an area of 0.16 nm^2 per molecule. Where the mineral has internal surfaces, as in montmorillonite, an additional measurement is required because the non-polar N_2 molecules do not penetrate into the interlayer space. However, small polar molecules such as water or ethylene glycol do penetrate so that monolayer adsorption of these compounds gives a measure of the internal and external surfaces. In the absence of an internal surface, as in kaolinite, the BET and ethylene glycol methods should give identical results. Adsorption is also influenced by electrostatic charge and atomic interactions between the surface (the *adsorbent*) and the adsorbing substance (the *adsorbate*) (Sections 7.1. and 7.3).

Box 2.5 Mineral surfaces and adsorption.

cmol charge per kilogram (Box 2.3). For the crystalline 1 : 1 and 2 : 1 phyllosilicates, permanent negative charge and hence *CEC* is the more important, and representative values of *CEC* for the main mineral groups as given in Table 2.3. Note that the *CEC* is quite variable within and between mineral groups, and not always the value expected from the moles of unbalanced negative charge per unit cell. Of the 2 : 1 clays, for example, the *CEC* of the illites is low because much of the charge is neutralized by interlayer K^+ which is non-exchangeable. On the other hand, the interlayer Ca^{2+} and Mg^{2+} ions of the soil vermiculites are exchangeable and the *CEC* is high.

The functional surface charge density (τ) of the clays is obtained from the ratio *CEC*/specific surface area (cmol $(-)$ m^{-2}). As shown in Table 2.3, despite the large range in charge per unit mass (the *CEC*), the values of τ are reasonably similar, suggesting that the clay crystals cannot acquire an indefinite number of

negative charges and remain stable. Kaolinite, which has very little isomorphous substitution, exists as large crystals and so has a much lower specific surface than montmorillonite; on the other hand, the large kaolinite crystals have a relatively greater edge area at which additional negative charges can develop and so augment the total surface charge density of the clay.

For minerals bearing predominantly pH-dependent charges (the sesquioxides, imogolite and allophane), both the *CEC* and *AEC* can be significant, depending on the prevailing pH. Imogolite and allophane have roughly comparable positive and negative charges in the pH range 6–7. The sesquioxides are positively charged up to pH 8 so in most soils they make no contribution to the *CEC*. Maximum *AEC* values (measured at pH 3.5) range from 30 to c. 50 cmol charge $(-)$ kg^{-1} for gibbsite and goethite, respectively. The development of charges on the surfaces of clays and oxides is discussed more fully in Chapter 7.

Table 2.3 Cation exchange capacities and surface charge densities of the clay mineral groups.

Clay mineral group	Cation exchange capacity (CEC)* (cmol (+) kg^{-1})	Surface charge density (τ) (cmol (–) m^{-2})
Kaolinites	3–20	2–6×10^{-4}
Illites	10–40	1–2×10^{-4}
Smectites	80–120	1–1.5×10^{-4}
Vermiculites	100–150	3×10^{-4}

*Although the charge on the surface is negative, it is measured by the number of moles of cation charge (+) adsorbed.

2.6 Summary

There is a continuous distribution of particle sizes in soil from boulders and stones down to clay minerals < 2 μm in equivalent diameter. The material that passes through a 2-mm sieve, the *fine earth*, is divided into *sand*, *silt* and *clay*, the relative proportions of which determine the *soil texture*. Texture is a property that changes only slowly with time, and is an important determinant of the soil's response to water – its stickiness, mouldability and permeability; also its capacity to retain cations and its rate of adjustment to ambient temperature changes.

The adsorption of water and solutes by clay particles (< 2 μm size) depends not only on their large *specific surface area* but also on the nature of the minerals present. Broadly, these are divided into the *crystalline clay minerals* (layer-lattice or phyllosilicates) and the *accessory minerals* which comprise various Fe and Al oxides and oxyhydroxides (*sesquioxides*), Mn and Ti oxides, silica and calcite. The basic crystal structure of the phyllosilicates consists of sheets of Si in tetrahedral coordination with O $[SiO_2]_n$ and sheets of Al in octahedral coordination with OH $[Al_2(OH)_6]_n$. Crystal layers are formed by the sharing of O atoms between contiguous silica and alumina sheets giving rise to minerals with Si/Al mol ratios $\leqslant 1$ (*imogolite* and the *allophanes*), the 1 : 1 layer lattice minerals (*kaolinites*) and the 2 : 1 layer lattice minerals (*illites*, *vermiculites* and *smectites*). Isomorphous substitution ($Al^{3+} \rightarrow Si^{4+}$ and Fe^{2+}, $Mg^{2+} \rightarrow Al^{3+}$) in the lattice results in an overall net layer charge ranging from < 0.005 to 2 mol of negative charge per unit cell for kaolinite to muscovite, respectively. The formation of *inner-sphere* complexes between the surface and unhydrated K^+ ions in the interlayer spaces of the micas effectively neutralizes the layer charge and permits large crystals to form. On weathering, K^+ is replaced by cations of higher ionic potential such as Ca^{2+} and Mg^{2+} and layers of water molecules intrude into the interlayer spaces – the *basal spacing* of the mineral increases. Ca^{2+} and Mg^{2+}, which form *outer-sphere* complexes with the surfaces, are freely exchangeable with other cations in the soil solution.

The edge faces of the clay crystals, especially kaolinite and the surfaces of iron and aluminium oxides (e.g. *goethite* and *gibbsite*) bear variable charges depending on the association or dissociation of protons at exposed O and OH groups. The surfaces are positively charged at low pH and negatively charged at high pH. The cation and anion charges adsorbed (in moles of charge per unit mass) measure the *CEC* and *AEC* respectively of the mineral. *CEC* ranges from 3 to 150 cmol charge (+) kg^{-1} for the clay minerals and *AEC* from 30 to 50 cmol charge (–) kg^{-1} for the sesquioxides. Imogolite and allophane have roughly comparable *CEC* and *AEC* at pH 6–7.

Clays in which the interlayer surfaces are freely accessible to water and solutes are called expanding-lattice clays. The total surface area is then the sum of the internal and external areas, reaching 750 m^2 g^{-1} for a fully dispersed Na-montmorillonite. Imogolite and allophane also have very large specific surfaces (c. 1000 m^2 g^{-1}) because of the internal and external surfaces associated with their respective tubular and spherical crystal structures. However, when the crystal layers are strongly bonded to one another, as in kaolinite, the surface area comprises only the external surface and may vary from 5 to 100 m^2 g^{-1} depending on the crystal size.

References

Bennett R. H. & Hulbert M. H. (1986) *Clay Microstructure*. Reidel, Boston.

Fitzpatrick E. A. (1971) *Pedology. A Systematic Approach to Soil Science*. Oliver & Boyd, Edinburgh.

Fripiat J. J. (1965) Surface chemistry and soil science, in *Experimental Pedology* (Eds E. G. Hallsworth and D. V. Crawford). Butterworth, London.

Hendricks S. B. (1945) Base exchange of crystalline silicates. *Industrial and Engineering Chemistry* **37**, 625–630.

Hodgson J. M. (1974) (Ed.) *Soil survey field handbook*. Soil Survey of England and Wales, Technical Monograph No. 5, Harpenden.

Klute A. (1986) (Ed.) *Methods of Soil Analysis. Part 1. Physical and Mineralogical Methods*, 2nd edn. Agronomy Monograph No. 9, American Society of Agronomy/Soil Science Society of America, Madison.

Loughnan F. C. (1969) *Chemical Weathering of the Silicate Minerals*. American Elsevier, New York.

McDonald R. C., Isbell R. F., Speight J. G., Walker J. & Hopkins M. S. (1990) *Australian Soil and Land Survey Field Handbook*, 2nd edn. Inkata Press, Melbourne.

Marshall C. E. (1964) *The Physical Chemistry and Mineralogy of Soils. 1. Soil Materials*. Wiley, New York.

Nahon D. B. (1991) *Introduction to the Petrology of Soils and Chemical Weathering*. Wiley, New York.

Rayment G. E. & Higginson F. R. (1992) *Australian Laboratory Handbook of Soil and Water Chemical Methods*. Australian Soil and Land Survey Handbook. Inkata Press, Melbourne.

Russell E. W. (1971) Soil structure: its maintenance and improvement. *Journal of Soil Science* **22**, 137–151.

Wada K. (1980) Mineralogical characteristics of Andisols, in *Soils with Variable Charge*. New Zealand Society of Soil Science, Lower Hutt.

Further reading

Dixon J. B. & Weed S. B. (1989) (Eds) *Minerals in Soil Environments*, 2nd edn. Soil Science Society of America Book Series No. 1. Soil Science Society of America, Madison, Wisconsin.

Mitchell J. K. (1993) *Fundamentals of Soil Behaviour*, 2nd edn. Wiley, New York.

Sposito G. (1989) *The Chemistry of Soils*. Oxford University Press, New York.

Taylor R. M., McKenzie R. M., Fordham A. W. & Gillman G. P. (1983) Oxide Minerals, in *Soils: an Australian Viewpoint*. Division of Soils CSIRO, CSIRO Melbourne/Academic Press, London, pp. 309–334.

Velde B. (1992) *Introduction to Clay Minerals: Chemistry, Origins, Uses and Environmental Significance*. Chapman & Hall, London.

Chapter 3
Soil Organisms and
Organic Matter

3.1 Origin of soil organic matter

The carbon cycle

The organic matter of the soil arises from the debris of green plants, animal residues and excreta that are deposited on the surface and mixed to a variable extent with the mineral component. The dead organic matter is colonized by a variety of *soil organisms*, most importantly micro-organisms (Section 3.2), which derive energy for growth from the oxidative decomposition of complex organic molecules. During decomposition, essential elements are converted from organic combination to simple inorganic forms, a process called *mineralization*. For example, organically combined N, P and S appear as NH_4^+, $H_2PO_4^-$ and SO_4^{2-}, ions, and about half the C is released as CO_2. Mineralization, especially the release of CO_2, is vital for the growth of succeeding generations of green plants. The remainder of the substrate C used by the micro-organisms is incorporated into their cell substance or *microbial biomass*, as is a variable proportion of the other essential elements N, P, S, etc. This incorporation renders these elements unavailable for plant growth until the organisms die and decay, so the process is called *immobilization*. The residues of the organisms, together with the more recalcitrant parts of the orginal substrate, accumulate in the soil. The various interlocking processes of synthesis and decomposition by which carbon is circulated through soil, plants, animals and air – collectively the *biosphere* – comprise the *carbon cycle* (Fig. 3.1).

For the past 200 years or so, the release of CO_2 from the combustion of fossil fuels, respiration by organisms, land clearing and burning has exceeded the sequestra-tion of C in living and dead organisms on land and in water. This has led to a steady rise in the CO_2 concentration in the atmosphere, currently at an annual rate of 0.4%, and to what has been described as the enhanced 'greenhouse effect' (Box 3.1). C in soil organic matter is a very large sink (*c.* 1500 Gt or 1500×10^9 t C), so that changes in the dynamic balance between the soil, vegetation and the atmosphere can significantly affect the

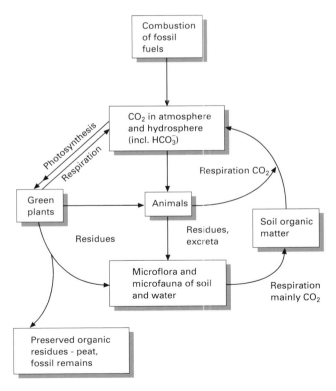

Fig. 3.1 The carbon cycle.

Box 3.1 Carbon cycling and the greenhouse effect.

The greenhouse effect is a natural phenomenon. Gases in relatively small concentrations in the atmosphere – H_2O vapour, CO_2, CH_4, N_2O and the chlorofluorocarbons (CFCs) – are transparent to incoming short-wave solar radiation, but absorb the long-wave radiation emitted from the Earth's surface. The absorption of this energy warms the lower atmosphere (the troposphere) and traps at the Earth's surface heat which would otherwise be radiated into space. The net effect is very beneficial for life on Earth because, in the absence of this insulating effect, the average temperature would be a chilly 30°C lower than it currently is. The process is very similar to that in a greenhouse except than in this case, the glass is transparent to incoming solar radiation and absorbs much of the long wave radiation emitted from the bodies inside.

Nevertheless, the increase in C gas emissions, mainly CO_2 but increasingly CH_4 as well, due to human activities since the Industrial Revolution in the 18th century, has *enhanced* the greenhouse effect. Over the past 160 years, the cumulative net release of CO_2 from agriculture is estimated to have been 264 Gt C compared to 200 Gt C from fossil fuel combustion and cement production. The release from agriculture was mainly associated with land clearing, especially of forests, burning and cultivation. Expansion of paddy rice cultivation has contributed to the increase in CH_4 release. Large-scale land clearing and deforestation are now occurring only in the tropics, and emissions from this source are currently *c.* 1.6 Gt C annually compared with *c.* 5.5 Gt C from fossil fuel combustion.

Importantly, the annual increase in CO_2-C in the atmosphere of *c.* 3.2 Gt is only about half the total annual anthropogenic emissions of *c.* 7.1 Gt, which means there is a significant sink for C in the terrestrial or marine environment. Identifying this sink is very difficult because of the uncertainties in calculating global net C fluxes and estimating the size of the known C sources and sinks.

net flux of CO_2 to the atmosphere. This makes the study of soil organic matter – its composition, accumulation and decomposition – a very important topic.

Inputs of plant and animals residues

The annual return of plant and animal residues to the soil varies greatly with the climatic region and the type of vegetation or land use. The amount of above-ground material – the annual litter fall – is easily measured and some average values, in terms of organic C, are given in Table 3.1. To this must be added timber fall in forests (although this does not provide a readily decomposable substrate) and the contribution from plant roots, which comprises dead root material and C compounds released from living roots into the surrounding soil (the rhizosphere C, Section 8.1). Root contributions are difficult to measure, but experiments in which the plant shoot has been placed in an atmosphere of CO_2 labelled with the radioactive isotope of carbon ^{14}C indicate that some 10–20% of the total plant C may be released into the rhizosphere, and the same proportion may remain

Table 3.1 Annual rate of litter return to the soil.

Land use of vegetation type	Organic C* (t ha^{-1})
Alpine and arctic forest	0.1–0.4
Arable farming (cereals)	1–2
Temperate grassland	2–4
Coniferous forest	1.5–3
Deciduous forest	1.5–4
Tropical rainforest (Colombia)	4–5
Tropical rainforest (West Africa)	10

* For an approximate conversion to organic matter, multiply these figures by 2.5.

as root residues. Thus, the below-ground return of C to the soil may range from 20 to 40% of the total photosynthate produced during a season of growth. The return of C in animal excreta is obviously highly variable but the figure for dung produced by dairy cows grazing highly productive pastures is 3.5–4.5 t ha^{-1} y^{-1}.

Composition of plant litter

Plant cells are made up primarily of carbohydrates, proteins and fats, plus smaller amounts of organic acids, lignin, pigments, waxes and resins. The bulk of the material is carbohydrate of which sugars and starch are rapidly decomposed, while hemicellulose, a pentose sugar polymer, and cellulose, a β(1–4) glucose polymer, are less readily decomposed. These and other decomposition reactions are catalysed by enzymes produced by specialized groups of micro-organisms, as identified in Section 3.2. Proteins are rapidly metabolized and *proteolysis* may begin in the senescing leaf before it has reached the soil. The ultimate products are amino acids (Fig. 3.2), some or all of which may be used in protein synthesis by the micro-organisms. Whether or not there is amino-N surplus to the needs of the micro-organisms, so that *net* mineralization can occur, depends on the C : N ratio of the substrate and on the properties of the decomposer organisms (Box 3.3). Generally, when the C : N ratio is > 25, net immobilization occurs, whereas at ratios < 25 net mineralization is likely.

As can be seen from Table 3.2, the C : N ratio of fresh litter is highly variable among plant species, but as the

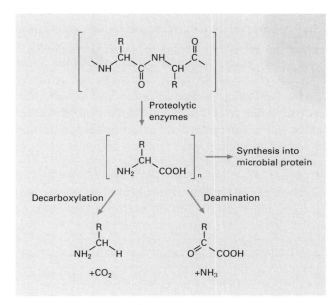

Fig. 3.2 Steps in proteolysis.

Table 3.2 C:N ratios of freshly fallen litter and soils.

	Range	Mean
Litter		
Herbaceous legumes	–	20
Cereal straw	40–120	80
Tropical forest species	27–32	30
Temperate hardwoods (elm, ash, lime, oak, birch)	25–44	35
Scots pine	–	91
Soil		
Soils from 63 sites in the USA	7–26	12.8
Rothamsted: old woodland	–	9.5
Old pastures, fertilized and unfertilized	11–12	11.8
Old arable, manured and unmanured	8–10	9.4

organic matter passes through successive cycles of decomposition in the soil the C : N ratio gradually narrows, and we find the C : N ratio of well drained soils of pH ≃ 7 to be very close to 10. Exceptions occur with poorly drained soils or those on which mor humus forms (Section 3.3).

N in complex heterocyclic ring compounds, such as

chlorophyll, is not easily mineralized, but the chitin of insect cuticles and of fungal cell walls, a glucosamine polymer, is eventually hydrolysed to glusose and NH_4^+. Most plant organic acids are readily decomposed; not so the fats, waxes and resins which can persist in the soil for some time. Similarly lignin, which is a complex phenolic polymer, is more stable than the carbohydrates and accumulates relative to these constituents during the decomposition of fresh residues. Lignin is a significant proportion of the dry matter of cereal straw (10–20%) and wood (20–30%). Its hydrolytic oxidation, catalysed by extracellular enzymes from various actino- mycetes and fungi, produces monocyclic phenols as shown in Fig. 3.3. These degradative products may serve as precursors for the synthesis of humic macro- molecules by the soil organisms.

3.2 The soil organisms

The soil organisms may be grouped, according to size, into:
- *macrofauna* – vertebrate animals mainly of the bur- rowing type, such as moles and rabbits, which live wholly or partly underground,
- *mesofauna* – small invertebrate animals representa- tive of the phyla Arthropoda, Annelida, Nematoda and Mollusca, and
- *micro-organisms*, comprising the *microfauna* (soil an- imals < 0.2 mm in length) and the *microflora* (members of the plant kingdom). In respect of their evolutionary development, the micro-organisms may be subdivided into:
- *prokaryotes* (organisms without a true nucleus), which include the bacteria, actinomycetes and Cyanophyceae, and
- *eukaryotes* (organisms with a membrane-bound nu- cleus), which include the fungi, algae and protozoa.

Soil organisms may also be classified on their mode of nutrition. Broadly, the *heterotrophs*, which include many species of bacteria and all the fungi, require C in the form of organic molecules for growth. However, the *autotrophs*, which include the remaining bacteria and most algae, can synthesize their cell substance from the

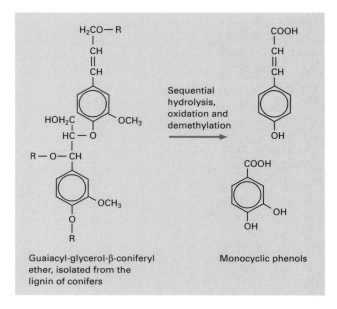

Fig. 3.3 Hydrolytic oxidation of a lignin-type compound (after Andreux, 1982).

C of CO_2, harnessing the energy of sunlight (in the case of the photosynthetic bacteria and algae), or chemical energy from the oxidation of inorganic compounds (the chemoautotrophs). Another way of subdividing the micro-organisms is on the basis of their requirement for molecular O_2; that is, into:
- *aerobes* – those requiring O_2 as the terminal acceptor of electrons in respiration,
- *facultative anaerobes* – those normally requiring O_2 but able to adapt to oxygen-free conditions by using NO_3^- and other inorganic compounds as electron accep- tors in respiration, and
- *obligate anaerobes* – those which grow only in the absence of O_2 because O_2 is toxic to them.

Biomass

Collectively, the mass of organisms in a given volume or mass of soil is referred to as *soil biomass*. Because the macro- and mesofauna can be physically separated from the soil, their mass can be measured directly and is usually expressed as kilograms (liveweight) per hectare to a certain depth. The macro- and mesofaunal biomass ranges from 2 to 5 t ha^{-1}, with earthworms making the largest single contribution (Table 3.3). For a review of the variety of methods used to extract invertebrates from soil, in the laboratory and field, the reader should consult Edwards (1991).

However, micro-organisms are intimately mixed with dead organic matter in soil and, being very small, are difficult to isolate for counting or weighing. This is particularly true of the microflora – bacteria, actinomycetes, fungi and algae – for which the individuals may be < 1 μm in size. For this group, collectively called the *soil microbial biomass*, the methods used to measure numbers and/or mass include:
• direct observations of organisms on agar plates or films, with or without dye staining, and
• physiological or biochemical methods such as the extraction of adenosine triphosphate (ATP), substrate-induced respiration (SIR), and chloroform fumigation–incubation.

Measurements in the second group tell nothing about the species of biomass, but give estimates of biomass size which is valuable in modelling C turnover in soil (Section 3.5). The most widely used method is CHCl$_3$ fumigation followed by incubation or extraction (Box 3.2).

Most of the organisms are concentrated in the top 15–25 cm of soil because C substrates are more plentiful there. Estimates of microbial biomass C range from 500 to 2000 kg ha^{-1} to 15-cm depth, with higher values being recorded on a per hectare basis when the soil is sampled to greater depths. Alternatively, biomass C can be expressed as a percentage of total soil C when it ranges from about 2% in arable soils to 3–4% in grassland or woodland soils.

Types of micro-organisms

Bacteria, including Actinomycetes

These organisms can be as small as 1 μm in length and 0.2 μm in breadth and therefore live in water films around soil particles in all but the smallest pores. They can be motile or non-motile, coccoid (round) or rod-shaped (Fig. 3.4), and can reproduce very rapidly in the soil under favourable conditions – as little as 8–24 h for the production of two daughter organisms from a single parent by fission. Accordingly, their number in the soil is enormous provided that living conditions, especially the food supply, are suitable. Bacterial numbers are estimated by observing the growth of colonies on special nutrient media which have been inoculated with drops of a very dilute soil suspension (the *dilution plate method*), or by direct microscopic observation of suitably stained thin soil sections, recording all the micro-organisms – living, dead and those which have formed spores. Estimates of between 1 and 4×10^9 organisms per gram of soil are obtained by the latter method.

Bacteria exhibit almost limitless variety in their metabolism and ability to decompose diverse substrates. Three major subgroups are recognized.
1 *Unicellular Eubacteria* – the most numerous group including both heterotrophs and autotrophs.
2 *Branched Eubacteria* or *Actinomycetes* – these are heterotrophs that form mycelial growths more delicate than the fungi (see below). They include such genera as *Streptomyces* which produce antibiotics and can degrade the

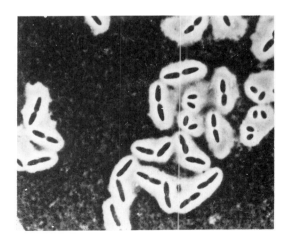

Fig. 3.4 Soil bacteria—*Azotobacter* sp. (after Hepper, 1975).

Box 3.2 Estimation of soil microbial biomass by $CHCl_3$ fumigation.

The method was developed for agricultural soils (generally < 5% organic C) which are not waterlogged nor very acidic (pH > 4.5). Measurements should not be made soon after large amounts of organic materials have been added to the soil. In the original method (Jenkinson and Powlson, 1976), a sample of soil (< 2 mm) was fumigated in $CHCl_3$ vapour to kill the organisms, the $CHCl_3$ evacuated, and the soil inoculated with fresh soil organisms before being incubated for 10 days at constant temperature. The flush of CO_2 evolved relative to a non-fumigated sample of the same soil was used to calculate the microbial biomass from the equation

$$\text{biomass C} = \frac{(CO_2 - \text{C from FS}) - (CO_2 - \text{C from UFS})}{K_c}, \qquad \text{(B3.2.1)}$$

where K_c is the empirically derived fraction of killed biomass C that is evolved as CO_2; FS is fumigated soil and UFS is unfumigated soil. The appropriate K_c value depends on the composition of the microbial population, being lower for fungi than bacteria. The most commonly used values are 0.45 for incubations at 25°C (Oades and Jenkinson, 1979) and 0.41 for incubations at 22°C (Anderson and Domsch, 1978).

The $CHCl_3$ fumigation–extraction method is a more rapid variant of the fumigation-incubation technique which can also be used on very acid soils. The fumigated soil is extracted in 0.5 M K_2SO_4 for 30 min and the dissolved organic C (DOC) is measured, relative to an unfumigated control soil. Biomass C is calculated from

$$\text{biomass C} = \frac{(\text{DOC extracted from FS}) - (\text{DOC extracted from UFS})}{K_{ec}}, \qquad \text{(B3.2.2)}$$

where K_{ec} is an empirical factor measuring the efficiency of C extraction; FS is fumigated soil and UFS is unfumigated soil. K_{ec} values show a similar range to K_c, depending on the soil type, but the generally accepted value is 0.45.

more recalcitrant C compounds such as lignin. They are aerobic sporeformers.

3 *Myxobacteria* (slime bacteria) – these unicellular organisms differ from the Eubacteria in the flexibility of their cell walls and their mode of locomotion. Many are specialized in degrading cellulose and chitin.

The activities of some of the specialized groups of bacteria are discussed in Chapters 8 and 10.

Fungi

Some fungi, such as the yeasts, are unicellular, but the majority produce long filamentous *hyphae*, 1–10 μm in diameter, which may be segmented and/or branched. The network of hyphae or *mycelium* develops fruiting bodies on which the spores formed are often highly coloured. Thus, members of the Basidiomycetes (white-rot fungi) and the abundant spore formers of the Fungi Imperfecti, such as *Penicillium* and *Aspergillus*, are often conspicuous on decaying wood and leaf litter in moist situations.

Fungi occur as *soil inhabitants*, those residing in the soil and feeding on dead organic matter (the saprophytes), and as *soil invaders*, those whose spores are normally deposited in the soil but which only germinate and grow when living tissue of a suitable host plant appears close by. Many of the latter fungi are pathogenic, growing parasitically on the host. Of special benefit, however, are the *mycorrhizal fungi* which live symbiotically in the host tissue, deriving C compounds from the host and in turn supplying it with mineral nutrients, especially P (Section 10.3).

Although soil fungal colonies are much less numerous than bacteria (1–4×10^5 organisms per gram), because of their filamentous growth their biomass is generally larger than the bacterial biomass (by as much as $3:1$, but the proportion of actively growing mycelium may be less than half the total biomass). The fungal population is also considerably less variable than bacterial numbers, for although fungi, excluding yeasts, are intolerant of anaerobic conditions, they grow better in acid soils (particularly at pH < 5.5) and tolerate variations in soil moisture better than bacteria. The fungi are all heterotrophic and are most abundant in the litter layer (Fig. 3.5) and the organically rich surface horizons of the soil, where their superior ability to decompose lignin confers a competitive advantage. Because of their slow growth rate relative to the bacteria, fungi cannot be reliably counted by the dilution plate method. Methods employing the transfer of fungal hyphae from the soil to nutrient agar plates give more accurate results.

An outline of methods for identifying and enumerating the many bacterial and fungal species in soil is given by Parkinson and Coleman (1991).

Algae

The algae are the major group of photosynthetic organisms in soil and are therefore confined to the soil surface, although many will grow heterotrophically in the

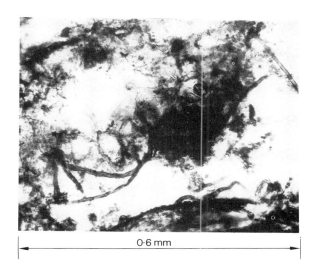

|← 0.6 mm →|

Fig. 3.5 Fungal mycelium in litter layer (courtesy of P. Bullock).

absence of light if simple organic solutes are provided. The two major subgroups are the prokaryotic Cyanophyceae (Cyanobacteria) or blue-green algae and the eukaryotic green algae. The 'blue-greens', as exemplified by the common genera *Nostoc* and *Anabaena*, are important because they can reduce atmospheric nitrogen and incorporate it into amino acids, thereby making a substantial contribution to the N status of wet soils (Section 10.2). The blue-greens prefer neutral to alkaline soils, whereas the green algae are common in acid soils. Algae in soil are much smaller than the aquatic or marine species and may number from $100\,000$ to 3×10^6 organisms per gram.

The prolific growth of aquatic algae during past eras in the Earth's history has produced large areas of sedimentary rocks such as the English Chalk (made up of calcified algal bodies – Fig. 5.6), and deposits of diatomaceous earth (silica skeletons of diatoms).

Protozoa

The smallest of the soil animals, ranging from 5 to $40\,\mu m$ in their longer dimension, the protozoa live in water films and move by means of cytoplasmic streaming, cilia, or flagellae, as in the case of *Euglena*. *Euglena* also contains chlorophyll and can live autotrophically, but nearly all of the protozoa prey upon other small organisms such as bacteria, algae, fungi and even nematodes. Protozoan biomass can be comparable to that of the earthworms so they are important in controlling bacterial and fungal numbers in the soil, and hence in nutrient cycling.

Soil enzymes

Many extracellular enzymes exist in soil independently of living organisms. They avoid denaturation and degradation to some extent by being adsorbed on to soil mineral or organic matter. The major types present are hydrolases, transferases, oxidases, reductases and decarboxylases. A very common and stable soil enzyme is *urease* which is important in the hydrolysis of urea (Section 10.2). Because of its biochemical function, the enzyme dehydrogenase is unlikely to occur in a free, extracellular state in soil. Thus, the assay of this enzyme has been used as an indirect method for measuring soil microbial activity.

The mesofauna

Next to the protozoa, the thread-like *nematodes* are

Fig. 3.6 Dung beetle (*Sisyphus rubripes*) rolling a 'dung ball' (after Waterhouse, 1973).

among the smallest of the soil fauna, ranging from a few tenths of a millimetre to a few millimetres in size. They are plentiful in soil and litter and may be saprophages or carnivores feeding on fungi, bacteria or other nematodes.

Other quite primitive soil animals are the *molluscs* (slugs and snails), many of which feed on living plants and are therefore pests, although some species feed on fungi and the faeces of other animals. As the mollusc biomass is only 200–300 kg ha^{-1} in most soils, their contribution to the turnover of organic matter is limited.

By far the most important groups involved in the turnover of organic matter are the *arthropods* and *annelids*.

Arthropods

This group includes the isopods (wood lice), arachnids (mainly mites), insects (winged and wingless) and myriapods (centipedes and millipedes). The more numerous are the mites and many of the insects, present as adults and larvae.

Mites and *springtails* (wingless Collembola) feed on plant remains and fungi in the litter, especially where thick mats build up under forest and undisturbed grassland. Their droppings appear as characteristic pellets in the litter (Fig. 3.10a). They may also be found at depth in the soil, living in the larger pores (0.3–5 mm). Predaceous adult mites and springtails feed on other mites and nematodes, while in their juvenile stage they feed on bacteria and fungi.

Many *beetles* and *insect larvae* live in the soil, some feeding saprophytically, whereas others feed on living tissues and can be serious pests of agricultural crops. Of particular value, however, are the coprophagous beetles of Africa and other tropical regions that feed upon the dung of large herbivores and greatly increase the rate at which such residues are comminuted and physically mixed with the soil (Fig. 3.6).

The *termite* (a member of the Isoptera) has been called the tropical analogue of the earthworm and indeed they are important comminuters of all forms of litter – tree trunks, branches and leaves – in the forest

Fig. 3.7 Termite mounds in New Queensland.

and especially the seasonal rainfall regions (savanna) of the tropics. Most species are surface feeders and build nests called termitaria by packing and cementing together soil particles with organic secretions and excrement. The small mounds only 30 cm or so high are the homes of *Cubitermes* species which feed under the cover of recent leaf fall (Fig. 3.7). The less frequent large mounds, some 4–6 m high, are the homes of wood-feeders and foraging termites, such as *Macrotermes* and *Odontotermes*, the latter usually colonizing the mounds of *Macrotermes* when they are abandoned. These species cultivate 'fungus gardens' within the termitaria to provide food for their larvae.

Some termite species or 'harvester ants' destroy living vegetation in Africa grasslands. In cool temperate regions, species of ant, such as *Formica fusca*, are of some importance in comminuting litter under pine trees, whereas *Lasius flavus* is active in grassland soils on chalk.

Centipedes are carnivorous whereas *millipedes* feed on vegetation, much of which is in the form of living roots, bulbs and tubers. Size comparisons between these and other soil invertebrates may be made by reference to the specimens in Fig. 3.8.

Annelids

These include the enchytracid worms and the lumbricids or earthworms.

The *enchytraeids* are small (a few millimetres to 2–3 cm long) and thread-like. Their biomass is rarely > 100 kg ha^{-1}. However, the *earthworms*, because of their size (several centimetres long) and physical activity are generally more important in the consumption of

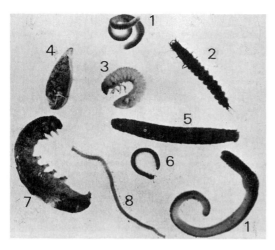

Fig. 3.8 The main litter-feeding soil animals (after Russell, 1973). 1 Earthworm, *Lumbricus* sp. (Oligochaeta); 2 Beetle larva, Carabidae (Coleoptera); 3 Chafer grub, *Phyllopertha* sp. (Coleoptera); 4 Slug, *Agriolimax* sp. (Mollusca); 5 Leatherjacket, Tipulidae (Diptera); 6 Millipede (Diplopoda); 7 Cutworm, Agrotidae (Lepidoptera); 8 Centipede (Chilopoda).

leaf litter than all the other invertebrates together. Exceptions occur in the drier savanna-region soils and soils of temperate regions on which mor humus forms (Section 3.3). An indication of the earthworm biomass under different systems of land use is given in Table 3.3.

Earthworms feed exclusively on dead organic matter. A few species, such as *Lumbricus rubellus* and *L. castaneous*, live mainly in the surface soil and litter layer, provided that temperature and moisture conditions are favourable. The majority burrow more deeply into the soil, and in the course of feeding ingest large quantities of clay- and silt-size particles. The large populations of earthworms in soils long down to grass may consume up to 90 t soil ha^{-1} annually. As a result, the organic and mineral matter are more homogeneously mixed when deposited in the worm faeces, which may appear as *casts* on the soil surface. The most useful species are *Aporrectodea caliginosa*, which is a very active topsoil-mixing species, *A. trapezoids* and the two deep-burrowing species *A. longa* and *Lumbricus terrestris*. When such earthworms are present, organic matter from the surface rapidly becomes incorporated throughout the upper part of the soil profile, in contrast to the sharp boundary between the litter layer and mineral soil proper in the absence of earthworms (Fig. 3.9a,b).

The consumption of dung and vegetable matter by earthworms can be prodigious. It has been estimated from feeding experiments that a population of 120 000 adult worms per hectare is capable of consuming 25–30 t of cow dung annually. The quantities of dung and leaf litter available in the field are normally much less than this, so that the size of the earthworm population is regulated by the amount of suitable organic matter available, as well as by soil temperature, moisture and pH. For example, few species of earthworms are found in soils of pH < 4.5. Of the common earthworms, *L. terrestris* is more tolerant of low pH than the *Aporrectodea* species, and most species prefer neutral to calcareous soils. In hot dry weather they burrow deeply into the soil and aestivate.

Grass and most herbaceous leaf litter are readily acceptable to earthworms, but there is great variability in the palatability of litter from temperate forest species. Litter from elm, lime and birch is more palatable than pine needles which must age considerably before being acceptable.

3.3 Changes in plant remains due to the activities of soil organisms

Types of humus

Decomposition begins with the invasion of ageing plant tissue by surface saprophytes, and proceeds in parallel

Table 3.3 Estimates of earthworm biomass in soils under different land use. After Edwards and Lofty, 1977.

Land use	Earthworm biomass
Hardwood and mixed woodland	370–680
Coniferous forest	50–170
Orchards (grassed)	640–287
Pasture	500–1500
Arable land	16–760

(a)

(b)

Fig. 3.9 (a) Litter accumulation on soil devoid of earthworms. (b) Uniform organic matter distribution in a soil with earthworms (after Edwards and Lofty, 1977).

with biochemical changes in the senescing tissue – the synthesis of protease enzymes, the rupture of cell membranes with consequent mixing of cellular constituents, and the auto-oxidation and polymerization of phenolic-type compounds. Decomposition accelerates, as evidenced by the rise in the rate of CO_2 production, when the plant material falls to the ground and is invaded by a host of soil organisms. The changes that occur in the plant residues lead not only to mineralization and immobilization of nutrient elements, but also to the synthesis of new compounds, less susceptible to decomposition, which collectively form a dark-brown to black, amorphous material called *humus*.

Mor and mull humus

Typically under deciduous forest species, there is a loose litter layer 2–5 cm deep under which the soil is well aggregated and porous, dark-brown in colour and changes only gradually with depth to the lighter colour of the mineral matrix. This deep (30–50 cm) organic A horizon of C : N ratio 10–15 and rich in animals, especially earthworms, is characteristic of *mull humus*. Alternatively, under coniferous species, the surface litter is thick (5–20 cm) and often ramified by plant roots and a tough fungal mycelium. It is sharply differentiated from the mineral soil which is capped by a thin band of blackish humus and is usually compact, poorly drained and devoid of earthworms. This superficial organic horizon, of C : N ratio > 30, is called *mor humus* and generally consists of three layers:
• L layer – undecomposed litter, primarily remains deposited during the previous year,
• F layer – the fermentation layer, partly decomposed, with abundant fungal growth and the faecal pellets of mites and springtails, and
• H layer – the humified layer, comprising completely altered plant residues and faecal pellets.
The contrasting microstructure of mor and mull humus is illustrated by the thin sections of Fig. 3.10.

Mor and mull humus, and intermediate types called *mor moder* and *mull moder*, occur under a variety of vegetation types. As well as the main deciduous trees

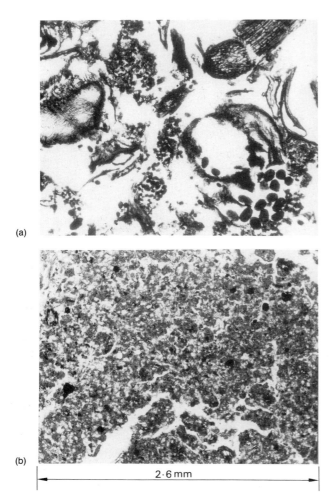

(a)

(b)

2·6 mm

Fig. 3.10 (a) Loose fabric of plant residues and mite droppings in mor humus. (b) Intimately mixed matrix of organic and mineral matter in mull humus (courtesy of P. Bullock).

mull on a fertile calcareous soil and mor on an infertile acid sand.

Colonization by soil micro-organisms

Polyphenolic compounds formed in the senescing leaf, which are not leached out by rainwater once it has fallen to the ground, have a considerable effect on the rate of decomposition of the leaf residue. This is well illustrated by the contrasting superficial layers and A horizons of soils under temperate forests (Fig. 3.10). The formation of mor humus is predisposed by high concentrations of phenols in the senescing leaf which precipitate cytoplasmic proteins on to the mesophyll cell walls, thereby making both the protein and cellulose more resistant to microbial decomposition. The process has been likened to the 'tanning' of leather, whereby plant tannins (polyphenols) are used to preserve the protein of animal skins. Where plant proteins and cellulose are not protected by tanning, these constituents and the sugars and storage carbohydrates are rapidly metabolized by the soil micro-organisms that colonize the litter in the sequence

sugar fungi and non-spore-forming bacteria $\rightarrow$ spore formers $\rightarrow$ cellulolytic myxobacteria $\rightarrow$ actinomycetes.

As indicated in Section 3.1, whether or not net mineralization occurs during decomposition depends on the C : N ratio of the organic substrate being below or above a critical value of *c.* 25. This critical value is set by the C : N ratio of the decomposing organisms and their growth yield (Box 3.3).

The role of soil animals

There is a great deal of interdependence between the activities of the microflora and small invertebrates in the decomposition of organic matter and formation of humus. Studies on the ecology of soil organisms suggest the existence of a 'food web' with several trophic (feeding) levels, and complex interactions between organisms within a trophic level and between levels. At the bottom of the web are plant litter, roots and detritus

(oak, elm, ash and beech), mull usually occurs under a well-drained grassland, whereas mor is found under heath vegetation (*Erica* and *Calluna* spp.) and under the major conifers (pine, spruce, larch and fir). However, the tendency towards mor or mull formation is determined not only by plant species, but also by the mineral status of the soil, particularly of Ca, N and P, so that the same species of deciduous tree can form

Box 3.3 The critical substrate C : N ratio for net mineralization of N.

For the N requirements of the decomposers to be satisfied, the C : N ratio of the substrate that is assimilated must be

$$\leqslant \frac{C_a}{N_m} \tag{B3.3.1}$$

where C_a, the C assimilated, is equal to the sum of C_m, the C incorporated into microbial cells, and C_r, the C respired as CO_2. N_m is the microbial N. Equation B3.3.1 may be written as

$$\frac{C_a}{C_m} \times \frac{C_m}{N_m} . \tag{B3.3.2}$$

Note that C_m/C_a is defined as the growth yield F of the micro-organisms. The critical C : N ratio of the substrate is therefore given by

$$\frac{C : N \text{ ratio of the microbial biomass}}{\text{growth yield, } F} \tag{B3.3.3}$$

C : N ratios range from 3.3 for the bacterium *Escherichia coli* to 12.9 for the fungus *Penicillium*, with the mean C : N ratio for bacteria being *c*. 4 and for fungi *c*. 10. Thus, the crotical C : N ratio will depend on the species composition of the microflora, but taking an average biomass C : N ratio of 5 and a growth yield of 0.2 gives

$$\left(\frac{C}{N}\right)\text{critical} = \frac{5}{0.2} = 25 . \tag{B3.3.4}$$

from organisms feeding at all the higher levels (including above ground). Successively, going up the web, there are:
• bacteria, saprophytic fungi and nematodes that feed on roots,
• protozoa, nematodes that feed on bacteria and fungi, and mites that feed on fungi,
• nematodes that feed on other nematodes and omnivorous nematodes, and
• top predators – mites and collembola – that feed on subordinate organisms.

A quantitative description of energy C and nutrient (N, P and S) flows through such food webs may help to explain the differences in productivity and physical and biological condition between various ecosystems such as grassland, arable land, or high input organic farms, and provide a pointer to their long-term sustainability. Although simulation models of food webs have been attempted, such is the complexity of the information required that it is currently impossible to test by direct measurement the validity of the many input variables and model parameters. Thus, C turnover and mineralization are usually described by more pragmatic models based on substrate properties and decomposition rates, as discussed in Section 3.5.

Turning to the larger invertebrates, we find that their digestive processes are relatively inefficient. Termites are an exception because they can digest cellulose and lignin by means of their intestinal microflora or the fungus gardens they cultivate in their termitaria. With the earthworms some microbial decomposition of the litter must occur before it is readily acceptable to these animals. Once ingested, the main effect of an earthworm's digestive activity is the comminution of the plant residues which vastly increases the surface area accessible to microbial attack. There is some chemical breakdown, usually less than 10%, which could be due to the activities of bacteria in the earthworm gut or to the effect of chitinase and cellulase which are secreted by the earthworms. The combined result of these processes is that earthworm casts have a higher pH, exchangeable Ca^{2+}, available phosphate and mineral nitrogen content than the surrounding soil. Whereas the partially humified residues in the L and F layers of mor humus retain a recognizable plant structure for many years, similar residues which have passed through the earthworm's digestive tract several times are reduced to a black, amorphous humus which is intimately associated with the mineral particles.

Biochemical changes and humus formation

The biochemical changes, which in total comprise *humification*, are complex because both degradative and synthetic processes occur simultaneously. Certainly, plant carbohydrates and proteins are decomposed and their microbial analogues synthesized, although the latter also become substrates to be decomposed in time. Important polymerization reactions involving aromatic* compounds also occur. For example, simple *o*- and *p*-hydroxy phenols occurring naturally in plants, or produced from the degradation of lignin and polyphenolic pigments (Fig. 3.3), are oxidatively polymerized to form humic precursors. The oxidation can be auto-catalytic or catalysed by the polyphenol oxidases. Nitrogen is incorporated into the polymers if amino acids condense with the phenols before polymerization occurs, as shown in Fig. 3.11. Some of the quinone rings formed during polymerization may be broken and additional carboxyl groups formed, as shown in Fig. 3.12,

---

* Unsaturated ring structures of C and H, of which the simplest is benzene C_6H_6.

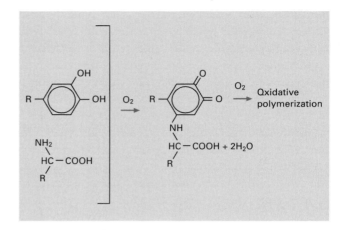

Fig. 3.11 Reaction between an amino acid and a simple phenols under oxidizing conditions (after Andreux, 1982).

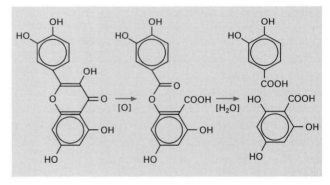

Fig. 3.12 Enzymatic decomposition of a quinine polymer, with increase in carboxyl groups (after Andreux, 1982).

which augment the acidity of the carboxyl groups in the non-aromatic (i.e. aliphatic) side chains. Carboxyls and the remaining reactive phenolic-OH groups form salts or chelate complexes (Section 3.4) with metal ions which increase the stability of the macromolecule. Their longevity in soil also depends on physical stabilization due to adsorption on mineral surfaces.

A typical elemental analysis of humic compounds extracted from soil is 54–58% C, 5–6% H, 3–6% N and 30–33% O. In view of the diversity of complex molecules that can be formed, it is not surprising that the identification of the chemical structure of these compounds has proved difficult. The C : H ratio is an index of aromaticity, the minimum value being 1, as for benzene. Values > 1 reflect the degree of condensation of the rings and the substitution of other elements for H in the structure. An accurate knowledge of the structure of humic compounds would help us greatly when assessing the validity of the various hypotheses of humus formation, and in explaining the properties of this material (Section 3.4).

3.4 Properties of soil organic matter

Fractionation of soil organic matter (SOM)

Physically, SOM may be subdivided into those components that are readily dispersed and separated from the mineral particles, and that which can be dispersed only by drastic chemical treatment. A suggested subdivision is as follows.

1 *Macro-organic matter.* This consists of the most recently added plant and animal debris, which, due to the entrapment of air, has a density near $1 \, g \, cm^{-3}$ and can be separated from the heavier mineral particles by flotation in water or aqueous salt solutions. It is then collected on a fine sieve (0.06 mm aperture).

2 *Light fraction.* This consists of partially humified and comminuted plant and faunal remains which can comprise up to 25% of the total organic matter in grassland soils. Separation from the mineral particles is achieved after ultrasonic vibration of the dry soil, to disrupt the soil aggregates, and flotation of the light fraction in liquids of high density ($\leqslant 2 \, g \, cm^{-3}$) with the aid of a surfactant.

3 *Humified fraction.* Much of the soil organic matter adheres strongly to the mineral particles, particularly the clay, to form a clay–humus complex. This organic matter cannot be separated by density flotation and is incompletely dissolved in metal-chelating solvents such as acetylacetone. It is most effectively dispersed by strong alkalis, such as sodium hydroxide which acts both by hydrolysing and depolymerizing the large organic molecules, or by a polar organic solvent such as

dimethylsulphoxide in acid solution. Commonly, humus is subdivided further according to its solubility in alkaline and acidic solutions, as summarized in Fig. 3.13. Some of the issues underlying the analysis of humus are outlined in Box 3.4.

Although progress in elucidating the structure of humic compounds in soil remains slow, because of the complexity of the problem, the following points seem to be generally accepted.

• The 'core' structure of the *humic acid* (HA) and *fulvic acid* (FA) fractions is composed of aliphatic and aromatic groups covalently bonded to form high molecular weight polymers (20 000–100 000) with much branching and folding.

• Polysaccharides and proteinaceous materials occupy the voids in the folded and branched macromolecules, but do not contribute more than 20% of the total mass. They may be covalently bonded or held by electrostatic charge effects or hydrogen bonding.

• Thirty-five to forty-five per cent of the HA is aromatic, consisting mainly of single ring structures that are highly substituted, particularly with carboxyl (—COOH) and phenolic hydroxyl groups of varying acid strength. Long-chain aliphatics, also with carboxyl groups, link the aromatic structures or exist as separate side chains attached to the groups. They confer a degree of hydrophobicity on the HA molecules.

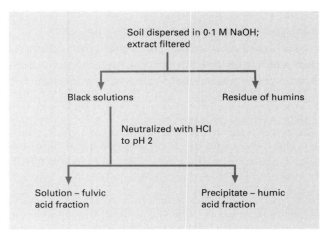

Fig. 3.13 Scheme for the fractionation of soil humus.

Box 3.4 Characterization of soil humus.

Central to the chemical characterization of humus is the problem that, because strong solvents and fairly drastic treatment are necessary to recover most of the humus from soils, it is unlikely that the compounds in the extract are the same as those originally in the soil. Further treatment with strong oxidizing agents such as $KMnO_4$ or an alkaline cupric oxide medium produces other changes in the compounds present. Nevertheless, powerful analytical techniques such as cross-polarization, magic angle spinning (CPMAS) ^{13}C nuclear magnetic resonance (NMR), Fourier transform infrared (FTIR) spectroscopy and pyrolysis–soft ionization mass spectrometry have been applied to the extracts of humic materials from soil. Reviews of the results of such studies are given by Hayes (1991) and Schnitzer and Schulten (1995). To avoid the use of strong solvents to extract humic compounds, other scientists have applied ^{13}C NMR to soil solids and attached organic matter, separated on the basis of both particle size and density (see the 'light fraction'). The solid-state NMR spectra have shown systematic differences in the organic matter in particles of different size, changing from material of plant origin to microbial products as the particle size decreases (and degree of humification increases).

- Approximately 25% of the FA is aromatic. FA molecules are smaller, more highly charged and more polar than HA. Charge repulsion causes the FA molecules to be more linear than the randomly coiled HA molecules. They do not precipitate in acid solutions and are susceptible to leaching.
- The type and steric arrangement of functional groups (—COOH, phenolic-OH and carbonyl C=O) facilitates the complexation of metal cations (see below).
- The non-extractable *humins* are thought to be HA-type compounds that are strongly adsorbed, or precipitated on mineral surfaces as metal salts or chelates.

Cation exchange capacity

Humification produces an organic colloid of high specific surface and high cation exchange capacity (*CEC*). Of the several functional groups containing oxygen, those which dissociate H^+ ions – the carboxylic and phenolic groups – are the most important. The former have p*K* values between 3 and 5 while the latter only begin to dissociate protons at pH values > 7. A few

basic–NH_3^+ groups may exist at pH values < 3. Thus, the *CEC* of soil organic matter is completely pH-dependent (Section 7.2) and buffered over a wide range of H^+ ion concentration: its value ranges between 150 and 300 cmol of charge (+) kg^{-1} dry matter. Organic matter can make a substantial contribution to the *CEC* of the whole soil, and hence to the retention of exchangeable cations, especially in soils of low clay content.

Chelation

Humic compounds can form coordination complexes with metallic cations by displacement of some of the water molecules from the cation's hydration shell. The resultant inner-sphere complex (Section 2.3) acquires additional stability through the 'pincer effect' of the coordinating groups, giving rise to a chelate compound (Fig. 3.14). Chelates formed with certain di- and polyvalent cations are the most stable, the stability constants falling in the order $Cu > Fe \approx Al > Mn \approx Co > Zn$. Metal complexes formed with the HA fraction are largely immobile, but the more ephemeral Fe^{2+}

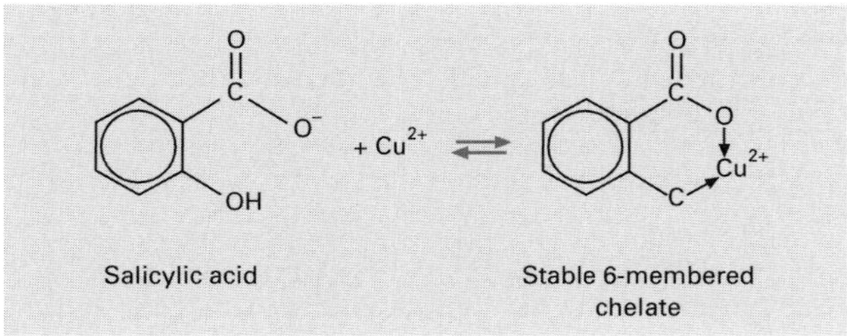

Salicylic acid   Stable 6-membered
                             chelate

Fig. 3.14 Organo-metal chelate formation.

polyphenolic complexes are soluble and their down-ward movement contributes to profile differentiation in podzols (Section 9.2).

Organic matter and soil physical properties

The presence of organic matter is of great importance in the formation and stabilization of soil structure. Polymers of the FA and HA fractions are adsorbed on mineral surfaces by a variety of mechanisms, primarily involving the functional groups – carboxyls, carbonyl ($-C=O$), alcoholic-OH, phenolic-OH, amino ($=NH$) and amine ($-NH_2$). These reactions are discussed in more detail in Sections 4.4 and 7.5. Uncharged non-polar groups on the humic polymers, and large un-charged polymers such as the polysaccharide and polyuronide gums synthesized by many bacteria, can be adsorbed at mineral surfaces by hydrogen-bonding and by van der Waals' forces (Table 4.2). They also function as bonding agents between mineral particles.

3.5 Factors affecting the rate of organic matter decomposition

Turnover

Grains of organic matter through litter decomposition and root death are offset by losses through decomposition and leaching. This is called *turnover*, which is defined as the flux of carbon through the organic C in a unit volume of soil. Given a constant environment, a soil will eventually attain a steady-state in which there is no measurable change in the organic matter content with time. The concept of turnover may be expressed mathematically by the equation

$$\frac{dC}{dt} = A - kC \qquad (3.1)$$

where C is the organic C content of the soil (t ha^{-1}); A is the annual addition of residues, assumed to be approximately constant, and k is a constant (y^{-1}), assuming that all the C in the soil decomposes at the same rate according to simple first-order kinetics. At equilibrium $dC/dt = 0$ and

$$A = kC. \qquad (3.2)$$

The *turnover time* is then given by

$$\frac{C}{A} = \frac{1}{k}. \qquad (3.3)$$

Annual inputs of C for old arable soils at Rothamsted in England have been estimated at 5–6% of the soil organic C, giving turnover times of 16–22 years. Theoretically, this is the time taken for all the organic C in the soil to be replaced, and in a sense is inversely related to the 'biological activity' of the soil. This picture is grossly over-simplified because the age of C in these soils as measured by radioactive carbon dating is 1290–3700 years, depending on depth. It is obvious that there must be a great range of decomposability of C compounds in the soil, with some of the C being much younger than 16–22 years and the bulk of it very much older.

This is confirmed by measurements of the rate of disappearances of ^{14}C from field soils to which ^{14}C-labelled residues have been added. For various residues (ryegrass shoots and roots, green maize, wheat straw), in a temperate climate, a surprisingly constant fraction (c. two-thirds) of the material decomposes and disappears in the first year (Fig. 3.15). In straw, the more resistant component is mainly lignin, but in the case of glucose the residue is mainly microbial metabolites produced during decomposition. After 5 years, approximately one-fifth of the added ^{14}C remains and its

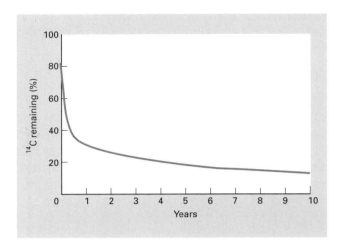

Fig. 3.15 Simplified time-course for the decomposition of ^{14}C-labelled C components in the field (after Jenkinson, 1981).

decomposition rate then still exceeds that of the unlabelled native soil C. Thus, although simple models such as Equation 3.2 can describe the C balance of soil–plant systems in steady state (attained only after many decades), more complex models are needed to fit data in the years following a change in the addition of residues to soil, or when changes in soil management are made.

Although a whole spectrum of k values must pertain in the field, changes in organic C in grassland and arable soils have been simulated quite successfully using models whose basic structure comprises three main modules:

- litter and other inputs of organic matter (manure, root exudates and dead roots),
- microbial biomass, and
- one or more components of SOM.

Within these modules, further subdivisions have been made based on the age and chemical composition of the residues, protection of biomass and its functional activity, and chemical or physical stabilization of SOM by soil minerals, giving rise to a wide range of k values. Decomposition rates are moderated by factors for temperature and moisture effects and sometimes for the effect of clay (see below). The predatory effect of soil fauna on bacteria and fungi is subsumed into the k

value for biomass decay. Most models postulate a component of very old or stable SOM which can be as much as half the total SOM, and which is excluded from the dynamics of the model. There may be justification for this in Australian soils where approximately one-quarter of the SOM has been identified as inert charcoal (presumed to be a relic from frequent bushfires before European settlement). But the inclusion of an 'inert SOM' term in C turnover models is indicative of the more general problem that few of the components in any such model can be experimentally verified. The components are convenient tools to explain the data used for model calibration.

For the reasons outlined in Section 3.1, there is much interest in SOM models in the context of modelling global climate change, or the sustainability of current land use systems (Chapter 15). Two of the more popular models are the CENTURY model (Parton *et al.*, 1988) and ROTH-26.3 (Coleman and Jenkinson, 1996). The two models are broadly similar in structure. ROTH-26.3 has evolved from the concepts set out by Jenkinson and Rayner (1977) and is briefly described in Box 3.5. It has been extensively tested on long-term field data, is modest in its input requirements and parsimonious in its parameters. Whereas the CENTURY model simulates C, N, P and S dynamics for several soil–plant systems (and includes a module for plant production and litter return), ROTH-26.3 deals only with soil processes affecting C turnover (although it has been linked to the dynamics of soil N turnover – see Box 11.3).

Substrate availability

Priming action

It was once thought that the stimulus to the resident soil microbial population provided by the addition of fresh residues was sufficient to accelerate the decomposition of soil humus. This effect was called *priming action*. Priming action, which can be positive or negative, has been demonstrated subsequently by following the decomposition of ^{14}C-labelled residues, as illustrated in

Box 3.5 ROTH-26.3, a model to simulate SOM turnover.

SOM is split into four active compartments and a small amount of inert organic matter (IOM), usually < 20% that does not turnover (age 50 000 years). The active compartments are decomposable (DPM) and resistant (RPM) plant material, microbial biomass (BIO), and humified organic matter (HUM). Each active compartment decomposes by first-order kinetics according to the rate constants

DPM $\quad k = 10\ \mathrm{y}^{-1}$

RPM $\quad k = 0.3\ \mathrm{y}^{-1}$

BIO $\quad k = 0.66\ \mathrm{y}^{-1}$

HUM $\quad k = 0.02\ \mathrm{y}^{-1}$.

These values were set by tuning the model to long-term field data at Rothamsted. Where possible the IOM fraction is adjusted so that the calculated average age of C in the soil matches the measured radiocarbon age. The compartments decay in a 'cascade' effect, as illustrated in Fig. B3.5.1 producing CO_2, BIO and HUM.

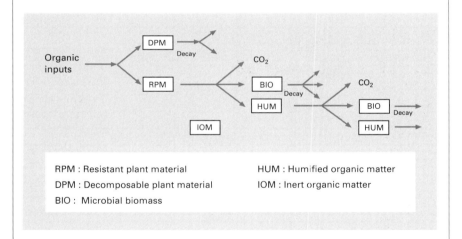

Fig. B3.5.1 Structure of the Rothamsted Carbon Model.

The ratio DPM/RPM in the input organic matter is adjusted for different vegetative cover and whether any pre-decomposition has occurred (as in manure or compost). The ratio of CO_2 to (BIO + HUM) decreases as the clay content of the soil increases to about 25%, reflecting the protection afforded organic matter when it is adsorbed on clays or occluded

Continued on p. 54

Box 3.5 *Continued.*

inside aggregates. The decay rates are modified for the effects of temperature, soil moisture deficit and 'plant retainment'; that is, the effect of plants in slowing the rate of organic matter decomposition relative to the rate in fallow soil.

The model has been tested on long-term field data in the UK, Africa and Australia and is continually undergoing revision as new information becomes available.

Fig. 3.16. The conclusion from this kind of experiment is that priming action generally has a negligible effect on the overall rate of decomposition. The addition of fresh residues does stimulate microbial activity, but the new biomass has little effect on the rate of humus decomposition. The Rothamsted model of SOM turnover ignores any priming effect.

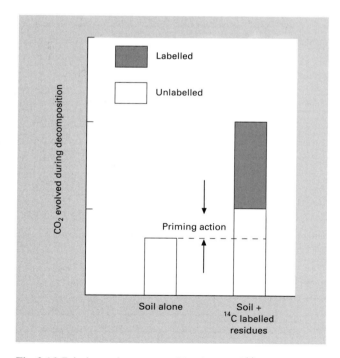

Fig. 3.16 Priming action measured by the use of ^{14}C-labelled residues (after Jenkinson, 1966).

Soil properties and environmental conditions

Soil organic matter levels are affected by moisture, O_2 supply, pH and temperature. The first two factors tend to counteract one another because when soil moisture is high, deficiency of O_2 may restrict decomposition, whereas when the soil is dry, moisture but not O_2 will be limiting. pH has little effect, except below 4 when the decomposition rate slows as in the case of mor humus and many upland peats (Section 9.5). On the other hand, temperature has a marked effect, not only on plant growth and hence on litter return, but also on litter decomposition through an effect on microbial respiration rate (Section 8.1). In a comparison of Nigerian (mean temperature 26.1°C) and English (mean temperature 8.9°C) soils, Jenkinson and Ayanaba (1977) found that the rate of disappearance of organic residues in both cases could be represented by the curve in Fig. 3.15 provided that 10 years in southern England were equated with 2.5 years in Nigeria.

The adsorption of protein by montmorillonite, especially in the interlamellar regions, protects the protein from microbial attack. Similarly the adsorption of various compounds by clays and sesquioxides generally serves to slow down their rate of decomposition. The organic matter held in the relatively stable pores in clay soils of diameter < 1 µm – the so-called 'sterile pores' – is also less accessible to microbial attack. Cultivation of the soil tends to break down the structure so that organic matter in sterile pores is exposed to microorganisms and its decomposition rate accelerated. Higher surface soil temperatures are also attained when the protection of the vegetative canopy and litter layer is lost.

The effect of clay content in slowing down the rate of SOM decomposition is moderated according to temperature. The part played by clay minerals and particularly sesquioxides in protecting soil organic matter under tropical conditions, where the potential for decomposition is high, is much greater than in temperate regions. Also in soils derived from volcanic ash, the considerable accumulation of organic matter is attributed to the stabilization of HA and FA fractions through their adsorption onto the Al sheets of imogolite and allophane clay minerals.

Activity of the soil biomass

Numbers of organisms

The effect of environmental and soil factors on the size and species diversity of the soil microbial population has been referred to briefly. Soil animals are also affected by types of litter, prevailing conditions of temperature, moisture and pH, and management practices, such as cultivation and the use of agricultural chemicals.

Physiological activity

Environmental and management factors can influence the physiological activity of the soil organisms. Striking effects occur, for example, when the soil is partially sterilized with toxic chemicals, such as chloroform or toluene. On the restoration of favourable conditions, the rapid multiplication of the few surviving organisms, feeding on the bodies of the killed organisms, produces a flush of decomposition as evidenced by a surge in the release of CO_2. Similar flushes of decomposition occur in soils subjected to extremes of wetting and drying (Section 10.2) and in soils of temperate regions on passing from a frozen to thawed state.

The dynamics of soil biomass changes in response to variations in soil and environmental factors are summarized graphically in Fig. 3.17.

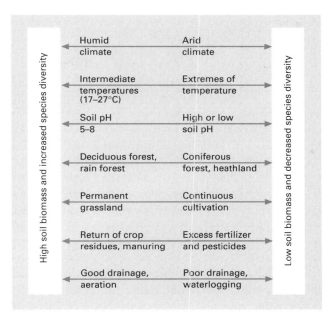

Fig. 3.17 Dynamics of soil biomass changes.

3.6 Summary

The input of C in the form of plant litter, dead roots, animal remains and excreta, all which are consumed by a heterogeneous population of soil organisms, is a very important part of the global C cycle. As the organic residues decompose, releasing a proportion of the C as CO_2 to the atmosphere, soil organic matter (SOM) accumulates and serves as a repository of C and other essential elements for successive generations of organisms. Release of these elements or *mineralization* depends on the decomposition rate and the species composition of the soil organisms.

The living organisms (*biomass*) vary in size from the *macrofauna*, vertebrate animals of the burrowing type, to the *mesofauna*, invertebrates such as mites, springtails, insects, earthworms and nematodes, to the *microorganisms* – broadly subdivided into the *prokaryotes* and *eukaryotes* and comprising the bacteria, Actinomycetes, fungi, algae and protozoa. The most important of the mesofauna are the *earthworms* in temperate soils, especially under grassland, and *termites* in many soils of the tropics and subtropics. Earthworm biomass normally ranges from 0.5 to 1.5 t ha^{-1} (0.04–0.12 t C ha^{-1}), whereas the microbial biomass makes up 2–4% of the total organic C in soil, or 0.5–2 t C ha^{-1}.

Annual rates of litter fall range from 0.1 t C ha^{-1} in alpine and Arctic forests to 10 t C ha^{-1} in tropical rain forests. Carbon substrates from rhizosphere deposition and the death of roots also make an important contribution that is difficult to quantify. The interaction of micro-organisms and mesofauna in decomposing litter leads not only to the release of mineral nutrients, but also to the synthesis of complex new organic compounds that are more resistant to attack. This is called *humification*. The chemical structure of humic compounds has proved difficult to elucidate, but analyses made of *fulvic acid* (FA) and *humic acid* (HA) fractions indicate that macromolecules consisting of a 'core' of substituted aromatic and aliphatic C groups, with a high density of —COOH and phenolic-OH groups, are the main constituents. Polysaccharides, probably of microbial origin, and proteinaceous material make-up about 20% of the humus polymer. Soil humus has a high cation exchange capacity (CEC) and is active in chelating metallic cations. Humic compounds are also important in the stabilization of soil structure (Chapter 4).

The organic matter content of a soil reflects the balance between gains of C from plant residues and excreta and the loss of C through decomposition by soil organisms. Experiments with ^{14}C-labelled residues show that under temperate conditions, about one-third of the C remains after 1 year and one-fifth after 5 years. In the humid tropics, decomposition is approximately four times faster than in temperate soils. Low pH retards decomposition, as does a high polyphenolic content in the senescing plant tissue, leading typically to the formation of *mor humus*. *Mull humus* forms on base-rich soils under herbaceous and deciduous forest species low in polyphenols. C : N ratios > 25 favour an increase in microbial biomass and *net immobilization* of N, whereas C : N ratios < 25 favour *net mineralization*.

References

Anderson J. P. E. & Domsch K. H. (1978) Mineralization of bacteria and fungi in chloroform-fumigated soil. *Soil Biology and Biochemistry* **10**, 207–223.

Andreux F. (1982) Genesis and properties of humic molecules, in *Constituents and Properties of Soils* (Eds M. Bonneau and B. Souchier). Academic Press, London, pp. 109–139.

Coleman K. & Jenkinson D. S. (1996) RothC-26.3 – A model for the turnover of carbon in soil, in *Evaluation of Soil Organic Matter Models* (Eds D. S. Powlson, P. Smith and J. U. Smith). NATO ASI Series 1: Global Environmental Change, Vol. 38. Springer, Berlin, pp. 237–246.

Edwards C. A. & Lofty J. R. (1977) *Biology of Earthworms*, 2nd ed. Chapman & Hall, London.

Edwards C. A. (1991) The assessment of populations of soil-inhabiting invertebrates. *Agriculture, Ecosystems and Environment* **34**, 145–176.

Hayes M. H. B. (1991) Concepts of the origins, composition and structures of humic substances, in *Advances in Soil Organic Matter Research. The Impact on Agriculture and the Environment* (Ed. W. S. Wilson). Special Publication No. 90. Royal Society of Chemistry, Cambridge, pp. 3–22.

Hepper C. M. (1975) Extracellular polysaccharides of soil bacteria, in *Soil Microbiology* (Ed. N. Walker). Butterworth, London.

Jenkinson D. S. (1966) The priming action, in *The Use of Isotopes in Soil Organic Matter Studies.* Proceedings of FAO/IAEA Technical Meeting, Brunswick, Volkenrode. Pergamon, Oxford, pp. 199–208.

Jenkinson D. S. (1981) The fate of plant and animal residues in soil, in *The Chemistry of Soil Processes* (Eds D. J. Greenland and M. H. B. Hayes). Wiley, Chichester, pp. 505–561.

Jenkinson D. S. & Ayanaba A. (1977) Decomposition of carbon-14 labelled plant material under tropical conditions. *Soil Science Society of America Journal* **41**, 912–915.

Jenkinson D. S. & Powlson D. S. (1976) The effects of biocidal treatments on metabolism in soil. 1. Fumigation with chloroform. *Soil Biology and Biochemistry* **8**, 167–177.

Jenkinson D. S. & Rayner J. H. (1977) The turnover of soil organic matter in some of the Rothamsted classical experiments. *Soil Science* **123**, 298–305.

Oades J. M. & Jenkinson D. S. (1979) Adenosine triphosphate content of the soil microbial biomass. *Soil Biology and Biochemistry* **11**, 201–204.

Parkinson D. & Coleman D. C. (1991) Microbial communities, activity and biomass. *Agriculture, Ecosystems and Environment* **34**, 3–33.

Parton W. J., Stewart J. W. B. & Cole C. V. (1988) Dynamics of C, N, P, and S in grassland soils: a model. *Biogeochemistry* **5**, 109–131.

Russell E. W. (1973) *Soil Conditions and Plant Growth*, 10th edn. Longman, London.

Schnitzer M. & Schulten H.-R. (1995) Analysis of organic matter in soil extracts and whole soils by pyrolysis-mass spectroscopy. *Advances in Agronomy* **55**, 168–217.

Waterhouse D. F. (1973) Pest management in Australia. *Nature New Biology* **246**, 269–271.

Further reading

Duxbury J. M. & Mosier A. R. (1993) Status and issues concerning agricultural emissions of greenhouse gases, in *Agricultural Dimensions of Global Climate Change* (Eds H. M. Kaiser & T. E. Drennen). St Lucie Press, Delray Beach, pp. 229–258.

Edwards C. A. & Bohlen P. J. (1996) *Biology and Ecology of Earthworms*, 3rd edn. Chapman & Hall, London.

Killham K. (1994) *Soil Ecology*. Cambridge University Press, Cambridge.

Stevenson F. J. (1994) *Humus Chemistry – Genesis, Composition, Reactions*, 3rd edn. Wiley, New York.

Wood M. (1989) *Soil Biology*. Blackie, Glasgow.

Chapter 4
Peds and Pores

4.1 Soil structure

As we have seen in Chapter 3, one reason why the soil supports a diversity of plant and animal life is the abundance of energy and nutrients supplied in organic remains. An equally important reason lies in the physical protection and favourable temperature, moisture and oxygen supply afforded by the structural organization of the soil. Weathering of parent material produces the primary soil particles of various sizes – the clay, silt, sand and stones. Without the intervention of external forces, these particles will simply 'pack' randomly, as for example in Fig. 4.1a, to attain a state of minimum potential energy. However, vital forces associated with plants, animals and micro-organisms, and physical forces associated with the change in state of water and its movement, act to arrange the soil particles into larger units of varied size called *aggregates* or *peds* (Fig. 4.1b). Work is done on the particles to create order out of disorder and increase the free energy of the system. Inherently, the structured soil is less stable than if the particles had packed randomly with no obvious structure, but the existence of structure confers many advantages on organisms living in the soil.

Aggregation of soil particles is the basis of *soil structure* for which several definitions have been offered: in the broadest sense, it includes:

• the size, shape and arrangement of particles and aggregates,
• the size, shape and arrangement of voids or spaces separating the particles and aggregates, and
• the combination of voids and aggregates into various types of structure.

Soil structure should be described only from a freshly exposed soil profile or a large undisturbed core. Samples from augers (Chapter 14) are unsuitable because of the disturbance, as are those from old road cuttings and gullies which have been exposed for some time.

4.2 Levels of structural organization

Many soils have structural features that are readily observed in the field – these features relate to the soil's *macrostructure*. One of the earliest systems for describing and classifying soil macrostructure was that of the USDA Soil Survey Manual (Soil Survey Staff, 1951). This has been the basis for similar systems developed in other countries, such as the one used by the Soil Survey of England and Wales (Hodgson, 1974), and that adopted in Australia (McDonald *et al.*, 1990). However, much of the fine detail of aggregation and the distribution of voids – the *microstructure* – can only be determined in the laboratory with the aid of some form of magnification, ranging from a hand lens ($\times 10$, also used in the field), to a light microscope ($\times 50$), to an electron microscope ($\times 10\,000$). Some of the macroscopic features have identifiable counterparts at the microscopic level, in which case a descriptive term common to both can be used. Other features are peculiar to the microstructure and require new descriptive terms and sophisticated analytical techniques, which have been devised by specialists in the study of *soil micromorphology* (Section 4.3). The new technique of *fractal geometry* is also being applied to the description of soil structure (Box 4.1).

Aggregation

Relatively permanent aggregates are recognizable in the field because they are separated by voids and natural planes of weakness. They should persist through cycles of wetting and drying, as distinct from the less permanent aggregates formed at or near the soil surface by the mechanical disturbance of digging and ploughing (*clods*), or the rupture of the soil mass across natural planes of weakness (*fragments*). A *concretion* is formed when localized accumulations of an insoluble compound irreversibly cement or enclose soil particles. *Nodules* are similarly formed but lack the symmetry and concentric internal structure of concretions.

Ped type

Four main *types* of ped have been recognized, as illustrated in Fig. 4.2. A brief description of each type follows.

Spheroidal. Peds that are roughly equidimensional and

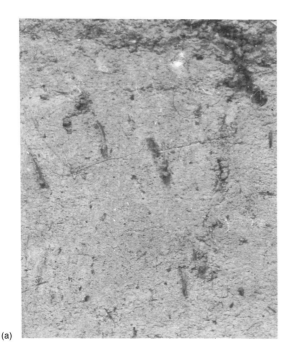

(a)

(b)

Fig. 4.1 (a) Random close packing of soil particles. (b) Structured arrangement of soil particles.

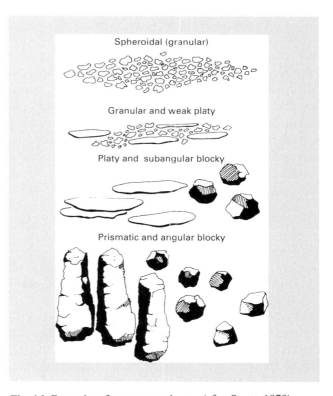

Fig. 4.2 Examples of common ped types (after Strutt, 1970).

The description of soil structure has traditionally been qualitative, whether it be at the microscopic or macroscopic level. This is because the shape, size and continuity of pores, and the enveloping peds, are so irregular. Attempts are being made, however, to use non-Euclidean geometry – the *fractal geometry* of Mandelbrot (1983) – not only to describe soil structure quantitatively, but also to model the physical, chemical and biological processes that operate over a range of scales to create soil structure (Section 4.4). Fractals are powerful tools because they have no characteristic scale: the morphological detail of a fractal object is the same at all scales (the property of self-similarity). Fractal characterization of soil structure requires the fractal dimensions for:

- irregular boundaries and surfaces,
- non-uniform mass distribution,
- fragmentation, and
- the dynamic properties of the fractal network (e.g. pore continuity and tortuosity)

to be measured. The can be done by computer-aided analysis of binary images of soil thin sections, or surfaces of undisturbed monoliths (Fig. 4.3). Some examples of this approach are given in Anderson *et al.* (1996).

Box 4.1 Towards a quantitative description of soil structure.

bounded by curved or irregular planar surfaces not accommodated by the faces of adjacent peds. There are two subtypes – *granules* which are relatively non-porous, and *crumbs* which are very porous. These peds are usually evenly coloured throughout, and arise primarily from the interaction of soil organisms, roots and mineral particles. They are therefore common in the A horizons of soils under grassland and deciduous forest.

Blocky. Peds bounded by curved or planar surfaces that are more or less mirror images of the faces of surrounding peds. Those with flat faces and sharply angular vertices are called *angular*; those with a mixture of rounded and flat faces and more subdued vertices are called *subangular*. Blocky peds are features of B horizons where their natural surfaces may be identified by their smoothness and often distinctive colours. The ped faces may be coated, for example, by organic matter, clay or sesquioxides (Section 4.3). The voids between peds often form paths for root penetration and water

flow. The ped faces may have live or dead roots attached, or show distinct root impressions.

Platy. The particles are arranged about a horizontal plane with limited vertical development; most of the ped faces are horizontal. When the ped is thicker in the middle than at the edges it is called *lenticular*. These peds occur in the A horizon of a soil immediately above an impermeable B horizon, or in the surface of a soil compacted by heavy machinery.

Prismatic. In these peds the horizontal development is limited when compared with the vertical; the peds are bounded by flat, vertical faces with sharp vertices. They are subdivided into prisms without caps (*prismatic*) and prisms with rounded caps (*columnar*). Prismatic structure is common in the subsoil of heavy clay soils subject to frequent waterlogging (Fig. 4.4). Columnar structure is associated particularly with the B horizon of sodium-affected soils (Fig. 4.5).

(a)

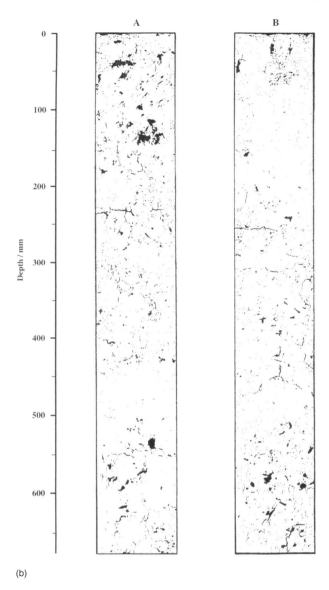

(b)

Fig. 4.3 (a) Thin section through a soil pore showing clay coatings (courtesy of P. Bullock). (b) Black and white images of soil monoliths of A, undisturbed soil under Eucalyptus forest, and B, same soil type but cultivated for 90 years. Black represents pore space; white represents soil solids (from McBratney *et al.*, 1992).

Compound peds

All types of ped shown in Fig. 4.2 are invariably compounded of smaller units which are clearly visible, and others which can only be seen under a microscope. For example, the large compound prismatic peds in Fig. 4.6 are composed of secondary subangular blocks which in

Fig. 4.4 Prismatic structure of a clay subsoil prone to water-logging.

from < 2 mm to > 500 mm, the average smallest dimension (vertical for platy peds, horizontal for prismatic and the diameter for spheroidal) being used to determine the class. The classes are described as 'fine', 'medium', 'coarse' and 'very coarse'. The reader is referred to the specialist soil survey handbooks for details.

Grade of structure may be described as follows.
• *Structureless* (or *apedal*). No observable aggregation nor definite arrangement of natural planes of weakness. *Massive* if coherent; *single grain* if non-coherent.
• *Weakly developed or weak.* Poorly formed, indistinct peds that are barely observable *in situ*. When disturbed, the soil breaks into a mixture of a few entire peds (< one-third), many broken peds and much unaggregated material.
• *Moderately developed or moderate.* Well-formed, moderately durable, distinct peds, but not sharply distinct in undisturbed soil. The soil, when disturbed, breaks down into a mixture of many entire peds (one-third to two-thirds), some broken peds and a little aggregated material.
• *Strongly developed or strong.* Durable, quite distinct peds in undisturbed soil which adhere weakly to one another or become separated when the soil is disturbed. The soil consists largely of entire peds (> two-thirds), and includes a few broken peds and little or no unaggregated material.

Grades of structure reflect differences in the strength of *intraped* cohesion relative to *interped* adhesion. Strong intraped cohesion is the basis of strong structure. The grade will depend on the moisture content at the time of sampling and the length of time the soil has been exposed to drying in the air, which is why it is recommended that structure be described only on freshly exposed soil (Section 4.1). Prolonged drying will markedly increase the structure grade, particularly of clay soils and soils high in iron oxides. Knowledge of how the strength of a soil's cohesion and adhesion changes with water content is important for many practical purposes, for example the use of soil as a foundation for roads and dam walls, its response to cultivation, and its suitability for under-drainage. These variations are described by the term *consistence* (Box 4.2).

turn break down to the primary structure of angular blocky peds < 10 mm in diameter. Each of these primary peds presents an intricate internal arrangement of granules and voids which can only be elucidated by careful appraisal of the soil in thin sections (Section 4.3).

Class and grade

In addition to type, the *class* of ped is judged according to its size, and the *grade* according to the distinctness and durability of the visible peds. Size classes range

Fig. 4.5 Columnar structure in the B horizon of a *solodized solonetz* (CSIRO photograph, courtesy of G.D. Hubble).

Fig. 4.6 Compound prismatic peds showing the breakdown into secondary structure.

4.3 Soil micromorphology

Micromorphology is concerned with the description, measurement and interpretation of soil components and pedological features at a microscopic to sub-macroscopic scale. The study of micromorphology began with Kubiena (1938) who was primarily interested in using a knowledge of microstructure to interpret soil genesis (Chapter 9). The advantage offered is that some pedogenic processes may not be sufficiently advanced to be seen macroscopically.

Incipient translocation down a soil profile of clay, organic matter or Fe is an example of a process which, although not obvious in the field, can be identified by the presence of microscopic-scale coatings on ped surfaces and pore walls. From this early emphasis on soil genesis, the scope of micromorphology has widened to cover many soil physical, chemical and biological properties, especially those involved in the quantitative de-

Box 4.2 Soil consistence.

Soil consistence depends on the grade of structure and the water content. It is measured by the resistance offered to breaking or deformation when a compressive, shear force is applied. In the field, consistence is assessed on samples 20–30 mm in diameter which are subjected to a compressive shearing force between thumb and forefinger. The sample may be an individual ped, compound ped or fragment, and the water content must be specified. Usually the test is done on both air-dry and moist soil (at the 'field capacity' as defined in Section 4.5). The resistance of the soil to rupture or deformation is expressed on a scale ranging from *loose* (no force required, separate particles present) to *rigid* (cannot be crushed under foot by the slow application of a person's bodyweight). Full details are given in the soil survey field handbooks cited in Section 4.1. A more quantitative relationship between consistence and soil water content, and its relevance for mole drainage, is given in Section 13.4.

scription of soil structure *in situ* (both solid and void phases). Some of the techniques of micromorphology are described in Box 4.3.

Micromorphologists have not yet agreed an unambiguous set of terms to describe soil features seen in thin sections. A 'foundation' set of terms was introduced by Brewer in 1964 and revised in 1976. Fitzpatrick (1984) provided a valuable glossary of terms and their meanings, and a group commissioned by the International Society of Soil Science prepared a *Handbook for Soil Thin Section Description* (Bullock *et al.*, 1985). Nevertheless, most of those working in the subject continue to use Brewer's terminology, a summary of which is presented in Table 4.1.

Reorientation of plasma *in situ*

During the repeated cycles of wetting and drying, and sometimes freezing and thawing, that accompany soil formation, the plate-like clay particles gradually become reorganized from a randomly arranged network of small bundles, the *asepic* S-matrix, into larger bundles that show a degree of preferred orientation, the *sepic* S-matrix. The bundles or plasma separations are called *domains*, which can be up to 5 µm long and 1–5 µm wide, depending on the dominant clay mineral present and their mode of formation.

As well as the reorganization of clay particles within peds, there is often a change in the orientation of clay particles at ped faces. In dry seasons, material from upper horizons sometimes falls into deep cracks in clay soils. On rewetting, the subsoil, which now contains more material than before, swells to a larger volume and exerts a pressure that displaces material upwards. Ped faces slide against each other to produce zones of prefered orientation called *slickensides* lying at an angle of about 45° to the vertical (Fig. 4.7).

Plasma translocation and concentration

One of the major processes of soil formation is the translocation of colloidal and soluble material from upper horizons into the subsoil. In the changed physical and chemical environment of the lower horizons, salts and oxides precipitate, clay particles flocculate and ped faces become coated with deposited materials. Brewer classified these *coatings* or *cutans* according to the nature of the surface to which they adhere and their mineralogical composition. Some of the more common examples are as follows.

• *Clay coatings* (argillans). Often different in colour from and with a higher reflectance than the S-matrix of the ped; they are easily recognized in sandy and loam soils, but difficult to distinguish from slickenside sur-

Box 4.3 Techniques and terminology in micromorphology.

Micromorphometric analysis is done on surfaces ranging in size from thin sections of a few cm^2 to monoliths of a few dm^2. Briefly, the method of thin sectioning involves taking an undisturbed soil sample in a Kubiena box (usual dimensions $10 \times 5 \times 3.5$ cm), removing the water in a way that minimizes soil shrinkage, and impregnating the voids with a resin under vacuum. Once the resin has cured, the soil block is sawn and polished to produce a section 25–30 µm in thickness which is stuck on to a glass slide. The section is examined under a petrological microscope to determine the characteristic optical properties of the minerals present. An example of a thin section is shown in Fig. 4.3a.

Large soil blocks (monoliths) up to $70 \times 5 \times 10$ cm can be taken from the field, usually after preliminary resin impregnation, followed by further resin impregnation, cutting and polishing in the laboratory. Such monoliths are most useful for studying the size, shape, distribution and frequency of different pore types in the void space. The polished surface is photographed or examined by computerized image analysing equipment, capable of resolving structural differences to a few nanometres. Resolution of the pore space is aided by fluorescence under ultraviolet light from the resin or dyes incorporated in the resin. Figure 4.3b shows an example of a two-dimensional image of the volume and geometry of pores in two monoliths – one from an undisturbed soil under forest (A); the other from the same soil type subjected to continuous cropping for 90 years. The effect of cultivation is seen in the loss of large pores and aggregate breakdown to produce a massive structure in the top 200 mm of soil B, relative to soil A.

Undisturbed soil cores have been used for the study of soil pores $> c.$ 0.5 mm using X-ray computed tomography.

faces in clay soils (Fig. 4.3a).

• *Sand or silt coatings* (skeletons). Formed by the illuvation of sand or silt, or by the residual concentration of these particles after clay removal.

• *Sesquioxidic coatings* (sesquans, mangans, ferrans). Formed by reduction and solution of Fe and Mn under anaerobic conditions and their subsequent oxidation and deposition in aerobic zones (Fig. 4.8). Colours range from black (mangans) through grey, red, yellow, blue and blue-green (ferrans), depending on the degree of oxidation and hydration of the iron compounds.

• *Salt coatings* (of gypsum, calcite, sodium chloride) and organic coatings (organans).

• Mixtures of these materials, particularly clay, sesqui-

oxides and organic matter, to form *compound coatings*.

The coatings may build up to such an extent that bridges of material form between soil particles and the smallest peds so that a cemented layer may develop in the soil. Such a layer or horizon is called a *pan* and is described by its thickness (if < 10 mm it is called 'thin') and its main chemical constituent. An example of a thin iron pan is shown in Fig. 4.9.

Compacted horizons that are weakly cemented by silica are called *duripans*. A *fragipan* is not cemented at all, but the size distribution of the soil particles is such that they pack to a very high bulk density (Section 4.5).

Cylindrical and spherical *voids*, usually old root channels or earthworm burrows, are called *pores* and those of roughly planar shape are called *fissures* or *cracks*.

Those of indeterminate shape are called *vughs*. Fissures start as cracks striking vertically downwards from the surface, especially in soils containing much of the expanding lattice-type clays (Fig. 4.10). Secondary cracks develop at right angles to the vertical fissures.

4.4 The creation and stabilization of soil structure

Formation of aggregates

The organization of soil material to form peds and pedological features requires the action of physical, chemical and biotic forces.

- Swelling pressures are generated through the osmotic effects of exchangeable cations adsorbed on clay surfaces.
- There are forces associated with the change in state of the soil water. Unlike any other liquid, water expands on cooling from 4 to 0°C. As the water in the largest pores freezes first, liquid water is drawn from the smaller pores thereby subjecting them to shrinkage forces; but the build-up of ice lenses in the large pores, and the expansion of the water on freezing, subject these regions to intense disruptive forces.
- Tension forces develop as water evaporates from soil pores, and the prevalence of evaporation over precipitation can lead to localized accumulations of salts and the formation of concretions anywhere in the soil profile.
- The mechanical impact of rain, animal treading and earth-moving machinery can reorganize soil materials and the energy of running water can move them.
- Burrowing animals reorganize soil materials, particularly the earthworm which grinds and mixes organic matter with clay and silt particles in its gut, adding extra calcium in the process to form a *cast*, which, on drying, has considerable strength. The faeces of other invertebrates such as mites and collembola also form stable pellets which may exist as microaggregates (Fig. 3.10a), or act as nuclei for aggregate formation.
- The physical binding effect of fine roots and fungal mycelia also contributes to aggregation. The metabolic

Table 4.1 Summary of descriptive terms in soil micromorphology introduced by Brewer (1964, 1976).

FABRIC – the spatial arrangement of particles, primary and compound (aggregated), and voids

Soil material with basic units or organization – the PEDS

S-MATRIX – comprising the plasma, skeleton grains and voids of primary peds, or comprising apedal soil material that does not occur as pedological features other than plasma separations. Thus we have:

PLASMA – colloidal (< 2 μm) and soluble materials not bound up in skeleton grains, which can be moved, reorganized and/or concentrated during soil formation: includes mineral and organic material

SKELETON GRAINS – larger than colloidal size and include detrital mineral fragments, secondary crystalline and amorphous bodies which are relatively stable and not usually translocated, concentrated or reorganized during soil formation

VOIDS – the spaces within the plasma and between plasma and skeleton grains

Soil material without obvious organization – APEDAL

PEDOLOGICAL FEATURES – those formed due to reorganization of matrix materials by pedogenic processes (*ORTHIC*) and those inherited from the parent material (*INHERITED*). Orthic features are subdivided into:

PLASMA CONCENTRATIONS – the concentration of any component of the plasma which has moved in solution or in suspension: e.g. carbonate and oxide nodules, clay coatings

PLASMA SEPARATIONS – the changed arrangement of the plasma components, as, for example, the orientation of clay minerals at the surface of peds, such as slickensides (see Fig. 4.7)

FOSSIL FORMATIONS – preserved root channels or burrows of the soil fauna: may be filled with various materials

energy expended by soil micro-organisms can maintain concentration gradients for the diffusion of salts, or induce a change in valency leading to the solution and subsequent reprecipitation of manganese and ferric oxides.

Interparticle forces

For a ped to persist, *interparticle* forces within the ped must be sufficient to prevent the particles separating

Fig. 4.7 Intersecting slickensides in a heavy clay subsoil (CSIRO photograph, courtesy of G.D. Hubble).

under the continuing impact of forces responsible for reorganization. Good structure for plant growth depends on the existence of peds between 1 and 10 mm in diameter (the primary peds mentioned in Section 4.2) that are stable when wetted. The reaction of aggregates to immersion in water is the basis of several tests of aggregate stability (Box 4.4).

It is pertinent here to consider ways in which the primary constituents – sand, silt, clay and organic matter – are organized to form peds. Some authors distinguish between *macroaggregates* (> 250 μm diameter) and *microaggregates* (< 250 μm). This division is made by taking account of the soil's response to disruptive treatments (sonic vibration, rapid wetting, or periodate oxidation), and the inferred mechanism of aggregate stabilization.

Several models of microaggregate organization have been proposed, all of which involve some combination of clay–clay and clay–organic matter interactions.

Microaggregates

Domains consisting of illite or vermiculite clay crystals up to 5 nm thick, stacked in roughly parallel alignment, and extending up to 5 μm in the *a–b* direction (Box 2.2), are thought to be the fundamental structural units of many soils. Similarly, single-layer crystals of montmorillonite (1 nm thick) can align themselves, with much overlapping, to form *quasi-crystals* several nanometres thick and up to 5 μm in the *a–b* direction. When Ca^{2+} is the dominant exchangeable cation, the *intercrystalline* spacing in the domains and the interlayer (*intracrystalline*) spacing in the quasi-crystals do not exceed 0.9 nm, even when the soil is placed in

Fig. 4.8 Manganese dioxide coatings on ped faces (courtesy of R. Brewer).

Fig. 4.9 A thin iron pan developed in ferruginous sand deposits.

distilled water, because of the stability of the trimolecular layer hydrate of Ca^{2+} ions between opposing clay surfaces (Section 2.4). Each unit – quasi-crystal or domain – acts as a stable entity in water, showing limited swelling, and is only likely to be disrupted if more than c. 15% of the Ca moles of charge is replaced by Na.

The generally accepted model of a microaggregate constitutes domains or quasi-crystals combined with organic polymers, through edge or face attractions, and also bonded to the oxygen surfaces of quartz grains through organic matter (Fig. 4.11). In the larger units (100–200 µm), the organic matter often consists of recognizable plant parts – leaf and root fragments. The

encrustation of these plant fragments with clay domains and other particles offers protection against microbial attack. Nevertheless, in smaller units (20–100 µm), the plant material is much degraded and only fragments of cell walls and xylem lignin fibres remain. At < 20 µm, all the plant material has been degraded and microbial products – mainly polysaccharides and polyuronide gums and mucilages – predominate. Humic polymers that complex polyvalent cations through carboxyl groups are also likely to be involved.

In more highly weathered soils in which 1 : 1 clays such as kaolinite may predominate, individual clay crystals can flocculate edge to face to form a 'cardhouse'

Fig. 4.10 Close cracking pattern in the surface of a montmorillonitic clay soil (scale in photo in 15 cm long) (CSIRO photograph, courtesy of G.D. Hubble).

type of microstructural unit (Section 7.4). Positively charged sesquioxide films on the planar surfaces also act as electrostatic bridges between crystals, or form complexes with negatively charged organic polymers. A summary of these various interparticle forces of attraction is given in Table 4.2.

Macroaggregates

The formation of macroaggregates (> 250 μm) depends primarily on the stabilization of microaggregates in larger structural units. Flocculated clay is a prerequisite

of microaggregate (< 250 μm) stability and the ability to resist disruption on wetting and mechanical disturbance. Clay–organic matter interactions, which may involve polyvalent cations (Fe^{3+}, Al^{3+} or Ca^{2+}) and humic polymers, or polysaccharides and clay surfaces, are also important. Inorganic compounds act as interparticle cements and stabilization agents; for example, calcium carbonate in calcareous soils, sesquioxides as films between planar clay surfaces and as discrete charged particles in many acid, highly weathered soils, and silica at depth in laterites (Section 9.4), where an indurated layer may form. The nature of microaggregate stabilization is such that it is relatively insensitive to changes in soil management, except when management induces:

• marked changes in the dominant exchangeable cat-

Box 4.4 Tests of aggregate stability.

A long established test is that of measuring *water-stable aggregates* (Klute, 1986). This involves sieving under water an air-dry sample of soil on a vertical array of sieves of different mesh size. After a standard time, the mass of aggregates remaining on individual sieves is measured and expressed as a percentage of the total. Since one of the original aims of this test was to assess a soil's susceptibility to water erosion, several variants of the test involving simulated rain falling on soil in the field have been developed. Another test of soil structure and aggregate stability involves placing small air-dry peds (0.2–0.5 g) in water of very low salt concentration (akin to rain water) (Emerson, 1991). If the ped quickly breaks down into subunits, which may themselves be aggregates of smaller particles, it is said to *slake*. This indicates that the intraped forces are not strong enough to withstand the pressure of air entrapped as the ped wets quickly, and the pressure generated by swelling. The subunits produced by slaking may also be unstable due to internal repulsive forces, in which case individual clay-size particles may separate – a process of *deflocculation* (sometimes called dispersion). Several undesirable effects may follow deflocculation such as:
- translocation of clay to form a dense clay B horizon (Section 9.2),
- blocking of drainage pores (Table 4.3) and, once the soil has dried,
- formation of a surface crust which impedes infiltration and seedling emergence.

Flocculation, deflocculation and swelling are surface phenomena which are discussed in Section 7.4. Soil management to maintain water-stable aggregates is discussed in Section 11.4.

---

ions (Na^+ replacing Ca^{2+}), or
- changes in soil pH (acid → alkaline pH decreases the positive charges on kaolinite edge faces and sesquioxide surfaces), or
- physical disruption of aggregates to expose 'protected' organic matter to further decomposition by soil organisms.

On the other hand, *macroaggregation* is much more dependent on management because it is stabilized primarily through the action of plant roots and fungal hyphae, particularly the vesicular arbuscular mycorrhizas (Section 10.3). Roots can penetrate pores > 10 μm diameter and fungal hyphae pores > 1 μm, and, in so doing, they enmesh soil particles to form stable macro-

aggregates. A diagrammatic model of a macroaggregate is shown in Fig. 4.12. Grass swards are especially effective because of the high density of roots in the surface horizon. Further to the physical binding effect of roots, rhizosphere C deposition (Section 3.1) provides substrate for the saprophytic micro-organisms producing the polysaccharide gums that bind soil particles together. These effects all depend on maintaining a high level of biotic activity – plants, animals (mainly earthworms) and micro-organisms – in the soil. They are most important in soils of pH 5.5–7, which derive little benefit through aggregate stabilization by either sesquioxides or $CaCO_3$. Substantial inputs of organic matter and minimal soil disturbance therefore favour macroag-

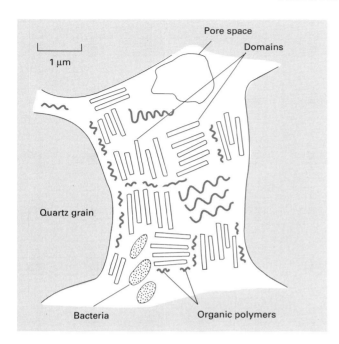

Fig. 4.11 Possible arrangements of clay domains, organic polymers and quartz grains in a microaggregate.

gregate stabilization. The stability of macro- and micro-aggregates in relation to the C content of soil, and the structural forms of that C, is summarized in Fig. 4.13.

4.5 Porosity

Pore volume

Clearly, the size, shape and arrangement of the peds determines the pore space or porosity of the soil. Pore volume is measured as the ratio

$$\text{pore space ratio } (PSR) = \frac{\text{volume of pores}}{\text{total soil volume}}. \qquad (4.1)$$

The *PSR* depends on the water content of the soil, since both the pore volume and total volume of an initially dry soil may change due to swelling as clay surfaces hydrate. The soil moisture content at which the porosity is measured therefore needs to be stated.

Table 4.2 Summary of interparticle forces of attraction involved in aggregate formation.

ELECTROSTATIC FORCES – these involve clay–clay, clay–oxide, and clay–organic interactions and include the following.
- Positively charged edge of a clay mineral attracted to the negative cleavage face of an adjacent clay mineral.
- Positively charged sesquioxide film 'sandwiched' between two negative clay mineral surfaces.
- Positively charged groups (e.g. amino groups) of an organic molecule attracted to a negative clay mineral surface. Organic molecules, in displacing the more highly hydrated inorganic cations from the clay, considerably modify its swelling properties.
- Clay–polyvalent cation–organic anion linkages or '*cation bridges*'. If the organic anion is very large (a polyanion), it may interact with several clay particles through cation bridges, positively charged edge faces or isolated sesquioxide films. Linear polyanions of this kind (e.g. Krilium) are excellent flocculants of clay minerals (Section 7.5).

VAN DER WAALS' FORCES – these involve clay–organic interactions such as the following.
- Specific dipole–dipole attraction between constituent groups of an uncharged organic molecule and a clay mineral surface, as in hydrogen bonding between a polyvinyl alcohol and O or OH surface.
- Non-specific forces between large molecules when in very close proximity, which are proportional to the number of atoms in each molecule. These forces can account for the strong adsorption of polysaccharide and polyuronide gums and mucilages on soil particles. These polymers of molecular weight 100 000–400 000 are produced in the soil by bacteria, such as *Pseudomonas* spp., which are abundant in the rhizosphere of many grasses. When the adsorption of such a large molecule results in the displacement of many water molecules, the increase in entropy of the system is large and the reaction is essentially irreversible. Large, flexible polymers of this kind can make contact with several clay crystals at many points and so help to hold them together in the manner of a 'coat of paint' (Section 7.5).

Total pore space indicates nothing about the actual *size distribution* of the pores. The distribution of pore sizes in non-swelling soils can be determined reasonably accurately from the volume of water released in response to known suctions, using the relationship between suction and the radius of pores that will just retain water at that suction (Section 6.1). In clay soils, however, where contraction of the soil volume as well

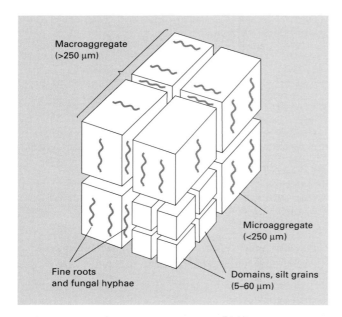

Fig. 4.12 Diagrammatic representation of a soil macroaggregate (after Greenland, 1979).

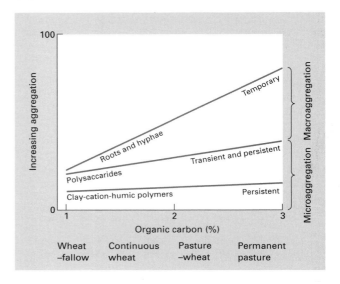

Fig. 4.13 The relative contributions of various forms of organic matter to the stabilization of aggregates in soil (after Tisdall and Oades, 1982).

as water release occurs as suction is applied, the use of the moisture-characteristic curve (Section 6.4) to determine pore-size distribution is unsatisfactory. Instead, water should be removed with the minimum of shrinkage, by critical-point drying for instance, and intrusion of mercury used to determine the pore-size distribution.

Some authors have subdivided the continuous pore-size distribution of a soil into classes based on presumed functions – aeration, drainage and water storage, as demonstrated in Table 4.3. The significance of these functions is explained below and in Chapter 6.

Bulk density

Porosity can be measured from the volume of a non-polar liquid, such as paraffin, that is absorbed into a dry ped under vacuum. Alternatively, porosity may be determined indirectly from the bulk density and particle density. Particle density (ρ_p) was introduced in Equation B2.1.1. *Bulk density* (*BD*) is defined as the mass of oven-dry (o.d.) soil per unit volume, and

depends on the densities of the constituent soil particles (clay, organic matter, etc.) and their packing arrangement. Thus, for a sample of o.d. soil, from Equation 4.1

$$PSR = \frac{\text{total soil volume} - \text{volume of solids}}{\text{total soil volume}}$$

$$= 1 - \frac{\text{mass of o.d. soil}}{\text{total soil volume}} \times \frac{\text{volume of solids}}{\text{mass of o.d. soil}}$$

$$= 1 - \frac{\text{bulk density}}{\text{particle density}}. \qquad (4.2)$$

Values of *BD* range from $< 1 \text{ g cm}^{-3}$ (or Mg m^{-3}) for soils high in organic matter, to 1.0–1.4 for well-aggregated loamy soils, to 1.2–1.8 for sands and compacted horizons in clay soils. Using Equation 4.2, it can easily be calculated that a sandy soil of bulk density 1.50 and particle density 2.65 has a *PSR* of 0.43 or 43%. Soil measurements made on a unit mass basis are translated to a unit volume basis by multiplying by the *BD* (Box 4.5). Where the *BD* is not known, it is common practice to assume an 'average' value of 1.33,

Table 4.3 Functional classifications of pores based on size. After Greenland, 1981 and De Leenheer, 1977.

Equivalent pore diameter (μm)	Descriptive function	Equivalent pore diameter (μm)	Descriptive function
> 500	Fissures	> 300	Aeration
500–50	Transmission pores	300–30	Normally draining pores
50–0.5	Storage pores	30–9	Slowly draining pores
		9–0.2	Useful water retention
0.5–0	Residual pores	0.2–0	Non-useful water retention

which corresponds to a soil mass of 2.0×10^6 kg ha^{-1} to a depth of 15 cm.

Water-filled porosity

The pores of a soil are partly or wholly occupied by water. Water is vital to life in the soil because of its existence as a liquid over the temperature range most suited to living organisms. It possesses several unique physical properties, namely the greatest specific heat capacity, latent heat of vaporization, surface tension and dielectric constant of any known liquid.

Soil *wetness* is characterized by the amount of water held in a certain mass or volume of soil; that is:

• *gravimetric water content* (θ_g), measured by drying the soil to a constant weight at a temperature of 105°C. θ_g is then calculated from

$$\theta_g = \frac{\text{mass of water}}{\text{mass of o.d. soil}}. \tag{4.3}$$

θ_g is often expressed as a percentage by multiplying by 100. Structural water, that is water of crystallization held within soil minerals, is not driven off at 105°C and is not measured as soil water.

• *volumetric water content* (θ_g), given by

$$\theta_v = \frac{\text{volume of water}}{\text{total soil volume}}. \tag{4.4}$$

The relationship between θ_g and θ_v is discussed in Box 4.5.

Box 4.5 The relationship between θ_g and θ_v.

Although θ_v is a dimensionless quantity, it is important to note it has the units of m^3 water m^{-3} soil. Thus, the concentration of a solute in the soil solution can be expressed on a soil volume basis by multiplying by θ_v.

θ_g and θ_v are related through the soil bulk density by the equation

$$\theta_v = \theta_g \times BD \tag{B4.5.1}$$

because 1 Mg water occupies 1 m^3 at normal temperatures. Because measurements of θ_v are more useful for most purposes than θ_g, soil water content will be expressed as a volumetric water content, using the symbol θ, throughout this book. For example, θ gives a direct measure of the 'equivalent depth' of water per unit area in the soil, which is useful for comparison with rainfall, evaporation or depth of irrigation water applied. An 'average' value for θ in a soil profile of 0.25 m^3 m^{-3} translates into 0.25 m^3 m^{-2} soil surface to 1 m depth, or 250 mm water per metre depth of soil.

Normally, water occupies less than the total pore volume and the soil is said to be *unsaturated* (although during a period of prolonged rain, the soil may temporarily become saturated to the surface; or permanently saturated by the rise of groundwater – Section 6.2).

Air-filled porosity

The pore space that does not contain water is filled with air. The air-filled porosity ε is defined by

$$\varepsilon = \frac{\text{volume of soil air}}{\text{total soil volume}} \qquad (4.5)$$

and it follows that

$$\varepsilon = PSR - \theta . \qquad (4.6)$$

Soils have been found to drain, after rain, to a water content in equilibrium with suctions between 5 and 10 kPa. This water content defines the *field capacity*, which is discussed in Chapter 6. For many well-structured British soils, the field capacity is attained at suctions close to 5 kPa when all pores > 60 µm diameter are drained of water. In Australia and the USA, the field capacity is set as the water content at a suction of 10 kPa, which corresponds to pores of equivalent diameter > 30 µm being drained. The pores drained at field capacity are called *macropores* and correspond to pores *between* soil peds, often created by faunal and root activity (the *biopores*). The value of ε for a soil in this state defines the *air capacity* C_a which, as Fig. 4.14 shows, may vary from 0.1 to 0.30 for topsoil, depending on the texture. A knowledge of the air capacity is important in:
- assessing soil aeration (Chapter 8),
- soil structural quality (Section 11.4), and
- designing soil underdrainage schemes (Section 13.4).

Composition of the soil air

The proportions of the major gases in the Earth's atmosphere are normally 79% N_2, 21% O_2 and 0.036% CO_2 by volume. According to Dalton's law of partial pressures, these concentrations correspond to partial pres-

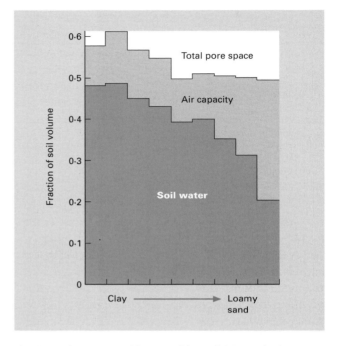

Fig. 4.14 Air and water-filled porosities at field capacity in top-soils of varying texture (after Hall *et al.* 1977).

sures of 0.79, 0.21 and 0.00036 of 1 atmosphere, which are equivalent to pressures of 0.802, 0.213 and 0.00037 bar, respectively.

The composition of the soil air may deviate from these proportions due to soil respiration (Section 8.1) which depletes the air of O_2 and raises the content of CO_2. The exchange of O_2 and CO_2 between the atmosphere and the soil air tends to establish an equilibrium that depends on the soil respiration rate and on the resistance to gas movement through the soil. Thus, for a soil in which the gases diffuse rapidly through the macropores, O_2 partial pressures are ~ 0.20 bar and CO_2 pressures ~ 0.002 bar. The normal gas compositions of a variety of soils put to different uses are given in Table 4.4.

However, if the macropores begin to fill with water and the soil becomes waterlogged, undesirable chemical and biological changes ensue as the soil becomes increasingly O_2 deficient or *anaerobic*: the O_2 partial pres-

Table 4.4. The composition of the gas phase of well-aerated soils. After Russell, 1973.

Soil and land use	Usual composition (% by volume)	
	O_2	CO_2
Arable land, fallow	20.7	0.1
Unmanured	20.4	0.2
Manured	20.3	0.4
Sandy soil, manured and cropped with potatoes	20.3	0.6
Serradella	20.7	0.2
Pasture land	18–20	0.5–1.5

sure may drop to zero and that of CO_2 rises far above 0.015 bar. In the case of soil placed as capping over refuse tips (landfill), the gas phase may consist almost entirely of CO_2 and methane (CH_4) in the approximate proportions of 40 : 60%. This is due to the proliferation of micro-organisms adapted to anaerobic conditions which produce, in addition to these gases and H_2, reduced forms of N, Mn, Fe and S. These reactions are discussed more fully in Section 8.3.

4.6 Summary

Physical, chemical and biotic forces act upon the primary clay, silt and sand particles, in intimate combination with organic materials, to form an organized arrangement of aggregates (*peds*) and intervening voids. The magnitude of the short-range interparticle forces within the peds must be sufficient to prevent the particles separating again under the influence of the same forces that are responsible for the reorganization. On a microscopic scale, individual crystals of montmorillonite (1 nm thick) can stack in roughly parallel alignment, with much overlapping, to form *quasi-crystals* which have long-range order in the *a–b* direction up to 5 μm. The somewhat larger crystals of illite and vermiculite can orient in a similar way to form larger units called *domains*. Both kinds of unit are stable in water when Ca^{2+} is the dominant exchangeable cation.

Domains, quasi-crystals, clay crystals, sesquioxides and organic polyanions are drawn together by the attraction of positively and negatively charged surfaces, reinforced by van der Waals forces and hydrogen-bonding to form *microaggregates* (< 250 μm diameter). The cohesion of microaggregates to form macroaggregates (> 250 μm) is enhanced by microbial polysaccharides and polyuronide gums which act like glue, and fungal hyphae and fine plant rootlets which enmesh the aggregates. Good structure for crop growth depends on the existence of water-stable aggregates between 1 and 10 mm in diameter.

The naturally formed visible peds, which comprise the soil's *macrostructure*, have spheroidal, blocky, platy or prismatic shapes. *Grade* of structure relates to the strength of intraped cohesion relative to interped adhesion, and is very dependent on water content. Variation in soil strength with respect to water content determines the soil's *consistence*. Soil also has a characteristic *microstructure* which is identifiable mainly under magnification (× 10 or more), and reflects quite strongly the predominant pedogenic processes of soil formation. Movement of materials in solution or suspension results in *plasma concentrations* (material < 2 μm) in the pores, on ped faces, and on sand particles (skeleton grains): these coatings are variously called clay coatings (*argillans*), organic coatings (*organans*), and iron oxide coatings (*ferrans*). Polished surfaces on ped faces formed under pressure are called *slickensides*.

The size, shape and arrangement of the peds determines the intervening volume of cylindrical and spherical *pores*, and planar *cracks* or *fissures*. The pore volume determines the size of the reservoir of water sustaining plant and animal life in the soil. It is usually expressed as a *pore space ratio* (*PSR*) and can be calculated approximately from the *bulk density* (dry mass per unit soil volume) and the mean particle specific gravity, that is

$$PSR = 1 - \frac{\text{bulk density}}{\text{particle density}}.$$

The *PSR* ranges from 0.5 to 0.6 irrespective of soil texture. Changes in soil wetness as measured by changes in the *volumetric water content* θ_v not only affect life processes, but also have a marked bearing on structure formation itself. Pores which drain at suctions up to 5–10 kPa are called *macropores* and exist mainly between peds: they include many of the pores created by burrowing soil fauna and plant roots (the biopores). The pore volume not occupied by water defines the *air-filled porosity* ε, for which in a well-aerated soil, the O_2 partial pressure is approximately the same as in the atmosphere, and that of CO_2 lies between 0.001 and 0.015 bar.

References

Anderson A. N., McBratney A. B. & Fitzpatrick E. A. (1996) Soil mass, surface and spectral fractal dimensions estimated from thin section photographs. *Soil Science Society of America Journal* **60**, 962–969.

Brewer R. (1964) *Fabric and Mineral Analysis of Soils.* Wiley, New York.

Brewer R. (1976) *Fabric and Mineral Analysis of Soils*, 2nd edn. Krieger, New York.

Bullock R., Fedoroff N., Jongerius A., Stoops G., Tursina T. & Babel U. (1985) *Handbook for Soil Thin Section Description.* Waine Research Publications, Wolverhampton.

DeLeenher L. (1977) Soil fertility study in maximizing crop yield on mechanized farms in Belgium: general and agricultural conclusions of a 15-year study (1961–1975). *Proceedings of the International Seminar on Soil Environment and Fertilizer Management in International Agriculture*, Tokyo, pp. 133–144.

Emerson W. W. (1991) Structural decline in soils, assessment and prevention. *Australian Journal of Soil Research* **29**, 905–921.

Fitzpatrick E. A. (1984) *Micromorphology of Soils.* Chapman & Hall, London.

Greenland D. J. (1981) Soil management and soil degradation. *Journal of Soil Science* **32**, 301–322.

Hall D. G. M., Reeve M. J., Thomasson A. J. & Wright V. F. (1977) (Eds) *Water Retention, Porosity and Density of Field Soils.* Soil Survey of England and Wales. Technical Monograph No. 9, Harpenden.

Hodgson J. M. (1974) (Ed.) *Soil Survey Field Handbook.* Soil Survey of England and Wales, Technical Monograph No. 5, Harpenden.

Klute A. (1986) (Ed.) *Methods of Soil Analysis. Part 1. Physical and Mineralogical Methods*, 2nd edn. Agronomy Monograph No. 9, American Society of Agronomy/Soil Science Society of America, Madison, Wisconsin.

Kubiena W. L. (1938) *Micropedology.* Collegiate Press, Ames.

McBratney A. B., Moran C. J., Stewart J. B., Cattle S. R. & Koppi A. J. (1992) Modifications to a method of rapid assessment of soil macropore structure by image analysis. *Geoderma* **53**, 255–274.

McDonald R. C., Isbell R. F., Speight J. G., Walker J. & Hopkins M. S. (1990) *Australian Soil and Land Survey Field Handbook*, 2nd edn. Inkata Press, Melbourne.

Mandelbrot B. B. (1983) *The Fractal Geometry of Nature.* Freeman, New York.

Russell E. W. (1973) *Soil Conditions and Plant Growth*, 10th edn. Longman, London.

Soil Survey Staff (1951) *Soil Survey Manual.* United States Department of Agriculture Handbook No. 18.

Strutt N. (1970) *Modern Farming and the Soil.* Report of the Agricultural Advisory Council on Soil Structure and Soil Fertility. HMSO, London.

Tisdall J. M. & Oades J. M. (1982) Organic matter and water-stable aggregates in soils. *Journal of Soil Science* **33**, 141–163.

Further reading

Fitzpatrick E. A. (1993) *Soil Microscopy and Micromorphology.* Wiley, Chichester.

Oades J. M. (1989) An introduction to organic matter in mineral soils, in *Minerals in Soil Environments* 2nd edn. (Eds J. B. Dixon & S. B. Weed). Soil Science Society of America, Madison, pp. 89–159.

Oades J. M. (1993) The role of biology in the formation, stabilization and degradation of soil structure. *Geoderma* **56**, 377–400.

Quirk J. P. (1978) Some physico-chemical aspects of soil structural stability – a review, in *Modification of Soil Structure* (Eds W. W. Emerson, R. D. Bond & A. R. Dexter). Wiley Interscience, Chichester.

Ringrose-Voase A. J. (1991) Micromorphology of soil structure: description, quantification, application. *Australian Journal of Soil Research* **29**, 777–813.

Part 2
Processes in the Soil Environment

'Do you think that to be an agronomist you must till the soil . . . with your own hands? No: you have to study the composition of various substances – geological strata, atmospheric phenomena, the properties of the various soils, minerals, types of water, the density of different bodies, their capillary attraction. And a hundred other things.'
From *Madame Bovary* by Gustave Flaubert (1856)

Chapter 5
Soil Formation

5.1 The soil-forming factors

The seminal book *Factors of Soil Formation* (Jenny, 1941) presented an hypothesis that drew together many of the current ideas on soil formation, the inspiration for which owed much to the earlier studies of Dokuchaev and the Russian school. The hypothesis was that soil is formed as a result of the interaction of many variables, the most important of which are:

- parent material,
- climate,
- organisms,
- relief,
- time.

Jenny called these variables *soil-forming factors* (Section 1.2) which controlled the direction and speed of soil formation. He attempted to define the relationship between any soil property *s* and the variables climate (*cl*), organisms (*o*), relief (*r*), parent material (*p*) and time (*t*), by a function of the form

$$s = f(cl, o, r, p, t \ldots). \tag{5.1}$$

The dots in the parentheses indicated that factors of lesser importance such as mineral accessions from the atmosphere, or fire, might need to be taken into account. An ensemble *S* of soil properties *s*, each being defined by a function such as Equation 5.1, makes up an individual soil type, of which there is an almost infinite number. Subsequently, Jenny (1980) redefined the soil-forming factors as 'state' variables and extended the terms on the LHS of Equation 5.1 to include total ecosystem properties, vegetation and animal properties, as well as soil properties.

To simplify the application of Equation 5.1, it has been the practice to express a soil property *s* as a function of only one controlling variable, under conditions where the others are constant or nearly so. If the controlling variable is climate, the relationship is then called a *climofunction*:

$$S = f(cl)_{o, r, p, t \ldots} \tag{5.2}$$

and the range of soils formed under different climates is called a *climosequence* (Fig. 5.1). Similarly, where other soil-forming factors are the predominant controlling variable, we have:

- parent material, a *lithosequence*,
- organisms, a *biosequence*,
- relief, a *toposequence*,
- time, a *chronosequence*.

The main virtue of Jenny's attempt to quantify the relationship between soil properties and soil-forming factors lies not in the prediction of the exact value of *s* or *S* at a particular site, but rather in identifying trends in properties and soil types that are associated with readily observable changes in climate, parent material, etc. The concept of a causal relationship between *S* and the factors *cl*, *o*, *r*, *p* and *t* has been implicit, and at times explicit (Section 5.3), in attempts at imposing order on the apparently chaotic variation of soils in the field; that is, *soil classification* (Chapter 14). Furthermore, factors such as climate and organisms determine the rate at which chemical and biological reactions occur in the soil (the *pedogenic processes*, Section 9.2); others such as parent material and relief define the initial state for soil development, and time measures the extent to which a reaction will have proceeded. In summary, there is a logical progression from

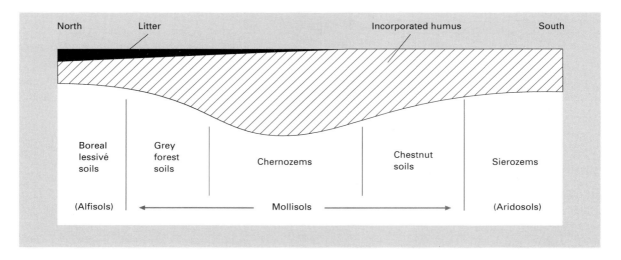

Fig. 5.1 A climosequence of soils found along a north-south transect through Russia showing soil depth and degree of humus incorporation (after Duchaufour, 1982).

the *environment* (an expression of soil-forming factors) → physical, chemical and biological *processes* → the development of characteristic *soil properties*.

This chapter deals briefly with the main factors affecting soil formation, and some of the outcomes expected for particular combinations of factors. The approach is essentially qualitative and intended to give the reader some insight as to why soils are so variable in the field. Subsequent to Jenny's pioneering work, attempts have been made to describe soil-forming processes more quantitatively through simulation modelling (e.g. Bryant and Arnold, 1994), but because of the complexity of the processes operating at a realistic scale in the field, progress has been limited.

Subsequent chapters (Chapters 6, 7 and 8) deal with important physical, chemical and biological processes (also covered in Chapters 2, 3 and 4), culminating in processes of profile development and nutrient cycling (Chapters 9 and 10). Soil properties and their management are discussed in Chapter 11 *et seq.*

5.2 Parent material

Soil may form by the weathering of:
• consolidated rock *in situ* (a residual soil), or
• unconsolidated superficial deposits which have been transported by ice, water, wind or gravity. These deposits originate from past denudation and geologic erosion of consolidated rock.

Weathering

Generally, weathering proceeds by physical disruption of the rock structure which exposes the constituent minerals to chemical alteration. Forces of expansion and contraction induced by diurnal temperature variations cause rock shattering and exfoliation, as illustrated by surface features in a desert soil (Fig. 5.2). This effect is most severe when water trapped in the rock repeatedly freezes and thaws, resulting in irresistible forces of expansion and contraction.

Water is the dominant agent in weathering, not only because it initiates solution and hydrolysis (Section 9.2), but also because it sustains plant life on the rock surface. Lichens play a special part in weathering, because they produce chelating agents which trap the elements of the decomposing rock in organo-metallic

Fig. 5.2 Stones of a desert soil surface fractured due to extreme variations in temperature (J.J. Harris Teall in Holmes, 1978).

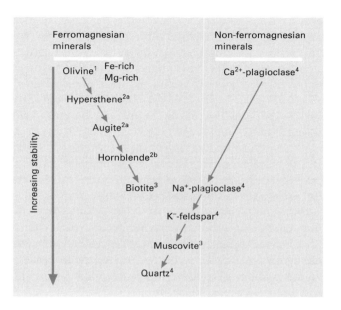

Fig. 5.3 Ranking of the common silicate minerals in order of increasing stability. Note that the superscripts indicate the number of oxygens shared between adjacent SiO_4^{4-} tetrahedra: 1 = none; 2a = two (single chain); 2b = 2 and 3 (double chain); 3 = three (sheet structure); 4 = four (three-dimensional structure) (after Goldich, 1938 and Birkeland, 1984).

complexes. In general, the growth, death and decay of plants and other organisms markedly enhances the solvent action of rainwater by the addition of carbon dioxide from respiration. Plant roots also contribute to the physical disintegration of rock.

Water carrying suspended rock fragments has a scouring action on surfaces. The suspended material can vary in size from the finest 'rock flour' formed by the grinding action of a glacier, to the gravel, pebbles and boulders moved along and constantly abraded by fast-flowing streams. Particles carried by wind also have a 'sand-blasting' effect, and the combination of wind and water to form waves produces powerful destructive forces along shorelines.

Mineral stability

The rate of weathering depends on:
• temperature,
• rate of water percolation,
• oxidation status of the weathering zone,
• surface area of rock exposed (largely determined by physical processes of fracturing and exfoliation),
• types of mineral present.

Much attention has been paid to the relationship between the crystalline structure of a rock mineral and its susceptibility to weathering. One scheme for the order of stability, or *weathering sequence*, of the more common silicate minerals in coarse rock fragments (> 2 mm diameter) is shown in Fig. 5.3. This sequence demonstrates that resistance of a mineral to weathering increases with the degree of sharing of oxygens between adjacent silicon tetrahedra in the crystal lattice (Section 2.3). The Si—O bond has the highest energy of formation followed by the Al—O bond, and the even weaker bonds formed between O and the basic metal cations* (Na^+, K^+, Ca^{2+}, Sr^{2+}, Ba^{2+}). Thus, olivine weathers rapidly because the silicon tetrahedra are only held together by O—metal cation bonds, whereas quartz is very resistant because it consists entirely of linked silicon tetrahedra. In the chain (amphiboles and pyroxenes) and sheet (phyllosilicate) structures, the weakest points are the O—metal cation bonds. Isomorphous substitution (Section 2.3) of Al^{3+} for Si^{4+} also contributes to instability because the proportion of Al—O to Si—O bonds increases and more O—metal cation bonds are necessary. This accounts for the decrease in stability of the calcium plagioclases when compared with the sodium plagioclases (Fig. 5.3).

Weathering continues in the finely comminuted materials (< 2 mm diameter) of soils and unconsolidated sedimentary deposits, with the least stable primary minerals appearing only in the early stages, and the more resistant minerals becoming dominant as weathering advances. The stages of weathering can be identified by certain type minerals, as shown in Table 5.1, which also includes some of the broad characteristics of the associated soils.

Rock types

Consolidated rocks are of igneous, sedimentary or metamorphic origin.

Igneous rocks

These are formed by the solidification of molten magma in or on the Earth's crust, and are the ultimate source of all other rocks. Material can be released explosively from the vents of volcanoes (volcanic ash and pyroclastic flows, collectively called *tephra*), or flow

---

* Cations of elements that form alkaline solutions such as NaOH and KOH when they react with water.

Table 5.1 Stages in the weathering of minerals (< 2 mm) in soils. After Jackson *et al.*, 1948.

Stage	Type mineral	Soil characteristics
Early weathering stages		
1	Gypsum	These minerals occur in the sand, silt and clay fractions of young soils all over the world, and in soils of arid regions where lack of water inhibits chemical weathering and leaching
2	Calcite	
3	Hornblende	
4	Biotite	
5	Albite	
Intermediate weathering stages		
6	Quartz	Found mainly in soils in temperate regions, frequently on parent materials of glacial or periglacial origin; generally fertile, with grass or forest as the natural vegetation
7	Muscovite (also illite)	
8	Vermiculite and mixed layer minerals	
9	Smectite	
Advanced weathering stages		
10	Kaolinite	Found in the clay fractions of many highly weathered soils on old land surfaces of humid and hot intertropical regions; often of low fertility
11	Gibbsite	
12	Hematite (also goethite)	
13	Anatase	

over the land surface as streams of lava that may cover many hundreds of square kilometres (extrusive formations). Other volcanic material wells up through lines of weakness in the Earth's crust to form *sills* and *dykes* (intrusive formations). Subsequent weathering of the surrounding rock may leave the hard, intrusive igneous rock remaining as a dominant feature in the landscape, as illustrated by the massive granite dome in Fig. 5.4.

Igneous rocks are broadly subdivided on mineral composition into:
• *acidic* rocks which are relatively rich in quartz and the light-coloured Ca or K/Na feldspars – mainly *intrusive*, and
• *basic* rocks which are low in quartz but high in the dark-coloured ferromagnesian minerals (hornblende, micas, pyroxene) – mainly *extrusive*.

The chemical composition of typical granite and basalt,

Fig. 5.4 Granite dome in Zimbabwe.

rocks representative of these two groupings, is given in Table 5.2. The higher the ratio of Si to (Ca + Mg + K), the more probable it is that some of the Si is not bound in the silicates, and may remain as residual quartz as the rock weathers. Soils formed from granite are therefore usually higher in quartz grains (sand- and silt-size) than those formed from basalt.

Table 5.2 Chemical composition (%) of typical granitic and basaltic rock. After Mohr and van Baren, 1954.

Element	Granite	Basalt
O	48.7	46.3
Si	33.1	24.7
Al	7.6	10.0
Fe	2.4	6.5
Mn	—	0.3
Mg	0.6	2.2
Ca	1.0	6.1
Na	2.1	3.0
K	4.4	0.8
H	0.1	0.1
Ti	0.2	0.6
P	—	0.1
Total	100.2	100.7

Sedimentary rocks

These are composed of the weathering products of igneous and metamorphic rocks and are formed after deposition by wind and water. Cycles of geologic uplift, weathering, erosion and subsequent deposition of eroded materials in rivers, lakes and seas have produced thick sequences of sediments. Under the weight of overlying sediments, deposits gradually consolidate and harden – the process of *diagenesis* – to form sedimentary rocks. Faulting, folding and tilting movements, followed by erosion, can cause the sequence of rock types in the geological column to be revealed by their surface outcrops (Fig. 5.5). *Clastic* sedimentary rocks are composed of fragments of the more resistant minerals. The size of fragments decreases in order from the conglomerates to sandstones to siltstones to mudstones. Other sedimentary rocks are formed by precipitation from solution. The most common are limestones and chalk (Fig. 5.6), which vary in composition from pure calcium carbonate to mixtures of calcium and magnesium carbonate (dolomite), or carbonates with detrital sand, silt and clay minerals.

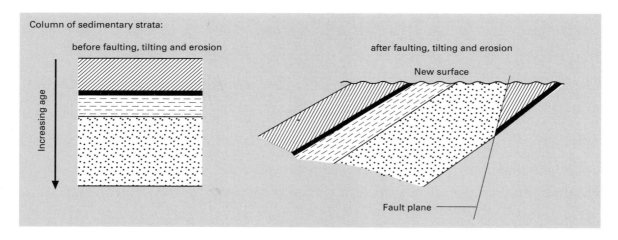

Column of sedimentary strata:

before faulting, tilting and erosion

after faulting, tilting and erosion

New surface

Increasing age

Fault plane

Fig. 5.5 Diagram to show how sedimentary rocks of different age and type can provide the parent materials for present-day soils.

Fig. 5.6 Electron micrograph of a sample of Middle Chalk (Berkshire, England) showing microstructure of calcified algae (courtesty of S.S. Foster).

Metamorphic rocks

Igneous and sedimentary rocks that are subjected to intense heat and great pressures are transformed into *metamorphic* rocks. Changes in mineralogy and rock structure generally make the metamorphosed rock more resistant to weathering.

Soils on transported material

Soil layers and buried profiles

Most parts of the Earth's surface have undergone several cycles of submergence, uplift, erosion and denudation over the hundreds of millions of years of geological time. During the mobile and depositional phases there is much opportunity for the mixing of materials from different rock formations. Thus, the interpretation of soil genesis on such heterogeneous parent material is often complex. In regions of active seismicity, such as along the crustal plate boundaries of the Pacific Ocean rim, volcanic eruptions and lava flows have, over millions of years, buried existing land surfaces and provided fresh parent material for soil formation (e.g. the North Island of New Zealand).

When two or more layers of transported material are superimposed they form a *lithological discontinuity*, and the resultant profile is often difficult to distinguish from the pedogenically differentiated profile of a soil formed on uniform parent material. In other cases, a soil may be buried by transported material during a period of prolonged erosion of an ancient landscape. The old soil is then preserved as a *fossil* soil, as illustrated by the buried 'mottled zone' of an old laterite (Oxisol (ST) or Ferrosol (A)) in Fig. 5.7.

Water

Transport by water on land produces alluvial, terrace and footslope deposits. Lacustrine deposits are laid down in lakes and marine deposits under the sea. During transport the rock material is sorted according to size and density and abraded, so that fluvial deposits characteristically have smooth, round pebbles. The larger fragments are moved by rolling or are lifted by the turbulence of water flowing over the irregular stream bed, and carried forward some distance by the force of the water. This 'bouncing' process is called *saltation* (*cf.* saltation of windblown particles, Section 11.5). The smallest particles are carried in suspension and salts move in solution. Colloidal material may not be deposited until the stream discharges into the sea, when flocculation occurs because of the high salt concentration. The modes of sediment transport in relation to the depth and velocity of the water are shown in Fig. 5.8.

Ice

Ice was an important agent in the transport of rock materials during the 2 million years of the Pleistocene era.

During the colder *glacial* phases, the ice cap advanced from polar and mountainous regions to cover a large part of the land surface, especially in the northern hemisphere (the ice reached a line only a few kilometres

Fig. 5.7 'Mottled zone' of a fossil laterite preserved under more recent transported material (alluvium) in southern New South Wales.

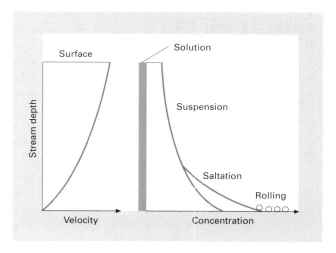

Fig. 5.8 Sediment transport in relation to depth and velocity of water (after Mitchell, 1993).

north of the present River Thames in southern Britain). In the southern hemisphere, where the Pleistocene ice sheets were much less extensive than in the northern hemisphere because of the greater area of sea relative to land, the influence of recent glaciations on soil formation is less pronounced. However, rock residues from much earlier glaciations do provide parent material for present-day soils in the southern hemisphere.

The moving Pleistocene ice ground down rock surfaces and the 'rock flour' became incorporated in the ice; rock debris also collected on the surface of valley glaciers by colluviation (see below). During the warmer *interglacial* phases, the ice retreated leaving extensive deposits of glacial drift or *till*, which is very heterogeneous in nature. Streams flowing out of the glaciers produced glacifluvial deposits of sands and gravels (Fig. 5.9). Repeated freezing and thawing in periglacial regions resulted in frost-heaving of the surface and much swelling and compaction in subsurface layers (forming fragipans, Section 4.3). Such deposits may often be distinguished from waterborne materials by the vertical orientation of many of the pebbles.

Wind

Wind moves rock fragments by the processes of rolling, saltation and aerial suspension (Section 11.5). Material swept from dry periglacial regions has formed aeolian deposits called *loess* in central USA, central Europe, northern China and South America. These deposits are many metres thick in places and form the parent material of highly productive soils, such as in the central Mississippi Basin of the USA (Fig. 5.10). Other characteristic wind deposits are the sand dunes of old sea and lake shores and desert areas.

Gravity

Gravity produces colluvial deposits, of which the scree formed by rock slides in mountainous regions is an obvious example. Less obvious, but widespread under periglacial conditions, are *solifluction* deposits at the foot of hillslopes. These are formed when a frozen soil surface thaws and the saturated soil mass slips over the frozen ground below, even on gentle hillslopes. *Creep* occurs when the surface mantle of soil moves by slow colluviation away from the original rock. The junction between creep layer and weathering rock is frequently delineated by a stone line. *Slips* and larger *earth flows* can occur on steep slopes on unstable parent materials when forest vegetation is cleared and replaced by short-rooted crop or pasture species, such as on the east coast of the North Island of New Zealand.

Fig. 5.9 Soil formation on glacifluvial sands and gravels of the Lleyn Pennisular, Wales.

Fig. 5.10 Deep loess near Mississippi, West Kentucky.

5.3 Climate

The importance of climate in soil formation is demonstrated by its explicit and implicit inclusion in criteria for the separation of soil classes at a high level in classifications such as *Soil Taxonomy* (Soil Survey Staff, 1975). The key components of climate are moisture and temperature.

Moisture

The effectiveness of *moisture* depends on:
- the form and intensity of the precipitation,
- its seasonal variability,
- the evaporation rate (from vegetation and soil, Section 6.5),
- land slope,
- permeability of the parent material.

Various functions of meteorological variables have been proposed to measure moisture effectiveness. Most focus on the balance between precipitation and evaporation, as shown in one of the earliest and simplest indices due to Thornthwaite (1931) (Box 5.1). For the British Isles, moisture effectiveness has been defined in terms of the maximum potential soil moisture deficit, abbreviated to *MD*. First, the potential water deficit (*PWD*) for each month in a period of years is obtained as the difference between mean monthly potential evaporation (Section 6.5) and rainfall (both in millimetres). These differences are summed to give the maximum *PWD* for each year and the yearly results are averaged to give the *MD* for that site.

Where moisture effectiveness is high, as in wet or humid climates, there is a net downward movement of water in the soil for most of the year, which usually results in greater *leaching* of soluble materials, sometimes out of the soil entirely, and the *translocation* of clay particles from upper to lower horizons. In arid climates, where moisture effectiveness is low, there is a net upward movement of water for most of the year, which carries salts to the soil surface. More is said about these processes, and profile differentiation, in Chapter 9.

Temperature

Temperature varies with latitude and altitude, and the extent of absorption and reflection of solar radiation by the atmosphere. Temperature affects the rate of mineral weathering and synthesis, and the biological processes of growth and decomposition. Reaction rates are

Box 5.1 Thornthwaite's climate indices.

As a measure of moisture effectiveness, Thornthwaite developed the *P–E* index, which is the summation over 1 year of the monthly ratios

$$\frac{\text{precipitation, } P}{\text{evaporation, } E} \times 10. \qquad\qquad (B5.1.1)$$

(The multiplier of 10 was introduced to eliminate fractions.) Based on the *P–E* index, five major categories of climate and associated vegetation types are recognized, namely

P–E index	Category	Vegetation
128 and above	Wet	Rainforest
64–127	Humid	Forest
32–63	Subhumid	Grassland
16–31	Semi-arid	Steppe
< 16	Arid	Desert

Thornthwaite also developed a temperature index, *T–E*, which is the sum for 1 year of the monthly values of

9/20 (mean monthly temperature in °C).   (B5.1.2)

This empirical expression was chosen so that the cold margin of the tundra had a *T–E* index of zero and the cooler margin of the rainforest an index of 128. The *T–E* indices and associated categories are:

T–E index	Category
128 and above	Tropical
64–127	Mesothermal
32–63	Microthermal
16–31	Taiga
1–15	Tundra
0	Frost

roughly doubled for each 10°C rise in temperature, although enzyme-catalysed reactions are sensitive to high temperatures and usually attain a maximum between 30 and 35°C.

One of the simplest indices of the thermal efficiency of the climate is Thornthwaite's *T–E* index (Box 5.1).

Another is a temperature index used in Britain, which is based on the accumulated degrees Celsius for days each year when the mean daily temperature exceeds 5.6°C. The combination of the *MD* and temperature indices produces a climatic classification for the British Isles, as used by Avery (1990) (Table 5.3).

Table 5.3 A climatic classification for the British Isles. After Avery, 1990.

Climatic regime	Mean annual accumulated temperature above 5.6°C (day-degrees)	Average maximum potential moisture deficit (*MD*) (mm)
Subhumid temperate	> 1375	> 125
Humid temperate	> 1375	50–125
Perhumid temperate	> 1375	< 50
Humid (oro)*boreal	675–1375	50–125
Perhumid (oro)*boreal	675–1375	< 50
Perhumid oroarctic*	< 675	< 50

* Altitude > 150 m.

The concept of soil zonality

From Dokuchaev (*c.* 1870) on, many pedologists in Europe and North America regarded climate as pre-eminent in soil formation. The obvious relationship between climatic zones, with their associated vegetation, and the broad belts of similar soils that stretched roughly east-west across Russia inspired the zonal concept of soils (Fig. 5.1).
• *Zonal* soils are those in which the influence of the regional climate factor and its associated vegetation, acting for a sufficient length of time, overrides the influence of any other factor. Zonal soils are *mature* soils (Section 5.6).
• *Intrazonal* soils, although showing well developed soil features, are those in which some local anomaly of relief, parent material or vegetation is strong enough to modify the influence of the regional climate.
• *Azonal* or immature soils have poorly differentiated profiles, either because of their youth or because some factor of the parent material or environment has arrested their development.

The zonal concept gave rise to systems of classification in which broad soil classes were related to climatic and vegetational zones on a continental scale, as in Russia, the USA and Australia. A key categorical level in such classifications was that of the *Great Soil Groups*, each of which embodied a central concept based on current understanding of soil genesis and environmental controls on genesis (Section 14.4). As these classifications evolved and higher and lower categories were developed, the definition of the Great Soil Groups became more 'factual'; that is, based on observable soil properties, their presence or absence and limiting values. Nevertheless, pedogenetic processes and their response to environmental factors continue to underpin many current classification systems, and the Great Soil Group names, as exemplified in Table 5.4, continue to have a wide currency in soil science.

Predictably, the concept of soil zonality is not very helpful when applied to soils of the subtropics and tropics. There, where land surfaces are generally much older than in Europe and North America, and have consequently undergone many cycles of erosion and deposition associated with climatic change, the age of the soil (Section 5.6) and its topographical relation to other soils in the landscape (Section 5.5) are factors of major importance. The zonal concept is also of little use in regions such as Britain and Scandinavia where much of the parent material of present-day soils is young (Pleistocene and more recent deposits) and its composition influences many soil properties: relief again plays a powerful role in soil formation.

5.4 Organisms

The soil, and the organisms living on and in it, comprise an *ecosystem*. The active components of the soil ecosystem are the plants, animals, micro-organisms and humans.

Vegetation

The primary succession of plants that colonize a weathering rock surface (Section 1.1) culminates in the development of a *climax community*, the species composition of which depends on the climate and parent material, but which, in turn, has a profound influence on the soil that is formed. For example, in the mid-West and Great Plains of the USA, deciduous forest seems to

Order	Suborder	Great Soil Groups
Zonal soils	1 Soils of the cold zone	Tundra soils
	2 Light-coloured soils of arid regions	Desert soils
		Red desert soils
		Sierozems
		Brown soils
		Reddish-brown soils
	3 Dark-coloured soils of semi-arid, subhumid, and humid grasslands	Chestnut soils
		Reddish chestnut soils
		Chernozems
		Prairie soils
		Reddish prairie soils
	4 Soils of the forest–grassland transition	Degraded chernozems
		Non-calcic brown or
		Shantung brown soils
	5 Light-coloured podzolized soils of the forest regions	Podzols
		Grey wooded or Grey podzolic soils
		Brown podzolic soils
		Grey-brown podzolic soils
		Red-yellow podzolic soils
	6 Lateritic soils of forested mesothermal and tropical regions	Reddish-brown lateritic soils
		Yellowish-brown lateritic soils
		Laterites
Intrazonal soils	1 Halomorphic (saline and alkali) soils of imperfectly drained arid regions and littoral deposits	Solonchak or saline soils
		Solonetz
		Soloth (solod)
	2 Hydromorphic soils of marshes, swamps, seep areas and flats	Humic gley soils
		Alpine meadow soils
		Bog soils
		Half-bog soils
		Low-humic gley soils
		Planosols
		Groundwater podzols
		Groundwater laterites
	3 Calcimorphic soils	Brown forest soils (Braunerde)
		Rendzinas
Azonal soils		Lithosols
		Regosols (includes Dry sands)
		Alluvial soils

Table 5.4 Higher soil categories according to the zonal concept. After Thorp and Smith, 1949.

accelerate soil formation compared to prairie (grass-land) on the same parent material and under similar climatic conditions. In parts of England, such as on the North York moors or heathlands of the Bagshot Sands,

the profiles preserved under the oldest barrows (burial mounds) of the Bronze Age people are typical of *brown forest soils* (Inceptisols (ST) or Tenosols (A)) formed under the deciduous forests before *c.* 4000 years ago. In contrast, podzols (Spodosols (ST) or Podosols (A)) have formed on the surfaces of the barrows, under the heathland vegetation which has developed since early man removed the forest. Differences in the chemical composition of leaf leachates can partly account for this divergent pattern of soil formation over time (Section 9.2).

Fauna

The species of invertebrate animals in the soil and the type of vegetation are closely related. The acid litter of pines, spruce and larch creates an environment unfavourable to earthworms, so that litter accumulates on the soil surface where it is only slowly comminuted by mites and Collembola and decomposed by fungi. The litter of elm and ash, and to a lesser extent oak and beech, is more readily ingested by earthworms and becomes incorporated into the soil as faecal material. Earthworms are the most important of the soil-forming fauna in temperate regions, being supported to a variable extent by the small arthropods and the larger burrowing animals (rabbits and moles). Earthworms are also important in tropical soils, under moist conditions; but in general the activities of termites, ants and coprophagous (dung-eating) beetles are of greater significance, particularly in the subhumid to semi-arid savanna of Africa and Asia.

Earthworms of the casting type build up a stone-free layer at the soil surface, as well as intimately mixing the litter with fine mineral particles they have ingested. The surface area of the organic matter that is accessible to microbial attack is then much greater. Termites are also important soil-formers in that they carry fine particles, water and dissolved salts from depths of 1 m or more to their termitaria on the surface (Fig. 3.7). When abandoned, the termitaria gradually disintegrate due to weathering and trampling by animals, to form a thin mantle on the soil surface.

Human influence

Humans influence soil formation through manipulation of the natural vegetation, and its associated animals, and the effects of agricultural practices and urban and industrial development. Indigenous populations of wild animals are regulated by natural controls; not so domestic stock which can, if too numerous, denude the vegetation so that the soil is exposed to severe erosion risk. Compaction by the traffic of domestic and wild animals decreases the rate of water infiltration into the soil, thereby increasing runoff and erosion. Human influences and the management of soil resources are discussed in Part III.

5.5 Relief

Relief has an important influence on the local climate, vegetation and drainage of a landscape. Major topographical features are easily recognized – mountains and valleys; escarpments, ridges and gorges; hills, plateaux and floodplains. Changes in elevation affect the temperature (a decrease of approximately 0.5°C per 100 m increase in height), the amount and form of the precipitation, and the intensity of storm events. These factors interact to influence the type of vegetation. More subtle changes in local climate and vegetation are associated with the *slope* and *aspect* of valley sides and escarpments.

Slope

Angle of slope and vegetative cover affect moisture effectiveness by governing the proportion of surface runoff to infiltration. Soils with impermeable B horizons, especially where they occur in an undulating landscape, can develop a perched watertable (Section 6.3) which predisposes to reducing conditions, and gleying and mottling in the profile. This is common, for example, during wet winters on the 'duplex' soils – red and yellow podzolics (Alfisols/Ultisols (ST) or Chromosols (A)) – on the lower slopes and foothills of the Great Dividing Range in southeastern Australia. Normally, however, in undulating landscapes on permeable materials, the soils at or near the top of a slope tend to be freely drained with the watertable at considerable depth, whereas at the valley bottom the soils are poorly

Box 5.2 Catenas and catenary complexes.

The importance of relief was highlighted by Milne (1935) in central East Africa, who recognized a recurring sequence of soils forming on slopes in a generally undulating landscape. He introduced the term *catena* (Latin for a chain) to describe the suite of contiguous soils extending from hill top to valley bottom. A catena is restricted to soils of the one climatic zone, that is, the members are variations within a zonal soil. Originally, there were two types of catenas:

• one in which the parent material is the same, and soil profile differences develop as a result of variations in drainage and aeration (as in Fig. 5.11). There can also be differential sorting of eroded material and accumulation of solutes according to slope.

• another in which the slope has been carved out of two or more superimposed parent rocks. The soils formed on the upper ridges and escarpment are old and denuded, whereas those formed on the lower slopes are younger, reflecting the rejuvenation of the *in situ* parent material by materials derived from above.

In the USA, the term has been restricted to sequences of soils on uniform parent material. In this sense, the catena is synonymous with Jenny's *toposequence*, which defines the suite formed when the influence of relief is dominant over that of the other soil-forming factors.

Although the catena was conceived as a sequence of contiguous soils on slopes, forming a convenient mapping unit for reconnaissance surveys (Section 14.3), in the USA a catena rarely consists of a geographical association of soils. There, when two or more members of a catena occur together in the landscape, they form a *catenary complex* which may be mapped as a *catenary soil association*. Other examples of catenas in the broader sense, as defined by Milne, are given in Chapter 9.

drained, with the watertable near or at the soil surface (Fig. 5.11a). The succession of soils forming under different drainage conditions on relatively uniform parent material comprises a *hydrological sequence*, an example of which is shown in Fig. 5.11b. As the drainability deteriorates, the oxidized soil profile, with its warm orange-red colours due to ferric oxides and oxyhydroxides (Section 2.4), is transformed into the mottled and gleyed profile of a waterlogged soil (iron in the ferrous state). Such a sequence of soils on a slope is an example of a *catena* (Box 5.2).

In addition to modified drainage, there is movement of soil on slopes. A mid-slope site is continually receiving solids and solutes from sites immediately up-slope by wash and creep, and continually losing material to sites below. Here the *form* of the slope is important whether smooth or uneven, convex or concave, or broken by old river terraces. Frost-heaving in very cold regions or faunal activity (rabbits, moles and termites) promotes the slow creep of soil down even slight slopes, which may create a deeper profile at the slope foot than on the slope above.

Small variations in elevation can also be important in flat lands, especially if the groundwater is saline and

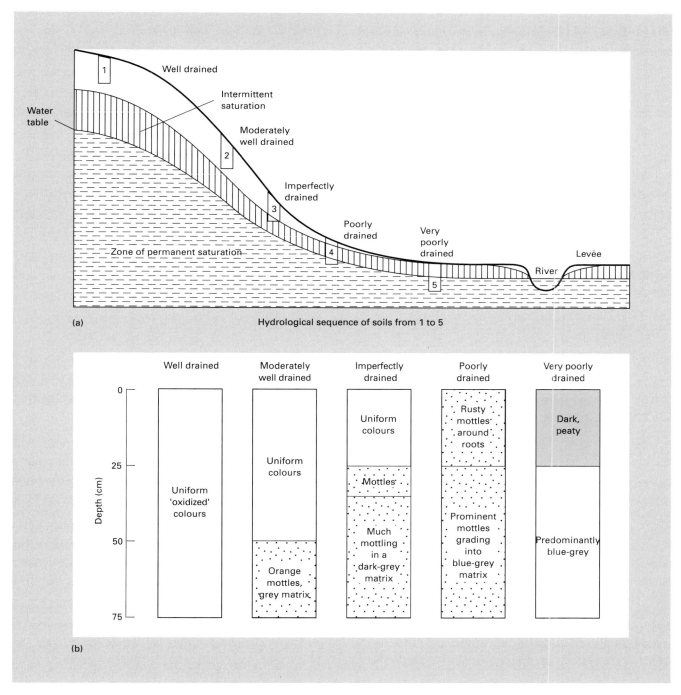

Fig. 5.11 (a) Section of slope and valley bottom showing a hydrological soil sequence. (b) Changes in profile morphology, principally colour, with the deterioration in soil drainage (after Batey, 1971).

rises close to the surface. A curious pattern of microrelief of this kind, called *gilgai**, has been observed in the black earths (Vertisols (ST) or Vertosols (A)) of semiarid regions in Australia (Fig. 5.12). The soils of the mound ('puff') and hollow ('shelf') are quite different, with calcium carbonate and gypsum occurring much closer to the surface in the yellow-brown crumbly clay of the puff than in the grey, compact cracking clay of the shelf. Comparable features are found in the black clay soils of Texas, the Free State of South Africa and in the Black Cotton soils of the Deccan, India. In the prairies of western Canada, glacial drift covers much of the landscape and forms a characteristic *knob* and *kettle* (depression) or *ridge* and *swale* topography. This markedly influences the local drainage and hence the type of soil formed, which ranges from being freely drained (even droughty) on the knobs to poorly drained (possibly saline) in the depressions.

5.6 Time

Time acts on soil formation in two ways.
- The *value* of a soil-forming factor may change with time (e.g. climatic change, new parent material, etc.)
- The *extent* of a pedogenic reaction depends on the time for which it has operated. *Monogenetic* soils are those that have formed under one set of factor values for a certain period of time; soils that have formed under more than one set of factor values are called *polygenetic*. The latter are much more common than the former.

It is well known that the world climate has changed over geological time: the most recent, large changes were associated with the alternating glacial and interglacial periods of the Pleistocene. Major climatic changes were accompanied by the raising and lowering of sea level, erosion and deposition, and induced isostatic processes in the Earth's crust, all of which produced radical changes in the distribution of parent material and vegetation, and in the shape of the landscape. On a much shorter timescale, covering a few thousand years,

* From the Aboriginal word for a small waterhole.

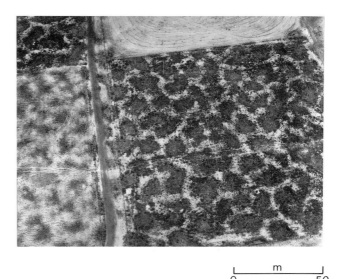

$$\begin{array}{ccc} & m & \\ 0 & & 50 \end{array}$$

Fig. 5.12 Aerial view of gilgai micro-relief (CSIRO photograph, courtesy of G.D. Hubble).

changes can occur in the biotic factor of soil formation, as witnessed by the primary succession of species on a weathering rock surface; and also in the relief factor through changes in slope form and the distribution of groundwater.

Rate of soil formation

Knowledge of a soil's age, together with information about any change in the factor values, enables rates of soil formation in different environments to be estimated. One criterion of a soil's age is its *profile morphology*, in particular, the number of horizons, their degree of differentiation and their thickness. The use of profile criteria is feasible for a soil derived from uniform parent material. Frequently, however, the past history of climatic, geologic and geomorphological change is so complex that an accurate assessment of a soil's age using these criteria is impossible.

The organic matter in individual horizons can be dated using the natural radioactivity of the C ($^{14}C/^{12}C$ ratio). Where the radio-C ages in a profile are not too dissimilar, an average age for the whole soil may be calculated. The relation between soil development and time can also be determined in special cases where the time that has elapsed since the exposure of parent material is known, and the subsequent influence of the other soil-forming factors has been reasonably steady. Such studies suggest that the rate of soil development is extremely variable, ranging from very rapid (several centimetres in 100 years or so) on volcanic ash in the tropics to very slow (1 cm per 5000 years) on chalk weathering in a cool temperate environment. Theoretically, the rate of soil formation sets the upper limit to the acceptable rate of soil loss by erosion (Section 11.5): in some cases, however, this may be unattainably low for a soil under agriculture so that the soil profile becomes thinner with time.

The concept of steady state

When the rate of change of a soil property with time is negligibly small, the soil is said to be in *steady state* with respect to that property. For example, for a soil forming on moderate slopes, the rate of natural erosion may just counteract the rate of rock weathering so that the soil, although immature, appears to be in steady state. In many cases, however, because the various pedogenic processes do not operate at the same rate, the properties that make up a soil do not all attain steady state at the same time. Soil formation is not linear with time, but tends to follow an exponential-type 'law of diminishing returns' (Section 11.2). This has been demonstrated for the accumulation of C, N and organic P in soils of temperate regions (Fig. 5.13). The time course of clay accumulation, on the other hand, is probably more sigmoidal in nature.

Because colonization by plants and animals of weathering parent material is an integral part of soil formation, it is more appropriate to consider steady state in the context of a soil and its climax vegetation (Section 5.4); that is, the soil–plant ecosystem. Studies of such ecosystems that have developed on parent material exposed by the last retreat of the Pleistocene ice (*c.* 10 000 y BP) suggest that a stable combination of soil

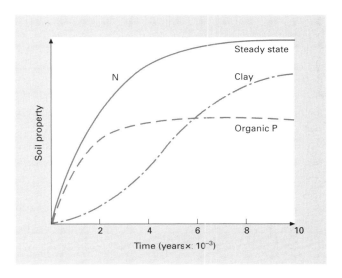

Fig. 5.13 The approach to steady state for selected soil properties under temperate conditions (after Jenny, 1980).

and vegetation (in the absence of human intervention) can be achieved in 1000 to 10 000 y. The system is stable in the sense that any changes occurring are immeasurably small within the time period since scientific observation began (*c.* 150 years). The soil component of such systems, the zonal soil, is sometimes referred to as *mature*.

By contrast, in many subtropical and tropical regions where the land surface has been exposed for much longer (10^5–10^6 y), the concept of a soil–plant ecosystem attaining steady state is much more problematic. The effect of time in the older landscapes is confounded by changes in other factors – climate, parent material and relief. Because the rate of mineral weathering is faster than in temperate regions, and leaching in humid climates potentially more severe, weathering occurs to much greater depths and the deep horizons are less influenced by surface organic matter. Crystalline clay minerals break down mainly to sesquioxides, with the release of silica and basic cations (Section 9.2). This trend in soil formation with time on a parent material of volcanic ash weathering in a wet tropical climate is illustrated in Fig. 5.14. Soil depth and vegetation bio-

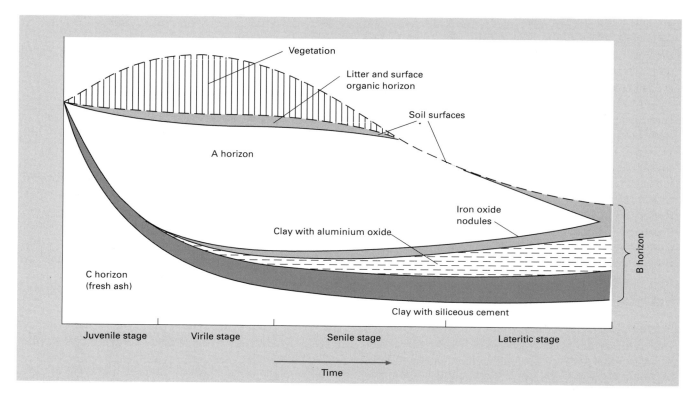

Fig. 5.14 Soil formation with time on volcanic ash under a wet tropical climate (after Mohr and van Baren, 1954).

mass attain maxima in the 'virile' stage, only to decline in the 'senile' and final stages. Climate exerts less effect than parent material and relief on the ultimate result of soil formation under such conditions.

5.7 Summary

Jenny (1941) sought to quantify the effects of *climate*, *organisms*, *relief*, *parent material* and *time* on soil formation by writing the equation

soil property, s = $\int (cl, o, r, p, t \ldots)$.

Lesser factors such as fire or atmospheric inputs are represented by the dots. Attempts have been made to solve this equation by examining the effect of one factor, such as climate, on *s* (or an ensemble of soil properties, *S*) when the other factors are constant or nearly so. Such conditions give rise to suites of soils called *climosequences*, *biosequences*, *toposequences*, *lithosequences* and *chronosequences*.

Although quantification of *clorpt* has not been very successful, recognition of the role of these factors, and their interactions, has helped scientists to understand why certain types of soil occur where they do. The concept has also been the basis of early attempts to impose order, through *soil classification*, on the apparently chaotic distribution of soils in the field. For example, the perceived dominant effect of climate and its associated vegetation on soil formation in continental regions, such as in Russia, the USA and Australia, gave rise to zonal systems of soil classification, the most durable hierarchical category of which is the *Great Soil*

Groups. Orginally developed around a central pedogenetic concept, Great Soil Groups have evolved in later classifications into classes with specified ranges of soil property values. Great Soil Group names continue to be used in soil science communication.

Where only one set of factor values has operated during soil formation, the resultant soil is said to be *monogenetic. Polygenetic* soils, which are much more common, arise when there is a significant change in the value of one or more of the factors during soil formation.

Parent material effects soil formation primarily through the type of rock minerals, their rate of weathering and the weathering products. These products are redistributed by the action of water, wind, ice and gravity, in many cases providing the parent material for another cycle of soil formation. Generally, the parent material of soils formed on transported material is more heterogeneous than that of soils formed on weathering rock *in situ* (residual soils). *Layering* of a soil profile occurs when new parent material is deposited on an existing soil, but is not of sufficient thickness to stop all pedogenesis in this soil.

It is difficult to separate cause and effect in the case of *organisms* because the range of colonizing species influences the pace and direction of soil formation, but the soil that forms affects the competition for survival amongst the plants and other organisms. Human activity can exert a profound effect on the soil, for good or bad, through the clearing of natural vegetation and the use of soil for agricultural and other purposes.

The importance of *relief* is obvious in the old weathered landscapes of the tropics, where the concept of the *catena* evolved to describe the recurring suites of soil that occur on slopes. Slope has an important effect on soil drainage, which in turn influences the balance between oxidation and reduction of Fe and Mn compounds in the soil profile. The effect of relief on soil formation can also be seen in the *gilgai* country of Australia and the *knob* and *kettle* country of the prairies in western Canada.

Time is important because the value of a soil-forming factor may change in time (e.g. climatic change), and the extent of a pedogenic reaction depends on the length of time for which it has operated. A soil profile may attain a *steady state* if the rate of natural erosion matches the rate of weathering of the parent material. Individual pedogenic processes generally approach steady state at different rates, in some cases following a 'law of diminishing returns', in others a sigmoidal response with respect to time. Soil–plant ecosystems developing on material exposed by the last retreat of Pleistocene ice sheets appear to have attained steady state within 1000–10 000 y. The zonal soils formed may be described as *mature*. On land surfaces which have been exposed for much longer (10^5–10^6 y), and where the rate of mineral weathering may be faster and leaching more severe, steady state soil–plant ecosystems are less likely to be achieved. Here parent material and relief determine the ultimate condition of the soil.

References

Avery B. W. (1990) *Soils of the British Isles.* CAB International, Wallingford.

Bryant R. B. & Arnold R. W. (Eds) (1994) *Quantitative Modeling of Soil Forming Processes.* Soil Science Society of America Special Publication No. 39. Soil Science Society of America, Madison, Wisconsin.

Batey T. (1971) *Soil profile drainage*, in ADAS Advisory Papers, No. 9, Soil Field Handbook. Ministry of Agriculture, Fisheries and Food, London.

Birkeland P. W. (1984) *Soils and Geomorphology.* Oxford University Press, New York.

Duchaufour P. (1982) *Pedology, Pedogenesis and Classification* (Translated by T. R. Paton). Allen & Unwin, London.

Goldich S. S. (1938) A study of rock weathering. *Journal of Geology* **46**, 17–58.

Holmes A. (1978) *Principles of Physical Geology*, 3rd edn. Nelson, London.

Jackson M. L., Tyler S. A., Willis A. L., Bourbeau G. A. & Pennington R. P. (1948) Weathering sequence of clay-size minerals in soils and sediments. 1. Fundamental generalizations. *Journal of Physical and Colloid Chemistry* **52**, 1237–1260.

Jenny H. (1941) *Factors of Soil Formation.* McGraw-Hill, New York.

Jenny H. (1980) *The Soil Resource, Origin and Behaviour.* Springer, New York.

Milne G. (1935) Some suggested units of classification and mapping, particularly for East African soils. *Soil Research* **4**, 183–198.

Mitchell J. K. (1993) *Fundamentals of Soil Behavior*, 2nd edn. Wiley, New York.

Mohr E. C. J. & van Baren F. A. (1954) *Tropical Soils*. N. V. Uitgeveriz, The Hague.

Soil Survey Staff (1975) *Soil Taxonomy. A basic system of soil classification for making and interpreting soil surveys*. Agriculture Handbook No. 436. United States Department of Agriculture, Washington DC.

Thornthwaite C. W. (1931) The climates of North America according to a new classification. *Geographical Review* **21**, 633–656.

Thorp J. & Smith G. D. (1949) Higher categories of soil classification. *Soil Science* **67**, 117–126.

Further reading

Amundson R., Harden J. & Singer M. (1994) *Factors of Soil Formation: A Fiftieth Anniversary Retrospective*. Soil Science Society of America Special Publication No. 33, Soil Science Society of America, Madison, Wisconsin.

Duff P. McL. D. & Smith A. J. (Eds) (1992) *Geology of England and Wales*. Geological Society, London.

Nahon D. B. (1991) *Introduction to the Petrology of Soils and Chemical Weathering*. Wiley, New York.

Paton T. R., Humphreys G. S. & Mitchell P. B. (1995) *Soils, a New Global View*. UCL Press, London.

Thomas M. J. (1994) *Geomorphology in the Tropics: A Study of Weathering and Denudation in Low Latitudes*. Wiley, Chichester.

Chapter 6
Hydrology, Soil Water and Temperature

6.1 Properties of soil water

Soil water potential

In Section 4.5 the concept of soil wetness measured by θ, the volumetric water content, was introduced. This is an *extensive* property of the soil water because it relates to the *quantity* of water in a defined volume of soil. However, to characterize the behaviour of water in soil, that is, its ability to move through the soil to plant roots, or to a watertable, it is necessary to know not only the quantity of water but also the several forces acting on the water which affect its *potential energy.* There is no absolute scale of potential energy, but we can measure *changes* in potential energy when useful work is done on a known quantity of water, or when the water does useful work. These changes are manifest as changes in the *free energy* of water, as demonstrated by a simple example.

Suppose a small quantity of water is added to a dry, non-saline soil in an isothermal (constant temperature) environment. This water is distributed so that the force of attraction between it and the soil is as large as possible. Work is done by the water to achieve a maximum reduction in free energy. Conversely, work would have to be done *on* the water to remove it from the soil and restore it to its original state. Additional water that is added to the soil is held by progressively weaker forces until, with the last drop of water necessary to saturate the soil, the free energy of the soil water approaches that of pure, free water at the same temperature, pressure and height above sea level. Thus, the energy status of soil water can be defined in terms of the *difference* in free energy between a known amount of water in the soil (say

1 mol) and the same amount of pure (no solutes), free (no external forces other than gravity) water at a standard temperature, pressure and a fixed, reference height above sea level. The free energy per mol of a substance defines its chemical potential – an intensive property of the system. So for water in soil, we define the *soil water potential* ψ by the equation

$$\psi = \mu_w(\text{soil}) - \mu_w^0(\text{standard state}) = RT \ln e/e^0 \qquad (6.1)$$

where μ_w is the chemical potential (free energy mol^{-1}) of water, e/e^0 is the ratio of vapour pressure of water in the soil to vapour pressure of pure water at a standard temperature (298 K) and pressure (1 atmosphere), R is the universal gas constant (8.314 J K^{-1} mol^{-1}), and T is absolute temperature (K).

The reference height used for soil water potential measurements is often the soil surface (Fig. 6.3). Note that for water at the standard temperature and pressure, and appropriate reference height, the maximum value of ψ is obtained at $e/e^0 = 1$, when $\psi = 0$, and also that ψ values will usually be negative.

The units of ψ are energy per mole (J mol^{-1}), but to be compatible with earlier pressure measurements, ψ is often expressed as energy per unit volume (J m^{-3}) by dividing equation 6.1 by V_w, the volume (m^3) per mole of water. The units of J m^{-3} are equivalent to N m^{-2} or pascals (Pa) which are units of pressure. The analogy with pressure measurements goes further with the use of the term *suction* (implying negative pressure) to describe the action of forces on water as it is drawn into a dry soil (see below).

Hydraulic head

The hydrostatic pressure due to a column of water h metres high (equivalent to ψ in units of energy per unit volume) is given by

$$\psi = h \rho g \tag{6.2}$$

where ρ is the density of water (1000 kg m^{-3}) and g is the acceleration due to gravity (9.8 m s^{-2}).

Thus, a pressure of 1 *bar* (10^5 Pa) is subtended at the base of a water column of height h, as calculated by rearranging Equation 6.2 to give

$$h = \frac{\psi}{\rho g} = \frac{10^5}{1000 \times 9.8} = 10.2 \text{ m}. \tag{6.3}$$

Hydrologists and soil scientists make use of this relationship and often express water potentials in terms of the height of an equivalent water column, referred to as *hydraulic head* or simply as *head*. Note that head h is an expression of total water potential in units of energy per unit weight; that is, J s^2/kg m = m, which is a very convenient unit for calculating head gradients in the field (Section 6.3). In summary, it can be seen that water potential ψ, expressed as energy per unit volume, is related to head h (energy per unit weight) by the equation

$$h = \frac{\psi}{\rho g}. \tag{6.4}$$

Comparative expressions for the free energy status of soil water are outlined in Box 6.1.

Components of the soil water potential

The free energy of water in the soil is reduced in the following ways.

(a) At very low relative humidities (RH < 20% or $e/e^0 < 0.2$), water is adsorbed on clay minerals of low permanent charge as a monolayer in which the molecules are hydrogen bonded to each other and to the surface. With an increase in RH more water is adsorbed and the influence of the surface on the structure of the liquid water extends further into the solution; yet the bulk of the effect occurs in the first three molecular layers within approximately 1 nm of the surface.

(b) At the charged planar surfaces of the 2 : 1 type clay minerals, water is also strongly attracted to the adsorbed cations, the strength of attraction depending on whether the cations form inner- or outer-sphere complexes with the surface (Section 2.4). The electric field of the cation orientates the polar water molecules to form a solvation or hydration shell containing up to six water molecules for a monovalent cation and up to 12 for a divalent cation such as Mg^{2+} (Fig. 6.1). The energy of hydration per mole of cation depends upon its ionic potential (Section 2.3) and polarizability, with en-

In dry soils, the suction of water, when expressed in head units in the old centimetre–gram–second system, was often a large and unwieldy number. It was therefore more convenient to use $\log_{10} h$ (cm), which defines the *pF value* of the soil water (compare the pH scale for H$^+$ ion activity). However, plant physiologists traditionally measured the energy status of plant water in pressure units (ψ values) which are not readily compatible with soil water *pF* values. Thus, for studies of water movement through the soil–plant–atmosphere continuum (SPAC), the energy status of water is best expressed in terms of water potentials ψ, and soil *pF* values are now rarely used. For water movement through soil as described by Darcy's law (Section 6.3), water potential is best expressed in units of head. Comparative values for the free energy status of water in soil at various states of wetness are given in Table 6.1.

Box 6.1 Expressions for soil water potential.

Table 6.1 Water potential, hydraulic head and *pF* of a soil of varying wetness.

Soil moisture condition	Water potential ψ (bar)	Head h (m)	pF	Equivalent radius of largest pores that would just hold water (μm)
Soil saturated or very nearly so	− 0.001*	0.0102†	0	1500‡
Moisture held after free drainage (field capacity)	− 0.05	0.51	1.7	30
Approximately the moisture content at which plants wilt	− 15	153	4.2	0.1
Soil at equilibrium with a relative vapour pressure of 0.85 (approaching air-dryness)	− 220	2244	5.4	0.007

* Calculated from Equation 6.1.
† Calculated from Equation 6.4.
‡ Calculated from Equation 6.6.

ergies for the common exchangeable cations falling in the order

$$Al^{3+} > Mg^{2+} > Ca^{2+} > Na^+ > K^+ \cong NH_4{}^+.$$

The greater the hydration energy of the cation, the more work done in attracting water molecules, and the larger the reduction in free energy of the water molecules trapped in its hydration shell. The relatively small amount of water present in dry soils (RH < 80% or e/e^0

Fig. 6.1 A hydrated exchangeable cation.

< 0.8) is therefore mainly associated with clay surfaces and their exchangeable cations.

(c) As water layers build up, the water begins to fill the smallest pores and interstices between mineral particles, forming curved air–water menisci (Fig. 6.2a). Equilibrium is attained when the surface free energy associated with the air–water–solid interface is at a minimum. The work done on the water by the surface

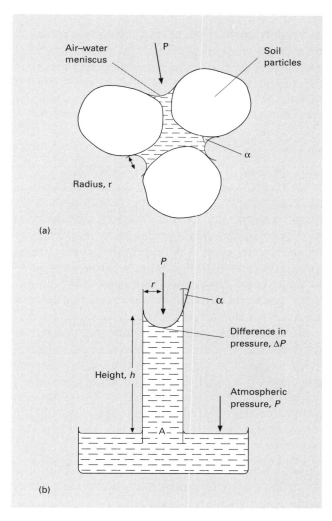

Fig. 6.2 (a) Water held under tension between soil particles. (b) Rise of water in a capillary tube.

tension forces is balanced by the increased potential energy of the water column, as for example in the rise of water in a capillary tube (Fig. 6.2b). It follows from the latter figure that

$$\pi r^2 \rho g h = 2\pi r \gamma \cos \alpha \qquad (6.5)$$

where π, r, ρ, g, h and α are as already defined, or indicated in Fig. 6.2, and γ is the surface tension of water. Note that atmospheric pressure acts on the *upper surface* of the water in the capillary tube, and on the free water surface outside the tube (Fig. 6.2b). The pressure in the water at the base of the column (point A) must also be atmospheric, and is greater than the pressure in the water at the *under surface* of the meniscus in the column. From Equation 6.2, we know that the difference in pressure ΔP across the meniscus must be given by $h\rho g$, and so from Equation 6.5 we have

$$\Delta p = \rho g h = \frac{2\gamma}{r} \qquad (6.6)$$

when the angle of wetting $\alpha = 0$ (see Box 6.2).

The combined effects of adsorption and capillarity on the free energy of soil water are expressed through the matric potential, ψ_m. Over the range of water contents for which capillary effects predominate, the matric potential ψ_m can be related directly to the pore radius through Equation 6.6. For example, when water in an initially wet soil has drained from pores > 100 μm diameter, the matric potential of water occupying the largest pores is – 0.03 bar or 3 kPa.

(d) Solutes (ions and organic molecules) also reduce the free energy of soil water, the extent of the reduction being measured by the osmotic potential, ψ_s. Osmotic potential is numerically equal to the osmotic pressure of the soil solution, which is defined as the hydrostatic pressure necessary to just stop the inflow of water when the solution is separated from pure water by a semi-permeable membrane. The osmotic pressure (π) in pascals of soil solutions can be calculated from the equation

$$OP = RTC \qquad (6.7)$$

where R and T have been defined and C is the concentration (mol m^{-3}) of solute particles (molecules and dissociated ions) in solution.

The effect of these forces may be summarized as follows. ψ_m and ψ_s are *component potentials* of the soil water potential ψ. In soil these potentials are always less than 0 so that their contribution to ψ is always negative. The contribution of ψ_s is negligible in non-saline soils. Positive contributions to ψ can be made by the gravitational potential ψ_g, if water is held at a height above that chosen for the reference level; or if soil water is subjected to a pressure greater than atmospheric, giving rise to a positive pressure potential, ψ_p.

The soil water potential ψ is therefore given by the sum

$$\psi = \psi_m + \psi_s + \psi_p + \psi_g. \qquad (6.8)$$

Although it is technically possible to increase the air pressure on a soil above normal atmospheric pressure, in practice variations in pressure potential due to changes in air pressure are negligible. The pressure potential ψ_p only becomes significant when the soil, or a zone in the soil, is saturated. In this case the pressure (p) at a depth below the upper surface of the saturated zone (the watertable) is directly proportional to that depth. The *piezometric head h* is given by

$$h = p + z \qquad (6.9)$$

where z is the depth in the soil below the reference level (the soil surface). By convention, z is measured positively upwards, so $h = p - z$ at point A in Fig. 6.3a. Obviously, ψ_m in the saturated soil must be zero so that ψ_m and ψ_p can be visualized as two subcomponents of a pressure–matric potential continuum for which:
• above the watertable, $\psi_p = 0$ (atmospheric pressure) and ψ_m is negative,
• at the watertable $\psi_m = \psi_p = 0$, and
• below the watertable, $\psi_m = 0$ and ψ_p is positive.

An example of water under negative pressure or *suction* ($\psi_m = -p$) in unsaturated soil is shown in Fig. 6.3b. Here the water in the manometer tube equilibrates with water in the soil surrounding the porous pot (at point B), such that the head h is given by

$$h = -p - z. \qquad (6.10)$$

The retention and movement of water in the soil is

The angle of wetting or *contact angle* is measured from the liquid–solid interface to the liquid–air interface. At most soil mineral surfaces, the angle is close to 0 and the water is said to 'wet' the surface (Fig. B6.2.1a). However, if these surfaces become coated with hydrophobic organic compounds, the cohesive forces between the water molecules become much stronger than the force of attraction between the water and the surface so that the contact angle increases markedly (Fig. B6.2.1b). The surface is then said to 'repel' the water and the soil as a whole may become *water repellent*. Such an effect can occur in sandy A horizons which undergo severe drying – water subsequently applied to the surface does not wet the surface, nor infiltrate, and tends to run off (Section 6.3). Many soil surfaces may exhibit some degree of water repellency when dried for a prolonged period, but the effect gradually disappears as the soil wets up again.

Box 6.2 Non-repellent and water-repellent soils.

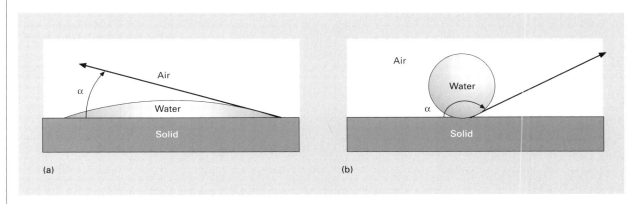

(a) (b)

Fig. B6.2.1 Illustration of air–water–solid interfaces (a) small contact angle, water 'wets' the surface (b) large contact angle, water is 'repelled' by the surface (after Jury *et al.* 1991).

discussed in terms of these potentials in the succeeding sections. Measurement of soil water potential is discussed in Box. 6.3.

6.2 The hydrological cycle

A global picture

Water vapour enters the atmosphere by evaporation from soil, water and plants; it precipitates as rain, hail, snow or dew. Precipitation *P* equals evaporation *E* over the entire Earth's surface; but for land surfaces only, the average annual *P* of 710 mm* exceeds the average annual *E* by 240 mm, the difference amounting to river discharge. If that part of the land surface which contributes little to evaporation – the deserts, ice caps and Tundra – is excluded from the calculation, the average rate of evaporation from the remainder is approximately 1000 mm per annum, which is not far short of

*Expressed as the 'equivalent depth' in m^3 water per m^2 surface × 1000 (Section 4.5).

Box 6.3 Field measurement of soil water potential.

A very useful instrument for measuring the combined expression of matric and gravitational potentials in the field is the *tensiometer*, which gives rise to a tensiometer potential ψ_t. A tensiometer consists of a porous ceramic cup (with an air-entry value of at least 85 kPa) sealed to a length of polyvinyl chloride (PVC) tubing, at the top of which is a pressure measuring device (Fig. B6.3.1a). The tube and cup are filled with water so that when the tensiometer is installed in the soil, the water in the tensiometer comes to equilibrium with water in the surrounding soil. The suction on the soil water (less the osmotic effect of any solutes which can diffuse freely through the ceramic cup walls) is transmitted through the water column, to be read with a Bourdon gauge or pressure transducer at the top. Tensiometers of different length installed at specific depths in the soil measure the gravitational potential of the soil water, with respect to a datum at the soil surface, as well as the matric potentials at those specific depths (Fig. B6.3.1b). This enables potential gradients for water movement to be measured (Section 6.3) and also reveals any zones of saturation in the soil ($\psi_m = 0$). Any depth where the tensiometer potential gradient is 0 defines a *zero flux plane* (Fig. B6.3.1b) where no net upward or downward movement of water would be expected.

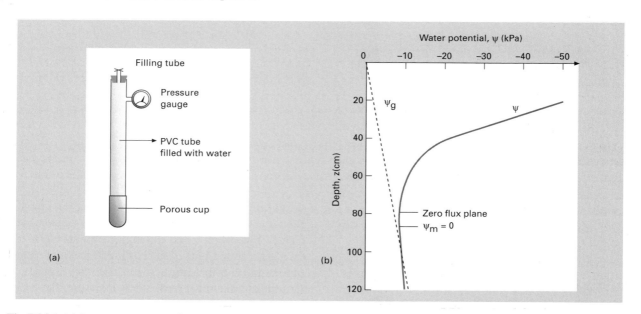

Fig. B6.3.1 (a) A vacuum gauge tensiometer. (b) A typical water potential profile in soil drying by evapotranspiration after being thoroughly wetted.

the 1200 mm annual rate of evaporation from the oceans. Evaporation depends on an input of energy, the energy available being much greater at the equator than at high latitudes. The capacity of the air to hold water also increases with temperature, so that the amount of precipitable water in the atmosphere is greatest at the equator (40–50 mm) and least at the poles (< 5 mm at the North Pole). Although relatively small, the atmospheric source replenishes the much larger reserves of surface water, and water in the soil and regolith which sustains terrestrial plant and animal life. Water in soil or the regolith exists in an unsaturated or *vadose* zone (above the watertable or *phreatic* surface – see Fig. 5.11a), or in a saturated zone (below the watertable). Estimates of the size of these reserves are subject to much uncertainty, but are approximately 250 mm for soil water (range 50–500 mm), 400–1000 mm for surface water (lakes and rivers), and 15–45 m for groundwater.

Water balance on a local scale

Water cycling at a local scale is illustrated by the processes occurring in a segment of a forested catchment (Fig. 6.4). Of the total precipitation on land, nearly all is intercepted by the vegetation with the remainder falling directly on the soil. Water in excess of that required to wet the leaves and branches, the *canopy storage capacity* (usually 0.5–1 mm), drips from the canopy or runs down stems to the soil. Canopy drip, stem flow and direct rainfall constitute the *net precipitation* or *net rainfall*. Water retained by the canopy is lost by evaporation and is referred to as *interception loss*: it may be as much as one-quarter of the gross rainfall for broadleaf trees of temperate regions in summer.

Of the rain that reaches the soil, the difference between that soaking in and that running off constitutes *surface runoff* or *overland flow*. The movement of water into the soil from above is called *infiltration*. When the soil is thoroughly wet down to the watertable, a steady rate of flow or percolation to the groundwater is maintained. Subsurface lateral flow or *interflow* also occurs through soils on slopes, or when vertical flow into the

subsoil is impeded as occurs in soils with impermeable B horizons (Section 5.5), or with indurated layers such as an iron pan (Section 4.3). Surface runoff generates a rapid reponse in streamflow in the catchment. This is augmented by flow from *contributing areas*, usually areas at the base of slopes and near streams that are fed by interflow from higher areas, which quickly reach saturation during or following periods of significant rainfall. Lateral flow of groundwater is much slower than surface runoff and interflow and provides the bulk of the stream's *baseflow*. Surface runoff, which may initiate soil erosion, and interflow make a much more variable contribution to streamflow than groundwater.

Slow upward movement of water occurs in reponse to evaporation from the soil surface. Vegetation, however, can offer a much larger surface for the evaporation of water. The resultant decrease in water potential in the leaves provides the driving force for water to be sucked into the roots and through the plant, to be lost by evaporation from the leaves. This is called *transpiration*. Some water flows into the water-depleted zones around absorbing roots while the remainder is intercepted by the roots as they grow through the soil. The combination of transpiration from vegetation and evaporation from bare soil surfaces is sometimes referred to as evapotranspiration, or simply as *evaporation*.

The interaction of all these processes in a catchment or isolated hydrological unit is expressed by the *water balance equation* (Box 6.4).

A knowledge of streamflow and deep drainage in a catchment is important, first because streamflow provides water that can be conserved for human and animal use; second, because water in deep aquifers (permeable rocks that hold groundwater) is a valuable reserve from which to supplement surface supplies when E is consistently greater than P. An example of equipment used for gauging streamflow in a small agricultural catchment is given in Fig. 6.5b (p. 109).

Precipitation is largely beyond human control, despite attempts in the USA and Australia to induce rainfall by 'cloud seeding' with dry ice or silver iodide. There are, however, opportunities for manipulating the other variables in the water balance which depend to a

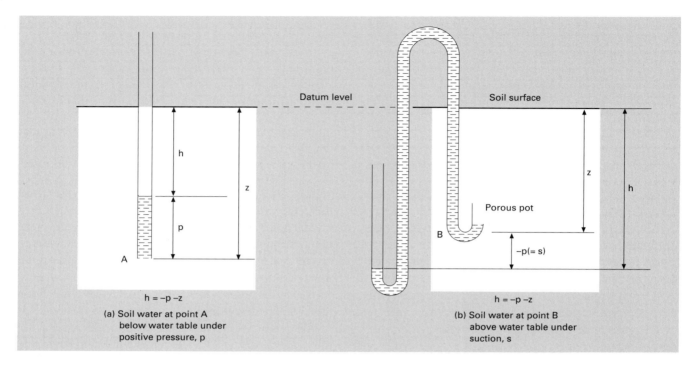

Fig. 6.3 Hydraulic head values for water is soil under a positive pressure, *p*, or suction, *s* (after Youngs and Thomasson, 1975).

large extent on soil properties and land management in the catchment. It is therefore relevant to discuss infiltration, soil–water movement and evaporation in some detail.

6.3 Infiltration, runoff and redistribution of soil water

Darcy's law

For water movement through porous materials, provided the velocity is low enough that flow is laminar and not turbulent, the rate of flow is directly proportional to the driving force and inversely proportional to the resistance, as expressed by Darcy's equation

$$J_w = -K \operatorname{grad} \psi \qquad (6.11)$$

where J_w is the volume per unit time crossing an area *A*, perpendicular to the flow, and is called the water flux density or *water flux*. Grad ψ denotes the *gradient* or change in water potential per unit distance in the direction of flow (which may be horizontal or vertical); and the proportionality coefficient *K* defines the *hydraulic conductivity* which is also the reciprocal of the flow resistance. (The negative sign accounts for the fact that flow occurs in the direction of decreasing potential so that the gradient in ψ is negative.)

The ideal condition for applying Darcy's law is when both grad ψ and the resistance to flow are constant over time, a condition attained during steady, saturated flow through a non-swelling soil (Box 6.5). Soils in the field usually do not satisfy this condition because they are in a dynamic cycle of wetting and drying, and the parame-

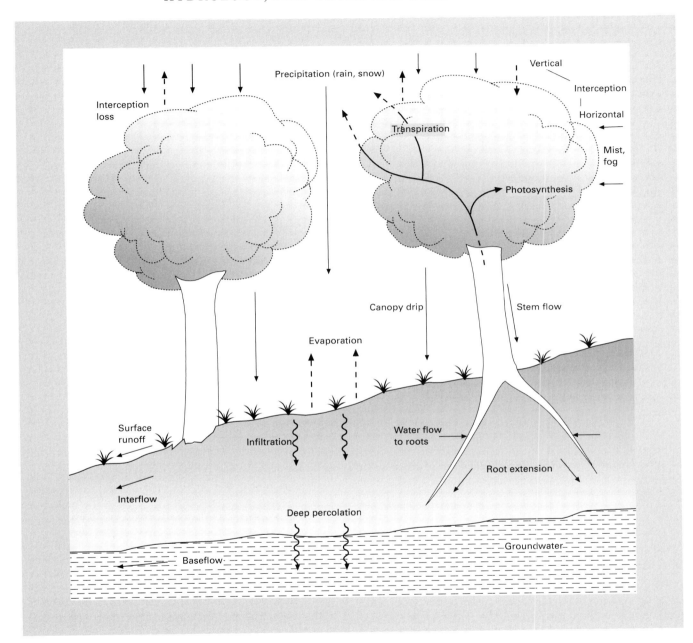

Fig. 6.4 Hydrological cycle in a forested catchment.

Box 6.4 The water balance equation.

The water balance equation is an equation of mass conservation, that is

$$P = E + \text{runoff}, R + \text{deep drainage}, D + \text{change in soil storage}, \Delta S. \quad \text{(B6.4.1)}$$

For catchments in which deep drainage losses are negligible, when periods of 1 year or more are considered, the changes in soil storage $\Delta S (+ \text{ or } -)$ tend to cancel out so that the difference between P and E is primarily due to runoff (surface and subsurface). Over shorter periods, values of ΔS are nevertheless important for plant growth, and if ΔS is persistently negative irrigation may be necessary to supplement rainfall. Changes in ΔS can be measured by inserting in the soil an aluminium tube, sealed at the base, into which a neutron probe is lowered. Fast neutrons are emitted from a radioactive source in the probe and are slowed down and reflected back to a detector, following collisions with hydrogen nuclei. By far the greatest source of variation in numbers of hydrogen nuclei in soil is due to changes in water content, so the neutron probe can be directly calibrated to give the volumetric water content θ. An example of a neutron probe is shown in Fig. 6.5a. Other non-destructive methods of measuring soil water content *in situ* rely upon measuring changes in the permittivity or dielectric number of the soil, using instruments such as a time domain reflectometer (TDR) or a capacitance probe. Water has a very high dielectric number compared to any other non-solid constituent in the soil, so changes in soil permittivity can be related to θ. The reader is referred to specialist texts such as Jury *et al.* (1991) for further details of these methods.

When P, E, R and ΔS are known, Equation B6.4.1 can be solved for various time intervals to estimate the drainage flux D, which represents water moving below the root zone that can augment the groundwater. The deep drainage flux is an important issue in many areas used for agriculture, particularly where the groundwater is saline and a rise in the watertable can cause problems for plant growth (Section 13.2)

ters in Equation 6.11 change in space and time. For all except saline soils, ψ_m and ψ_g are the main component potentials governing water movement. Written in head units, the water potential becomes the sum of the pressure head component p, where $-p$ is the suction s for soil water under tension, and the gravitational head component z, measured positively upwards from an arbitrary reference level. That is

$$h = s + z. \quad (6.12)$$

Noting that for unsaturated conditions, K is a func-

tion of matric potential or suction s, Darcy's law for describing vertical flow, written in head units, becomes

$$J_w = -K(s) \text{ grad } h$$
$$= -K(s) \text{ grad } (s + z). \quad (6.13)$$

Put in partial differential notation, Equation 6.13 becomes

$$J_w = -K(s) \left(\frac{\partial s}{\partial z} + 1 \right). \quad (6.14)$$

If the solution to Equation 6.14 is negative, this means

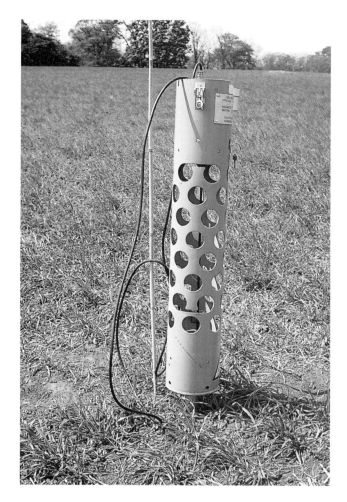

Fig. 6.5 (a) Neutron probe in spring barley. (b) V-notch weir for stream gauging.

that net water flow is downwards (drainage); if positive, net water flow is upwards (e.g. in response to evaporative demand). Furthermore, because potential has been expressed in head units, K has the same dimensions as the water flux J_w (L T^{-1}).

Stages of infiltration

Rain falling on the soil surface is drawn into the pores under the influence of both a suction and gravitational head gradient, and if the rainfall intensity is less than the initial *infiltration rate* (*IR*) all the water is absorbed. In the early stages of wetting a dry soil, the suction gradient is predominant. However, as the depth of wet soil increases, the suction gradient decreases (the same difference in suction is spread over an ever-increasing depth interval Δz) and the gravitational head gradient becomes the main driving force. The flux density J_w then approaches the saturated hydraulic conductivity K_s, under which circumstances the soil may not be able to accept water rapidly enough to prevent ponding on

Box 6.5 Factors affecting hydraulic conductivity K.

The value of K depends on the amount of water in the pores and its viscosity, as well as the pore size distribution, tortuosity and surface roughness. Some insight into the effect of pore size (radius) on K is obtained from Poiseuille's law which describes the rate of water flow Q through a full cylindrical tube under a constant pressure gradient, that is

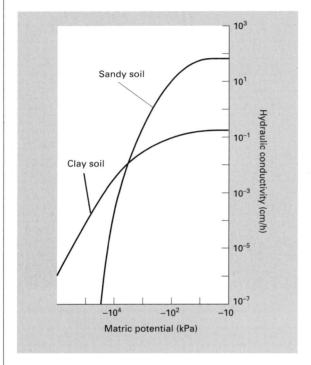

Fig. B6.5.1 Typical hydraulic conductivity–matric potential curves for a sandy and a clay soil.

$$Q = \frac{\pi r^4 \Delta P}{8L\eta} \tag{B6.5.1}$$

where r is the tube radius, η the dynamic viscosity and $\Delta P/L$ is the change in pressure over the length of tube L. Equation B6.5.1 shows that the volume of water flowing per unit time varies with the fourth power of the radius, and the water flux or velocity J_w varies with the radius squared. Hence in a soil, other factors being equal, water would move 100 times faster through water-filled pores of 1-mm radius compared to 0.1-mm radius.

As a soil desaturates ($\psi_m < 0$), water drains out of the largest pores first (Section 6.4) and the average size of pores still conducting water decreases. Water columns in the soil may become discontinuous so the average pathway for water movement becomes more tortuous. The overall result is that hydraulic conductivity K decreases markedly from its maximum value at saturation (K_s). Examples of the change in K with matric potential ψ_m are shown in Fig. B6.5.1.

K_s can be measured in the laboratory by first saturating an intact core (usually from the bottom upwards to displace entrapped air), then ponding water on the core surface and measuring the rate at which water flows out of the base of the core. The gradient in ψ or head through the core will be unity so that at steady state $J_w = K = K_s$ in Equation 6.11. Field K_s and K values at very low suctions (up to -150 mm head) can be measured in the field using a disc permeameter or tension infiltrometer (Green *et al.*, 1986).

the surface. For bare soil, structural changes in the surface following raindrop impact may also reduce the rate of acceptance so that the *IR* falls below the rainfall intensity within a few minutes to a few hours. Initially, ponding may be confined to small depressions and this

creates favourable conditions for preferential flow down macropores (see below). But once the surface detention reaches a critical value, runoff occurs. This sequence of events for rainfall of moderately high intensity is illustrated in Fig. 6.6.

The wetting front

If rain continues to fall on the soil surface, as in Fig. 6.6, and the soil is structurally homogeneous, water travels downwards at a constant rate J_w which determines the *infiltration capacity* or *infiltrability* of the soil. As water penetrates more deeply a zone of uniform water content, the *transmission zone*, develops behind a narrow wetting zone and well defined *wetting front*. The plot of water content against depth for such a soil, illustrated in Fig. 6.7, shows a sharp change in water content at the wetting front. This occurs because water at the boundary takes up a preferred position of minimum potential energy in the narrowest pores, for which the hydraulic conductivity is very low, and does not move at an appreciable rate until the large pores begin to fill.

For a constant rainfall intensity J_w, and soil water content θ, the mean *pore water velocity* $\bar{v}$ in the wet soil zone is given by

$$\bar{v} = \frac{J_w}{\theta},$$ (6.15)

and the distance z travelled by the wetting front in time t is given by

$$z(t) = \frac{J_w t}{\theta}.$$ (6.16)

In practice, field soils are rarely structurally homogeneous due to the presence of old root channels, worm holes and planes of weakness between aggregates, so the downward movement of water is more erratic than Fig. 6.7 would imply. A wetting front in a natural soil profile is shown in Fig. 6.8. Although the change in water content at the wetting front is sharp in this soil, there is also evidence of water at some points moving ahead of the main transmission zone. At a local scale there will be a range of pore water velocities because some of the infiltrating water flows rapidly down channels and cracks (sometimes referred to as 'preferential' or 'bypass' flow), whereas the bulk of the water pene-

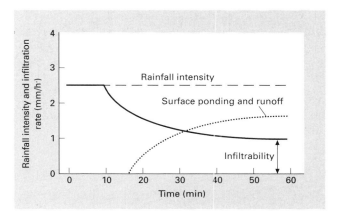

Fig. 6.6 Infiltration, surface ponding and runoff during a storm.

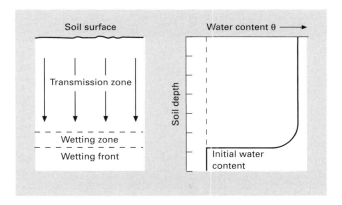

Fig. 6.7 Vertical infiltration into a homogeneous column of dry soil.

trates slowly into the micropores within aggregates ('matrix' flow).

Within a well-developed transmission zone, the soil is near saturated (allowing for some trapped air pockets) and the suction gradient $\partial s/\partial z$ is negligible. Therefore, Equation 6.14 simplifies to

$$J_w = -K$$ (6.17)

where K approaches the saturated hydraulic conductivity of the soil (c. 0.95 K_s). Values for K_s range from 1 to

Fig. 6.8 Vertical infiltration into a soil in the field.

500 mm d^{-1}, depending on the texture and structure of the soil. A method of measuring K_s in the field, or the infiltration capacity, is described in Box. 6.6.

Field capacity

When infiltration ceases, redistribution of water occurs at the expense of the initially saturated zone of soil. If the soil is completely wet to the watertable, or an artificial outlet such as a drain, drainage ceases when the suction gradient acting upwards balances the gravitational head gradient acting downwards. Even when the

soil is not completely wet depthwise, the drainage rate often becomes very small 1–2 days after rain or irrigation. This is possible because of the sharp fall in hydraulic conductivity when most of the macropores have drained of water. The water content then defines the *field capacity* (*FC*), which for well-structured British soils is in equilibrium with a matric suction of about 5 kPa (Section 4.5). In Australia, the water content at field capacity is set at a suction of 10 kPa. As is seen in Section 6.4, the concept of field capacity is useful for setting an upper limit to the amount of 'available' water in the soil; but it is imprecise for two reasons:
• the soil may remain above the *FC* due to frequent showers of rain for several days, and
• soils with a high proportion of micropores continue to drain slowly for several days after rain.

Steady state and transient flow

Flow under saturated or near saturated conditions is approximately in *steady state*; that is, the water flux into a given volume is equal to the water flux out. However, as the larger, more continuous pores drain and the soil becomes unsaturated, the resistance to flow increases and the value of K decreases accordingly. Flow then becomes non-steady or *transient*. Darcy's Equation 6.14 can still be used to calculate water fluxes, provided that

---

Water is ponded to a known depth inside a ring infiltrometer – a metal circle 30–60 cm in diameter that is driven a short distance into the soil surface. When quasi-steady state infiltration is achieved, the rate of water loss from the ring is taken as an estimate of the *infiltration capacity*. In the previous discussion one-dimensional infiltration (vertical only) was assumed. This is not necessarily the case with a ring infiltrometer, particularly in dry and heavy textured soils where considerable lateral flow will occur under a suction gradient. The infiltrometer will therefore overestimate the one-dimensional infiltration capacity, or field K_s, unless an outer 'buffer ring' filled with water is used, as in the *double-ring infiltrometer*. The infiltration rate is measured from the inner ring only.

Box 6.6 Measuring field infiltration capacity.

the suction gradient can be measured over short time intervals, and the $K(s)$ function is known.

For drainage to a very deep watertable $\partial s / \partial z$ is much smaller than 1, and the water content throughout the profile adjusts so that $J_w \simeq K(s)$ (compare with near saturated flow through a transmission zone during infiltration). Unsaturated flow of this kind occurs in deep permeable soils below the root zone. Alternatively, flow is sometimes impeded by a layer of less permeable soil or parent material deep in the profile, resulting in localized saturation and the development of a *perched watertable* (Fig. 6.9). The increase in hydraulic head immediately above the impeding layer induces lateral flow which may eventually appear as a line of springs at another position in the landscape.

Non-steady state conditions

Soil water is rarely in steady state in the root zone. Intermittent rainfall and surface evaporation combined with the variable plant uptake of water cause the water content θ to fluctuate with time at any particular point in the soil. Flow can then only be described by combining Darcy's equation with an equation for the conservation of mass – *the continuity equation*. The latter relates the rate of change of θ in a small soil volume to the change in water flux into and out of that volume, allowing for any water removed by evaporation and transpiration (ET) (Fig. 6.10). Assuming that flow is vertical only and that ET is negligible, the continuity equation for the soil volume in Fig. 6.10 is written as

$$\frac{\partial \theta}{\partial t} = -\frac{J_w}{\partial z}.$$ (6.18)

Substituting for J_w from Equation 6.14 we obtain the *Richards' equation*

$$\frac{\partial \theta}{\partial t} = \frac{\partial}{\partial z}\left\{ K(s)\left(\frac{\partial s}{\partial z} + 1 \right) \right\}.$$ (6.19)

The equation cannot be solved in this form because it contains two unknowns, s and θ. This problem is overcome by making use of the moisture characteristic function $s(\theta)$ (Section 6.4) to substitute for either θ or s in Equation 6.19. Recognizing that, since K is a function of s and s is a function of θ, K may be written as a function of θ, that is, $K(\theta)$, we have

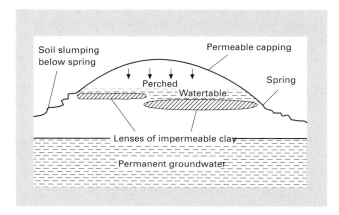

Fig. 6.9 A landscape profile showing perched watertables and spring lines.

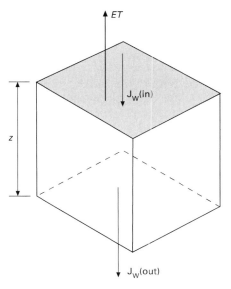

Fig. 6.10 Water flow through a soil volume under non-steady state conditions.

$$\frac{\partial s}{\partial z} = \frac{\partial s}{\partial \theta}\frac{\partial \theta}{\partial z} \qquad\qquad (6.20)$$

where $\partial s/\partial \theta$ is obtained from the moisture characteristic curve. On substitution in (6.14) we have

$$J_{\mathrm{w}} = -K(\theta)\frac{\partial s}{\partial \theta}\frac{\partial \theta}{\partial z} - K(\theta)\,. \qquad\qquad (6.21)$$

The expression $K(\theta)\dfrac{\partial s}{\partial \theta}$ is called the *soil water diffusivity* $D_{\mathrm{w}}(\theta)$. On substitution for J_{w} in Equation 6.18 we obtain the water content form of the Richard's equation

$$\frac{\partial \theta}{\partial t} = \frac{\partial}{\partial z}\left(D_{\mathrm{w}}(\theta)\frac{\partial \theta}{\partial z}\right) + \frac{\partial K(\theta)}{\partial z}\,. \qquad (6.22)$$

This equation generally can only be solved numerically for specified initial and boundary conditions; examples are given by Klute and Dirksen (1986). Another example of the use of the continuity equation is in the mathematical description of solute leaching through soil (Box 6.7).

6.4 The moisture characteristic curve

Pore-size distribution

If a saturated soil core is placed on a porous plate attached to a hanging water column and allowed to attain equilibrium (Fig. 6.11a), the soil water content at the applied suction can be measured. For suctions > 85 kPa, the soil is usually brought to equilibrium with an air pressure which is then equal but opposite to the water suction in the soil pores. A graph of θ against soil suction is constructed, which is called the *moisture characteristic* or *moisture retentivity* curve, illustrated by curve A in Fig. 6.11b. The rate of change of θ with s ($\mathrm{d}\theta/\mathrm{d}s$) is the *specific moisture capacity.*

Since s is inversely proportional to the radius of the largest pores holding water at that suction (Equation 6.6), the volume of pores having radii between r and $r - \delta r$, where δr is a very small decrement in r corresponding to an increase in suction δs, can be deter-

mined from the slope of the moisture characteristic curve A, provided that shrinkage of the soil on drying is negligible. Furthermore, the larger the value of $\mathrm{d}\theta/\mathrm{d}s$, the greater is the volume of pores holding water within the size class defined by δr. Therefore, the value of s when $\mathrm{d}\theta/\mathrm{d}s$ is at a maximum defines (through the relationship $s \propto 1/r$) the most frequent pore-size class, as illustrated by curve B in Fig. 6.11b. The example chosen is typical of a well-structured soil showing an adequate volume of large pores between peds (macropores), which promote drainage and aeration, relative to fine pores within peds (micropores) that hold water in the so-called available range (see below).

Hysteresis

The shape of the moisture characteristic curve depends on whether the soil is drying or wetting – a phenomenon known as *hysteresis* which occurs for two main reasons.

(a) Large pores do not empty at the same suction as they fill, if access to the pore is restricted by narrow necks. Take, for example, a large pore of diameter D, which is connected to two smaller pores of diameter d_1 and d_2. On drying, the large pore will not empty until the suction is great enough to break the meniscus across the larger of the two connecting pores, diameter d_2 (Fig. 6.12a). On wetting, water will advance into the large pore through pore d_1 or d_2 at roughly the same suction because they are of similar size. However, the large pore will not fill completely until the suction is small enough to sustain a meniscus across the distance D (Fig. 6.12b). Because $D > d_1$ or d_2, the suction at which the large pore fills on wetting is less than the suction at which it empties on drying.

(b) For a given pore, the contact angle between the water meniscus and the pore walls is greater when the meniscus is advancing (wetting) than when it is retreating (drying). This means that the radius of curvature is greater for the advancing meniscus and that the suction developed is correspondingly less for the same water content.

The combined effect of (a) and (b) is illustrated in

Box 6.7 Modelling solute transport.

Infiltration of rain and subsequent drainage of water in the soil can lead to the redistribution of solutes. The net downward movement of solutes is called *leaching* (Section 5.3). On the assumption that the mechanisms involved are mass flow of solution (convection) and diffusion between regions of different concentration in adjacent pores, the solute flux density J_s moving vertically downwards through the soil can be written as

$$J_s = J_v + J_d \qquad (B6.7.1)$$

where the convective flux J_v is given by

$$J_v = J_w C \qquad (B6.7.2)$$

and the diffusive flux J_d is given by Fick's first law of diffusion

$$J_d = -D\theta \frac{\partial C}{\partial z}. \qquad (B6.7.3)$$

C is the concentration of the solute in the soil solution and D is its diffusion coefficient in the porous medium. Recalling the principle of mass conservation embodied in the continuity Equation 6.18, for solute flux through a defined volume of soil we may write

$$\left\{ \frac{\partial(\theta C)}{\partial t} \right\} = -\frac{\partial J_s}{\partial z}. \qquad (B6.7.4)$$

Substituting for J_s from Equations B6.7.1, B6.7.2 and B6.7.3 we have

$$\theta \frac{\partial C}{\partial t} = -\frac{\partial}{\partial z} \left\{ J_w C - D\theta \frac{\partial C}{\partial z} \right\}$$

and rearranging terms gives

$$\frac{\partial C}{\partial t} = D \frac{\partial^2 C}{\partial z^2} - \frac{J_w}{\theta} \frac{\partial C}{\partial z}. \qquad (B6.7.5)$$

In practice, if the substitution $J_w/\theta = \bar{v}$ (Equation 6.15) is made in Equation B6.7.5, the value of D obtained by solving Equation B6.7.5 is invariably found to be greater than the porous medium diffusion coefficient of the solute. This is due to the coupled effects of molecular diffusion and hydrodynamic dispersion on the solute as water flows through the heterogeneous pore space, a subject outside the scope of this book. For this reason, Equation B6.7.5 is written in the form

$$\frac{\partial C}{\partial t} = D_h \frac{\partial^2 C}{\partial z^2} - \bar{v} \frac{\partial C}{\partial z} \qquad (B6.7.6)$$

where D_h is called the *hydrodynamic dispersion coefficient* and (B6.7.6) is known as the *convection–dispersion* equation for solute transport.

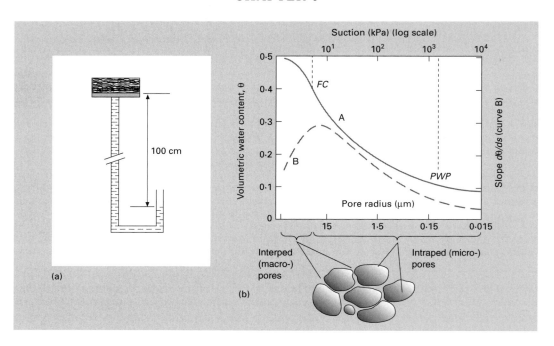

Fig. 6.11 (a) Soil water under a negative hydrostatic pressure of 100 kPa. (b) Graphs of the moisture characteristic (A) and specific moisture capacity (B) of a clay soil.

Fig. 6.12c which shows that the soil come to equilibrium at a higher water content for a given suction on the drying than the wetting phase of the cycle. A minor cause of hysteresis applies to clays that hold appreciable water at high suctions. It is thought to be due to changes in the size and arrangement of the smallest pores as water films contract and clay domains become re-oriented. Such changes may not be reversible. Similarly, the water content attained at zero suction at the end of the wetting phase may differ slightly from that at the start of the original drying phase because of air entrapment in the soil (Fig. 6.12c).

Available water capacity

The available water capacity (AWC) defines the amount of water in a soil that is nominally available for plant growth. The upper limit is set by the field capacity (FC) and the lower limit as the value of θ at which plants lose turgor and wilt; that is, the *permanent wilting point* (*PWP*). When related to the moisture characteristic curve, these water contents are found to correspond to suctions of between 5 and 10 kPa for FC and 1500 kPa for *PWP* for a number of soil–plant associations. It must be recalled from Section 6.3, however, that FC is an inexact concept for many soils. Further, the onset of permanent wilting depends as much on the plant's ability to lower the water potential in its tissues (to maintain a favourable gradient for water inflow from the soil) as it does on the soil water potential. For this reason, the choice of 1500 kPa as the suction at the *PWP* is arbitrary. Nevertheless, the use of AWC as a measure of the soil's reserve of plant-available water is widespread, particularly for assessing the soil water available to crops under dryland farming, and for calculating the amount of water to be applied to crops under irrigation (Section 13.2). As the soil water content falls below the FC, the difference between FC and θ defines the *soil moisture deficit* (*SMD*), and the

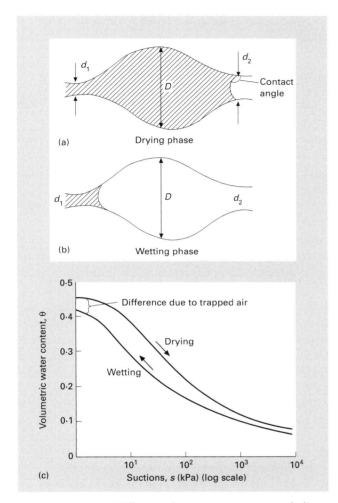

Fig. 6.12 (a) and (b) Differences in pore water content at similar suctions during wetting and drying. (c) Hysteresis in the soil moisture characteristic curve.

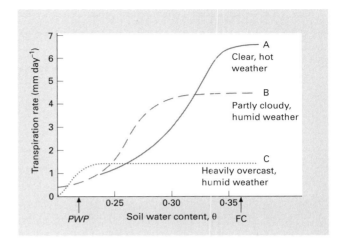

Fig. 6.13 The effect of atmospheric conditions and soil water content on plant transpiration rate (after Denmead and Shaw, 1962).

critical *SMD* for reapplication of irrigation water is often set as a fraction of the *AWC*.

It is sometimes assumed that water held between *FC* and *PWP* is equally available for growth, in the sense that a plant will transpire at a constant rate over the whole range. This is only approximately true if:
• the soil moisture characteristic curve is of a type that the bulk of the available water is held at low suctions at which the hydraulic conductivity of the soil is relatively high – this is true of many sandy and sandy loam soils but not of clay loams and clays;
• the evaporative power of the atmosphere is low enough that the rate of water movement from soil to root can satisfy the plant's transpiration. This is dem-onstrated by Fig. 6.13, which shows that when the evaporative power of the atmosphere is low (curve C), almost all the available water is transpired before the plant begins to suffer moisture stress and the transpira-tion rate falls. By contrast, when the evaporative power of the atmosphere is high (curve A), the transpiration rate drops markedly after only a small fraction of the *AWC* has been consumed. The plant then shows signs of wilting at θ values well above the *PWP*, but turgor is restored if cool, humid weather conditions supervene. This is a case of temporary wilting.

These observations illustrate the complexity of the interaction between soil properties (θ–suction and K–s relationships), plant properties (rooting depth, root morphology, salt accumulation, osmotic adjustment) and atmospheric factors affecting the plant's transpira-tion rate. The strong link between atmospheric factors

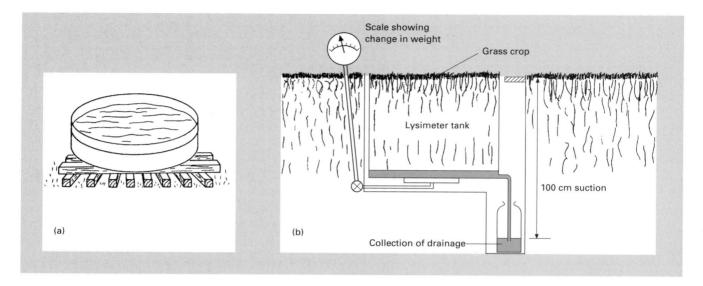

Fig. 6.14 (a) Class A evaporation pan. (b) A weighing lysimeter (after Pereira, 1973).

and transpiration rate, which determines the rate at which the plant must withdraw water from the soil to remain turgid, focuses attention on the physically based measurements of evaporation that are discussed in Section 6.5.

6.5 Evaporation

The energy balance

Evaporation from an open water surface E_o is measured using evaporation pans, which are shallow tanks of water exposed to wind and sun. An example of the widely used Class A pan is shown in Fig. 6.14a. Note that the pan should be raised above the soil surface to allow air to circulate and minimize any overheating of the water. Evapotranspiration ET can be measured directly using weighing *lysimeters*, which are large tanks filled with soil and vegetation (Fig. 6.14b). Where such a lysimeter is installed in a field and planted to the same crop, pasture or even trees as in the surrounding area, the evaporative conditions should be the same. Prefera-

bly the soil in the lysimeter should be undisturbed and have a small suction applied to simulate the condition of the natural soil. The ET (mm) for the crop and soil is calculated from the water balance equation (B6.4.1) for the lysimeter. When direct measurement is not possible, and especially for the estimation of ET losses on a regional scale, the *potential* rate of evaporation from any surface – open water, plant or wet soil – can be calculated from meteorological data. The basic information required is:
• the net input of energy into the system, and
• the capacity of the air to take up and remove water vapour.

The main source of energy is solar radiation of which a fraction is directly reflected, depending on the *albedo* of the surface (ranging from 0.15 for forests to 0.30 for agricultural crops), and some is lost as a return of long-wave radiation from the surface to the atmosphere. Of the *net radiation* absorbed R_N, part is dissipated by evaporation (component LE, where L is the latent heat of vaporization and E the rate of evaporation), part by the transfer of sensible heat to the air (Q) and part is transferred as heat to the soil, J_H, that is,

$$R_N = LE + Q + J_H . \tag{6.23}$$

The transport capacity of the air is estimated from the vapour pressure gradient over the surface and the average wind speed, with an added empirical factor to account for surface roughness when evaporation from vegetation, rather than a free water surface, is considered (Box 6.8).

For a fully vegetated surface, over a 24-h period, J_H in Equation 6.23 can be ignored (not so for bare soil and much longer periods – Section 6.6). The ratio of the components Q (sensible heat) to LE (latent heat) – Q/LE – defines the *Bowen ratio*. Under humid temperate conditions, Q/LE is small so to a reasonable approximation, Q may be neglected in Equation 6.23 and we may write

$$E \simeq \frac{R_N}{L}. \qquad (6.24)$$

Penman (1970) suggested that approximately 40% of the incoming radiation R_1 is dissipated in the evaporation of water from a short grass crop completely covering the soil surface and for which moisture is not limiting. Thus, substituting $R_N = 0.4\ R_1 = 219\ \text{J m}^{-2}\ \text{s}^{-1}$ and $L = 2.4 \times 10^9\ \text{J m}^{-3}$ in Equation 6.24 gives an evap-

oration rate E of 3 mm day^{-1}. This is an estimate of potential ET (E_p), which for the conditions specified, is comparable with the actual rates of ET under mild cloudy conditions shown in Fig. 6.13. However, when the soil moisture supply becomes limiting, actual rates of ET (ET_a) fall progressively below potential rates and formulas based on the soil's water-holding and supply capacity need to be used. Under drier conditions when plants may suffer water stress, the leaf surfaces will warm up and a much greater share of R_1 is lost to the air as sensible heat. In this case the Bowen ratio increases and the approximation used in Equation 6.24 is no longer valid. E_p then has to be calculated from the complete energy balance and the transport characteristics of the air (Box 6.8).

Evaporation from soil

Stages of drying

While the soil surface remains wet, the rate of evaporation is determined by the energy balance between the soil and the atmosphere and the transport capacity of

Fig. 6.15 Meteorological measuring instruments.

One of the most successful approaches to estimating evaporation from moist surfaces under non-limiting water supply conditions is due to Penman (1948), who developed an equation based on energy balance (Equation 6.23) and aerodynamic formulas for the turbulent transport of heat and water vapour from the evaporating surface to the air above. The Penman equation is

$$E_p = \frac{y}{(y + \gamma)L} \{R_N - J_H + (au + b)\Delta P_v\} \tag{B6.8.1}$$

where y is the rate of change of the saturation vapour pressure with temperature, γ is the psychrometer constant, ΔP_v is the vapour pressure saturation deficit above the canopy, u is the horizontal wind speed at 2 m height, and a and b are constants for a particular region. Doorenbos and Pruitt (1977) provide tables for calculating net radiation from solar radiation, temperature and other factors, and for selecting appropriate values for the constants a and b.

Another widely used equation, similar to Penman but which does not include a variable term for the vapour pressure deficit and wind speed, is that of Priestley and Taylor (1972)

$$E_p = \left[\alpha \frac{y}{y + \gamma} (R_N - J_H)\right]/L \tag{B6.8.2}$$

where α is a dimensionless constant and the other terms are defined above. α needs to be chosen for a particular region, but a value of 1.26 has been found to have wide applicability.

Examples of field instruments for measuring the meteorological variables used in the calculation of E_p are shown in Fig. 6.18.

Box 6.8 Calculation of potential rates of evaporation.

the air. Depending on the albedo, the rate can approach that of evaporation from an open water surface (commonly for wet soil $E = E_p = 0.8 E_0$). This stage is called *constant-rate* drying and may last for 2–14 days, depending on the soil type, latitude and season of the year.

With high evaporation rates from a soil, unaffected by a watertable, the rate of water loss soon exceeds the rate of supply from below and a surface layer, 1–2 cm deep, becomes air-dry. As the macropores empty of water first, the unsaturated conductivity for water flow across the dry zone from below decreases sharply and more than offsets the steep rise in suction gradient. The rate of evaporation is now controlled by soil properties and decreases with time, a stage known as *falling-rate* drying. The dry surface layer has a *self-mulching* effect because it helps to conserve moisture at depth, especially under conditions of high incident radiation in soils with a fine-granular or crumb structure.

Water movement in dry soil

Despite the self-mulching effect of an air-dry surface, if evaporation is prolonged and uninterrupted by rainfall, the zone of dry soil gradually extends downwards. This is because there is a slow flow of liquid water through soil wetter than the wilting point, to the drying surface, and there is also some diffusion of water vapour.

The flow of liquid water through relatively dry soil can be demonstrated by the simultaneous transfer of water and its dissolved salts in soil subjected to controlled, isothermal evaporative conditions. As Fig. 6.16 illustrates, the movement of water and chloride from the lower half of a series of soil blocks at initially uniform water contents is identical down to a water content only slightly above the wilting point. Below that it is mainly water that moves, by vapour diffusion, in response to vapour pressure gradients created by differences in matric suction, salt concentration or temperature. The effect of matric or osmotic suction on vapour pressure is normally small because, over the range 0–1500 kPa, the relative vapour pressure (e/e^0 in Equation 6.1) merely changes from 1.0 to 0.989. Only when the soil becomes air-dry and the relative vapour pressure of the soil water drops below 0.85 does the vapour pressure gradient under isothermal conditions become appreciable. Temperature, however, has a large effect on vapour pressure, producing a threefold increase between 10 and 30°C.

An exaggerated diurnal variation in temperature can cause redistribution of salt within the top 15 cm or so of a dry soil through the alteration of vapour and liquid phase transfer of water (Fig. 6.17). As the surface heats up during the day, water vapour diffuses into the cooler layers and condenses. Transfer and condensation of water means heat transfer so that the temperature gradient, and hence the vapour pressure gradient, gradually diminishes until the evening when, with concurrent surface cooling, the temperature differential disappears. At this point, the increased water content and matric potential in the condensation zone induces a return flow of water to the surface, carrying dissolved salts that are deposited when the water subsequently evaporates.

Capillary rise

Soil water above the watertable is drawn upwards in continuous pores until the suction gradient acting upwards is balanced by the gravitational head gradient. This rise of groundwater, known as *capillary rise*, depends very much on the pore-size distribution of the soil, being usually not greater than 1 m in sandy soils, but as much as 2 m in some silt loams. Thus, a watertable within 1–2 m of the soil surface can lead to excessive evaporative losses. The actual rate of loss is governed by the external atmospheric conditions up to the point where the conductivity of the soil becomes

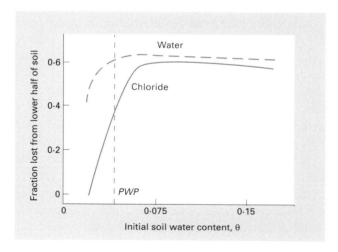

Fig. 6.16 The movement of water and chloride through soil blocks exposed to constant evaporative conditions (after Marshall, 1959).

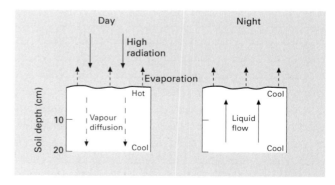

Fig. 6.17 Diurnal alternation of vapour and liquid water movement at the surface of a drying soil.

limiting. Accumulation of salts at the surface, as occurs in mismanaged irrigated soils, encourages capillary rise because the high osmotic suction maintains an upward head gradient even in moist soils, when the unsaturated hydraulic conductivity is high and appreciable water movement can therefore occur. The rise of groundwater often exacerbates the problem of *soil salinity*, which is discussed in Chapter 13.

6.6 Soil temperature

The effects of temperature on the physicochemical and biological reactions in soil are discussed in Sections 5.2, 5.3 and 8.1. This section deals with the causes and nature of soil temperature variations.

Diurnal and annual temperature variations

Clearly the net radiant energy R_N (Equation 6.23), and hence the soil heat flux J_H, will depend on the albedo a of the surface, the time of day and season of the year. Not only does the type of surface affect a (Section 6.5),

but a also varies with latitude (from *c.* 7% at the equator to 56% at the North Pole), and with aspect and slope in the landscape. J_H is negative during the day (heat flowing into the soil) and positive at night (heat flowing upwards from the soil to the air). This variation in flux causes the soil temperature to change in a sinusoidal way over a 24-h period (*diurnal* variation); the longer term variation in heat flux between summer and winter causes a similar sinusoidal temperature variation of much longer period τ (*annual* variation). These diurnal and annual variations can be described reasonably well by the equation

$$T(t) = T_A + A \sin\left(\frac{2\pi t}{\tau}\right) \tag{6.25}$$

where t is time, T_A is the average temperature, τ is the period of the wave (1 day or 1 year) and A the amplitude.

Variations in soil temperature are greatest at the surface and of decreasing amplitude with depth (Fig. 6.18). There is also a time lag between the attainment of either

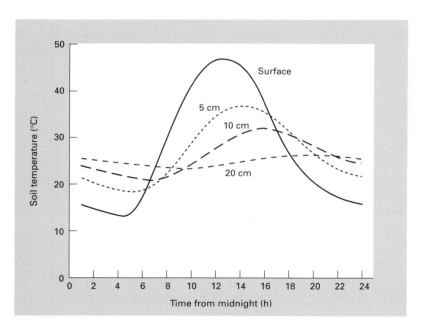

Fig. 6.18 Diurnal variations in soil temperature measured at different depths (after Jury *et al.* 1991).

the maximum or the minimum temperature which increases with depth. These effects are due to the nature of heat transport from the surface, given the time-dependent input as described by Equation 6.25 and the properties of the soil – in particular, its thermal conductivity λ_e and its heat capacity per unit volume C_h. The main processes involved in heat transport are conduction (involving molecular collisions) and convection of latent heat (involving the movement of water vapour – see Fig. 6.17), which can be described by the continuity equation for heat flow (*cf.* Equation 6.16)

$$\frac{\partial T}{\partial t} = K_T \frac{\partial^2 T}{\partial z^2}, \tag{6.26}$$

where z is depth and $K_T (= \lambda_e/C_h)$ is the *thermal diffusivity*.

The heat flux equation (6.26) can be solved using Equation 6.25 as an upper boundary condition, and the assumption that the amplitude of the temperature wave far below the surface is 0 (as $z \to \infty$). The solution of Equation 6.26 for these conditions is

$$T(z, t) = T_A + A \exp\left(-\frac{z}{d}\right) \sin\left(\frac{2\pi t}{\tau} + \frac{z}{d}\right) \tag{6.27}$$

where the damping depth d is given by

$$d = \left(\frac{K_T \tau}{\pi}\right)^{1/2}. \tag{6.28}$$

These results show that:
• for a given value of τ the time lag for temperature change is proportional to z^2 and inversely proportional to K_T,
• for given thermal properties, the damping depth $d \propto \sqrt{\tau}$, and
• all depths have the same average annual temperature.

The analysis confirms the observation that a rapidly changing surface temperature variation (diurnal) will not penetrate as deeply as a slowly changing temperature (annual period) of the same amplitude. Generally speaking, diurnal variations are small below 30 cm so this depth is commonly used to measure average soil temperatures. The amplitude of diurnal variations in surface temperature is much reduced by vegetation and surface mulches through their effects on J_H and the thermal conductivity of the surface. As well as having low thermal conductivity, light-coloured granular plastic mulches transmit incoming short-wave radiation but prevent the loss of long-wave thermal radiation, thereby creating a 'greenhouse effect' at the soil surface (Section 3.1).

6.7 Summary

The interaction between soil water and organo-mineral surfaces, exchangeable cations and dissolved salts reduces its potential energy. There is no absolute scale of potential energy, but we can measure *changes* in potential energy when useful work is done on a known quantity of water, or when the water does useful work. These changes are manifest as changes in the *free energy* per mole of the water (i.e. its chemical potential ψ_w) relative to pure water outside the soil at the same temperature and pressure and at a fixed reference height. The *soil water potential ψ* is therefore defined as

$$\psi = \mu_w(\text{soil}) - \mu_w^0(\text{standard state}) = RT \ln e/e^0.$$

where e/e^0 is the relative vapour pressure of the soil water at temperature T. ψ can be expressed as energy per unit volume, equivalent to pressure in bars or pascals; or as energy per unit weight, which is equivalent to *hydraulic head* in units of length. Head h and potential ψ are related through

$$h = \psi/\rho g.$$

The value of ψ may be partitioned between several component potentials attributable to matric, osmotic, pressure and gravitational forces, respectively, as

$$\psi = \psi_m + \psi_s + \psi_p + \psi_g.$$

Usually, ψ_m and ψ_p are regarded as parts of a continuum such that at normal atmospheric pressure in an unsaturated soil $\psi_p = 0$ and ψ_m is negative, whereas in saturated soil ψ_p is positive and $\psi_m = 0$.

The interchange of water between the atmosphere

and Earth's surface by precipitation P and evaporation E from soil, water and plants comprises the *hydrological cycle*. Within a defined catchment, $P = E$ + streamflow + deep drainage + change in soil storage, ΔS. Over a year or longer, ΔS is negligible so that streamflow and deep drainage determine the potential water storage, either in surface dams or as groundwater. Nevertheless, changes in ΔS are crucial in the short term and may decide crop success or failure.

Changes in soil volumetric water content θ can be measured by a neutron probe; changes in soil water suction s (equivalent to $-\psi_m$), up to $c.$ 85 kPa are measured by a tensiometer. Under static equilibrium conditions, the relationship between θ and s defines the *moisture characteristic curve*, the slope of which ($\partial\theta/\partial s$) depends on the pore-size distribution in non-swelling soils. Two points identified on this curve, the field capacity FC corresponding to suctions between 5 and 10 kPa and the permanent wilting point PWP corresponding to a suction of 1500 kPa, set the approximate upper and lower limits for the *available water capacity* (AWC) of the soil. However, the rate of water uptake by plants also depends on plant properties and atmospheric factors, the latter exerting an overriding effect on evaporation and transpiration (*evapotranspiration*) rates when soil moisture is non-limiting.

The flux of water J_w through the soil is given by *Darcy's equation*

$$J_w = -K \operatorname{grad} h$$

where K is the soil *hydraulic conductivity* (the reciprocal of the resistance to flow). For vertical flow in non-saline soils, the main driving forces are the gradients in matric and gravitational potentials, represented in terms of head by s and depth z, respectively. During saturated flow, $\partial s/\partial z = 0$ and gravity is the driving force, so that $J_w = -K_s$, the *saturated hydraulic conductivity*. As the soil desaturates K becomes a function of s, and once the large pores have drained, K falls very rapidly with increases in s. Even in unsaturated soil, flow to a deep watertable can be in steady state if $\partial s/\partial z$ is constant, but normally flow under unsaturated conditions is transient and both $K(s)$ and $\partial s/\partial z$ must be known to calculate J_w.

The change in water content in an unsaturated profile can be calculated by combining Darcy's equation with an equation for the conservation of water mass to give the *Richards' equation*

$$\frac{\partial\theta}{\partial t} = \frac{\partial}{\partial z}\left\{ K(s)\left(\frac{\partial s}{\partial z} + 1 \right) \right\}.$$

Solute transport through structurally homogeneous soils can be described by combining equations for mass flow and diffusion, which give rise to the *convection–dispersion equation*.

Evaporation from wet soil surfaces or vegetation well supplied with water proceeds at a potential rate E_p, determined by the net energy input R_N and the meteorological conditions, and can be estimated using the Penman equation. When the supply of water is limiting, actual evaporation falls below E_p. For bare soil, the evaporation rate falls to a very low level once the top 1–2 cm has become air-dry. Vapour phase transport is insignificant in non-saline soils wetter than the PWP unless a temperature gradient exists between different layers in the soil.

Soil temperature is determined by the net heat flux J_H, a component of R_N. The temperature at the surface fluctuates in an approximately sinusoidal wave both diurnally (period 1 day) and annually (period 1 year). The phase lag in the wave increases with depth, but the amplitude of the temperature variation decreases.

References

Denmead O. T. & Shaw R. H. (1962) Availability of soil water to plants as affected by soil moisture content and meteorological conditions. *Agronomy Journal* **54**, 385–390.

Doorenbos J. & Pruitt W. O. (1977) *Guidelines for predicting crop water requirements*. FAO Irrigation and Drainage Paper No. 24. FAO, Rome.

Green R. E., Ahuja L. R. & Chong S. K. (1986) Hydraulic conductivity, diffusivity and sorptivity of unsaturated soils: field methods, in *Methods of Soil Analysis. Part 1. Physical and Mineralogical Methods*, 2nd edn (Ed. A. Klute). Agronomy Monograph No. 9. American Society of Agronomy/Soil Science Society of America, Madison, pp. 771–798.

Jury W. A., Gardner W. R. & Gardner W. H. (1991) *Soil Physics*, 5th edn. Wiley, New York.

Klute A. & Dirksen C. (1986) Hydraulic conductivity and diffusivity: laboratory methods, in *Methods of Soil Analysis. Part 1. Physical and Mineralogical Methods*, 2nd edn (Ed. A Klute). Agronomy Monograph No. 9. American Society of Agronomy/Soil Science Society of America, Madison, pp. 687–734.

Marshall T. J. (1959) *Relations between Water and Soil.* Commonwealth Bureau of Soils, Technical Communication No. 50.

Penman H. L. (1948) Natural evaporation from open water, bare soil and grass. *Proceedings of the Royal Society (London) Series A* **193**, 120–145.

Penman H. L. (1970) The water cycle. *Scientific American* **223**, 98–108.

Pereira H. C. (1973) *Land Use and Water Resources in Temperate and Tropical Climates.* Cambridge University Press, Cambridge.

Priestley C. H. B. & Taylor R. J. (1972) On the assessment of surface heat flux and evaporation using large-scale parameters. *Monthly Weather Review* **100**, 81–92.

Youngs E. G. & Thomasson A. (1975) Water movement in soil, in *Soil Physical Conditions and Crop Production.* MAFF Bulletin No. 29, pp. 228–239.

Further reading

Hillel D. (1982) *Introduction to Soil Physics.* Academic Press, New York.

Kutilek M. & Nielsen D. R. (1994) *Soil Hydrology.* Catena, Cremlingen-Destedt, Germany.

Marshall T. J., Holmes J. W. & Rose C. W. (1996) *Soil Physics*, 3rd edn. Cambridge University Press, Cambridge.

Monteith J. L. & Unsworth M. H. (1990) *Principles of Environmental Physics*, 2nd edn. Edward Arnold, London.

Chapter 7
Reactions at Surfaces

7.1 Charges on soil particles

Some definitions

Chapter 2 introduced the concept of:
- *permanent charge* on soil minerals arising out of the substitution of elements of similar size but different valency within the crystal lattice, and
- *pH-dependent charge* arising through the reversible dissociation of protons from surface groups according to their acid strength and the activity of H^+ ions in the ambient solution.

The net permanent charge created by isomorphous substitution in clay minerals (mainly the 2 : 1 clays) is invariably negative and is unaffected by the concentration and type of ions in the soil solution, unless the pH is so low that it induces decomposition of the crystal lattice. With some of the hydrated oxides, there can be a small positive permanent charge due to the substitution of cations of higher valency within the crystal lattice. The result is a mineral of constant surface charge density (σ_0) and variable electrical potential ϕ. On the other hand, the reversible dissociation of protons from carboxyl and phenolic groups in organic polymers, or at O and OH groups in the surfaces of oxides and edge faces of kaolinite crystals, results in a *variable* surface charge density (σ_H). H^+ (and OH^-) are called *potential-determining* ions for such surfaces, and σ_H measures the difference between the moles H^+ and OH^- complexed by the surface. The electrical potential of the surface remains constant provided the pH of the solution does not change (see Equation B7.2.2). Hence the expression 'constant potential' surface which is sometimes used. When surfaces of constant and variable

charge occur on the one mineral, such as kaolinite and some allophanes (Section 2.4), the resultant surface charge density is called the *intrinsic* surface charge density (σ_{in}), defined by

$$\sigma_{in} = \sigma_0 + \sigma_H . \tag{7.1}$$

Methods of measuring or calculating σ_0, σ_H and σ_{in} are discussed by Sposito (1989). σ_{in} is an operationally defined parameter because its value will depend on the experimental conditions under which it is measured.

The charged particle–solution interface

The existence of immobile charges on a mineral surface produces a change in the distribution of mobile charges (ions) in the soil solution in contact with the surface. A simple example is provided by a group of aluminosilicates called the *zeolites*, in which the permanent negative charge is distributed throughout the crystal lattice to give a uniform volume charge. Ions dissolved in the solution permeating the mineral can diffuse through a network of interconnecting pores, undergoing frequent changes in electrical potential as they move from the influence of one lattice charge to another. However, when equilibrium is reached between the exchanger and the bathing salt solution, each ion suffers an average change in electrical potential on passing into or out of the exchanger, the value of which measures the *Donnan membrane potential* or simply the *Donnan potential*. Cations are retained, or positively adsorbed, within the exchanger and anions are excluded or negatively absorbed. The volume from which anions are effectively excluded is called the *Donnan free space* (Fig. 7.1). This model applies equally well to soil humic polymers and

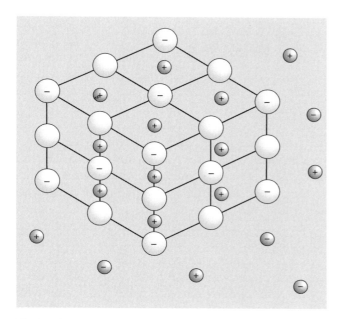

Fig. 7.1 Concentration of mobile cations (+) within the Donnan free space and exclusion of mobile anions (–).

plant constituents, cell walls and cytoplasmic proteins, except that the charges on these are not permanent but pH dependent.

Non-uniform volume charge

Unlike the zeolites, the crystalline clay minerals have thin laminar structures (see Section 2.4 and Box 2.2) in which the permanent negative charge acts as if it were spread over the planar surfaces of the crystals. The negative charge creates a surplus of cations (the *counterions*) and deficit of anions (the *co-ions*) in the solution immediately adjacent to the surface. The simplest case is one in which each surface charge is neutralized by the close proximity of a mobile charge of opposite sign – the parallel alignment of charges in two planes is called a *Helmholtz double layer* (Fig. 7.2).

However, the Helmholtz model is unrealistic in aqueous solution. The high dielectric constant of water diminishes the electrostatic force of attraction between the fixed and mobile charges, which is also opposed by

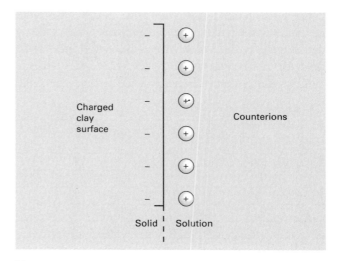

Fig. 7.2 The Helmholtz double layer.

a diffusive force tending to draw the cations out of the region of high concentration near the charged surface. At equilibrium the electrostatic force across a plane of unit area near to and parallel to the surface is just balanced by the difference in osmotic pressure between the plane and the bulk solution far removed from the surface. The result is a diffuse distribution of cations and anions in solution (Fig. 7.3a), which together with the surface charge comprises the *Gouy–Chapman dou-*

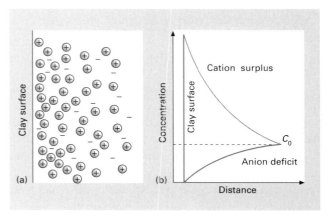

Fig. 7.3 (a) and (b) The ionic distribution at a negatively charged clay surface.

ble layer, so named after the two scientists who independently developed the mathematical description of the diffuse space charge. Such a diffuse double layer (DDL) can develop at surfaces of constant charge or variable charge.

Characteristics of the Gouy–Chapman double layer

The distribution of ions in solution in the direction normal to the charged planar surface (Fig. 7.3a) is given by the Boltzmann equation

$$C(x) = C_0 \exp[-zF\phi(x)/RT] \qquad (7.2)$$

where $C(x)$ is the concentration of an ion of valency z (cation or anion) at a distance x from the surface where the potential is $\phi(x)$, C_0 is the ion concentration at a plane in the bulk solution, F is Faraday's constant and R and T have their usual meaning (see Equation 6.1). It is implicit in Equation 7.2 that the potential in the bulk solution is zero. Thus, for a negatively charged surface, as x becomes smaller and the absolute value of $\phi(x)$ increases, the concentration $C(x)$ of cations increases exponentially relative to C_0. At the same time, the concentration of anions decreases exponentially, becoming vanishingly small as the surface is approached. The concentration profiles for cations and anions in solution near a negatively charged surface are illustrated in Fig. 7.3b.

At a clay mineral planar surface, the volume density of charge in the diffuse layer $\sigma(x)$ must remain constant irrespective of changes in the ionic composition of the solution. For each square centimetre of surface, the solid phase charge is neutralized by the surplus of cations over anions in solution, such that

$$\sigma(x) = \sigma(+) + \sigma(-) \qquad (7.3)$$

where $\sigma(+)$ and $\sigma(-)$ are the cationic and anionic charges in the diffusive layer (cmol($+$) and ($-$) cm^{-2}, respectively). (Note that when $x = 0$, $\sigma(x) = \sigma_0$ the density of permanent negative charge on the planar clay surfaces.) As in a parallel plate condenser, the capacity of the DDL increases as the thickness of the diffuse

Table 7.1 Calculated diffuse layer thickness for different concentrations of 1 : 1 and 2 : 1 electrolytes.

Electrolyte concentration C_0 (mmol cm^{-3})	Effective diffuse layer thickness, d_{ex} (nm)	
	NaCl	CaCl$_2$
0.1	1.94	1.0
0.01	6.2	3.2
0.001	19.4	10.1

layer decreases. The effective thickness of the diffuse layer is measured by the parameter d_{ex}, which is the mean distance over which a co-ion is depleted at the charged surface. It can be shown that d_{ex} depends on the valency of the counterion and the concentration C_0 of the bulk solution, according to the equation

$$d_{ex} \simeq \frac{2}{z\sqrt{\beta C_0}} \qquad (7.4)$$

where β, the DDL constant, has the value 1.084×10^{16} cm cmol^{-1}. Calculated values of d_{ex} for different concentrations of NaCl and CaCl$_2$ solutions in contact with clay surface are given in Table 7.1. The figures illustrate how, as the valency of the cation increases or the solution concentration increases, the diffuse layer is compressed.

Modifications to the Gouy–Chapman double layer

The Gouy–Chapman theory works well for cation solutions of intermediate concentrations (0.1–0.0001 M) and surface charge densities of $1–4 \times 10^{-4}$ cmols m^{-2}. It breaks down once certain limits of cation valency and solution concentration are reached because:
• the ions are assumed to be point charges, which leads to absurdly high concentrations of cations in the inner region of the diffuse layer, and
• the theory ignores forces between the surface and counterions other than simple electrostatic forces. It predicts that cations of the same valency will be adsorbed with the same energy, whereas it is known from

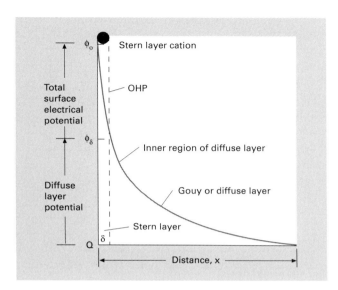

Fig. 7.4 Electrical potential gradients at a planar clay surface (after van Olphen, 1977).

cation exchange measurements that Li^+ and K^+, or Ca^{2+} and Ba^{2+}, for example, are not held with the same tenacity by clays in contact with solutions of identical normality.*

These factors are accommodated in the *Stern model* – essentially a combination of the Helmholtz and Gouy–Chapman concepts – which splits the solution component of the double layer into two parts.

(a) A plane of counterions (cations) of finite size located within a few Ångströms† of the surface, the Stern layer, across which the electrical potential decays linearly.

(b) A diffuse layer of cations, across which the potential falls almost exponentially, the inner surface of which abuts onto the Stern layer and is called the outer Helmholtz plane (*OHP*) (Fig. 7.4). Allowance is made in the Stern model for the reduction in the dielectric constant of water due to its ordered structure very near the surface, and for specific adsorption forces between

---

* A normal (N) solution is one containing 1 mmol of cationic or anionic charge per cm^3.
† 1 Ångström (Å) = 10^{-10} m.

the surface and Stern layer cations (Box 7.1).

The Stern model predicts that an increase in solution concentration not only compresses the double layer but also increases the proportion of ions in the Stern layer. The diffuse layer potential ϕ_δ decreases relative to the total surface potential ϕ_0 and the repulsion of anions from the inner region of the diffuse layer is diminished. Cations in the diffuse part of the DDL are said to be non-specifically adsorbed.

Even the Stern model of the DDL is incomplete, because although inner-sphere complexes (Stern layer cations) are accounted for, outer-sphere complexes are not. This shortcoming has led to the development of models based on the molecular properties of the interface (surface complexation models), as described by Sposito (1994).

Surfaces of variable charge

Oxide surfaces

Free oxides in soil (Section 2.4) exhibit pH-dependent charges due to the reversible adsorption of potential-determining H^+ ions. Of these oxides, the Fe and Al oxides and oxyhydroxides are quantitatively the most important. The change in surface charge with pH change is illustrated for hydrated Al_2O_3 in Fig. 7.5. The O^-, OH and OH_2^+ groups of the oxide surface lie in the inner Helmholtz plane (*IHP*). Naturally, when the surface has a net positive charge, anions are attracted from the solution to form a double layer. If the charge on the surface is all balanced by counterions forming inner- and outer-sphere complexes, so that the diffuse layer potential is zero, then the solution pH defines the *point of zero charge* (*PZC*) of the solid (Box 7.2).

Clay mineral edge faces

Aluminium atoms coordinated to O and OH groups are also exposed at the edges of clay crystals. This structure differs from that of the oxides only in that the oxygens are also coordinated to Si atoms in at least one contiguous silica layer. The full negative charge of two electrons

Box 7.1 Specific adsorption of cations.

At a distance from the surface less than 4 Å the assumption of a plane of uniform surface charge does not hold and cations in the Stern layer experience a polarization force in the vicinity of each charged site. Depending on the balance of forces – the polarizing effect of the cation on the water molecules in its hydration shell compared with the polarization of the cation by the surface charge – the cation may shed its water of hydration to enter the Stern layer where it forms an *inner-sphere* complex (Section 2.3) with surface groups of a siloxane trigonal cavity. (The plane of O atoms in the surface of a 2 : 1 layer silicate is called a siloxane surface.) The cations most likely to do this belong to elements of Group I in the Periodic Table. Lithium, Na, K, Rb and Cs form a series in which the ionic potential and hence hydration energy of the monovalent cation decreases with increasing atomic size. The small cation Li^+ tends to remain hydrated at the surface whereas the large Cs^+ ion tends to dehydrate and become tightly adsorbed.

The attractive force between the charged surface and the solution cations, over and above the simple electrostatic force, is called a *specific adsorption* force and the cations are said to be specifically adsorbed. Cations held in the Stern layer are continually exchanging with cations in the diffuse layer, but at any instant the proportion held in the Stern layer increases from about 16% for Li^+, to 36% for Na^+ and 49% for K^+, and the overall strength of adsorption increases in the sequence Li < Na < K < Rb < Cs.

Divalent cations, such as those formed by the Group II elements Mg, Ca, Sr, Ba and Ra, can also form inner-sphere complexes at clay surfaces, but because of their much higher ionic potentials (and hydration energies) they tend to retain some of their hydration water. For example, when Ba^{2+} is adsorbed on vermiculite, a monolayer of water is retained between layers of the clay crystal. However, K^+ which has an almost identical ionic radius to Ba^{2+}, is adsorbed without any hydration water. This retention of hydration water results in the divalent cations forming *outer-space* complexes with siloxane surface (one layer of water molecules between the surface and the cation), which is the basis for clay quasi-crystal formation when Ca^{2+} is the predominant cation (Section 4.4). The strength of adsorption of the divalent cations of Group II increases in the order Mg < Ca < Sr < Ba < Ra.

Following from Equation 7.1, the density of net charge on a clay particle (σ_p) may be written as

$$\sigma_p = \sigma_0 + \sigma_H + \sigma_{IS} + \sigma_{OS} \tag{B7.1.1}$$

Continued

Box 7.1 *Continued*

where σ_{IS} is the charge density due to ions forming inner-sphere complexes (other than H^+ and OH^-) and σ_{OS} is the charge density due to outer-sphere complexes. For clays, σ_p is almost invariably negative and the balance of charge for the particle and its whole ionic atmosphere is made up by σ_d, so that

$$\sigma_p + \sigma_d = 0 \qquad\qquad (B7.1.2)$$

where σ_d is the diffuse layer charge density comprising the surplus of cations *not* complexed with the surface in any way, over the uncomplexed anions, according to Equation 7.3. Equation B7.1.2 is a statement of the *charge balance* for the surface and its surrounding solution.

Box 7.2 Points of zero charge for mineral surfaces.

Points of zero charge are pH values at which one or more of the surface charge components in Equation B7.1.2 is zero. The most general value – the *PZC* as defined – occurs when $\sigma_d = 0$, so that σ_p must equal zero too. It follows from Equation B7.1.1 that when $\sigma_p = 0$

$$\sigma_H = -(\sigma_0 + \sigma_{IS} + \sigma_{OS}) \qquad (pH) = (PZC) \qquad (B7.2.1)$$

The value of σ_H changes inversely with the pH of the solution in contact with the surface (pH_s), and the potential of such a surface (H^+ is the potential-determining ion) is related to the difference between *PZC* and pH_s by the Nernst equation

$$\phi_0 = \frac{2.303RT}{zF}(PZC - pH_s) \qquad\qquad (B7.2.2)$$

Thus as pH_s decreases, σ_H increases (+) and ϕ_0 becomes more positive.

In soil, various inorganic cations and anions can form inner- and outer-sphere complexes with oxide surfaces. If $(\sigma_{IS} + \sigma_{OS})$ in Equation B7.2.1 increases through net cation adsorption, σ_H must decrease and pH = *PZC* will increase (the acid strength of the surface weakens).

Conversely, if $(\sigma_{IS} + \sigma_{OS})$ decreases through net anion adsorption, σ_H must increase and the pH = *PZC* decreases. Thus, the formation of surface complexes changes the *PZC* of the pure mineral in the same direction (+ or –) as the change in net surface charge $(\sigma_{IS} + \sigma_{OS})$. This has implications for the retention of nutrient ions and for particle to particle interactions (Section 7.4). The *PZC* for common soil oxides and oxyhydroxides ranges from 2–3 for quartz and silica to 7–8 for goethite and 8–9 for gibbsite.

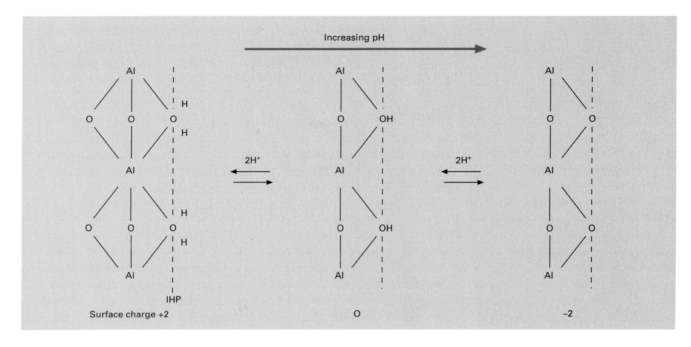

Fig. 7.5 Changes in surface charge on an aluminium oxide.

per 0.4 nm^2 of edge area is only developed above pH 9 when the $\equiv$Si—OH groups dissociate (Fig. 2.16). The association of H$^+$ with the O and OH coordinated to the aluminium increases as the pH decreases from 9 to 5 in the manner illustrated in Fig. 7.6, the *PZC* of the edge face occurring *c*. pH 6 for kaolinite. Thus, the edge faces of the clay minerals show a pH-dependent charge.

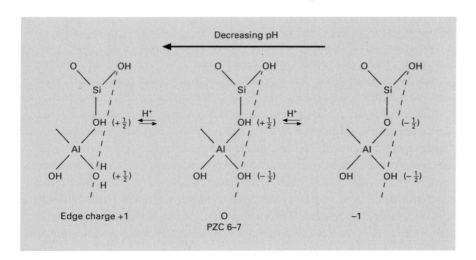

Fig. 7.6 Charge development at the edge face of kaolinite with change in pH.

Organic matter

Although the charge of humified organic matter is completely dependent on pH due to the dissociation of carboxyl and phenolic groups (Section 3.4), the charge is negative from pH 3 upwards and so augments the permanent negative charge of the clay minerals. In soils of high organic content, however, it has the effect of considerably increasing the pH-dependence of the soil's CEC.

The significance of pH-dependent charges in soil

Edge charges are of greatest significance in the kaolinites which have low permanent charges (2–5 cmol kg^{-1}) and high edge : planar area ratios (1 : 10–1 : 5). The development of one positive charge per 0.4 nm^2 of edge area amounts to 4×10^{-4} cmol m^{-2}, or 0.4–4 cmol kg^{-1} of kaolinite ranging in edge face area from 1 to 10 m^2 g^{-1}. Conversely, edge charges are unimportant in the 2 : 1 clay minerals which have high permanent negative charges (40–150 cmol kg^{-1}) and exist as smaller crystals, especially in the direction of the c axis, thereby offering a smaller edge: planar ratio than the kaolinites. The charge characteristics of soil clays may, however, be considerably modified by the presence of sesquioxides in variable amounts and spatial distribution.

The formation and occurrence of iron and aluminium oxides is discussed in Section 2.4. With *PZC* values in the range 7–9, they are normally positively charged in soil and can occur as thin films adsorbed on the planar surfaces of clays, as illustrated by the Fe(OH)$_3$ film precipitated on kaolinite at pH 3 (Fig. 7.7). Al(OH)$_3$ films occur preferentially within the interlayer spaces of mica-type clay minerals to form mixed layer minerals. Positive charge densities of 30–50 cmol kg^{-1} have been recorded for soil oxides, so that soils low in organic matter which contain mainly kaolinite and sesquioxides can developed a *net positive* charge at low pH. Such behaviour is prevalent in highly weathered soils of the tropics, exemplified by an acid subsoil from South Africa, the charge vs. pH curve of

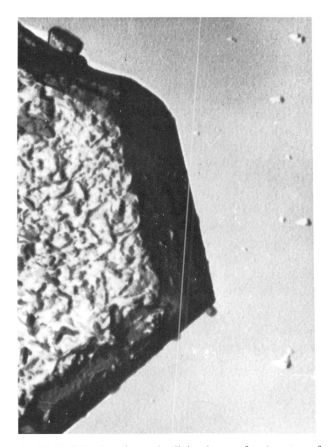

Fig. 7.7 Fe(OH)$_3$ deposit on a kaolinite cleavage face (courtesy of D.J. Greenland).

which is illustrated in Fig. 7.8. The presence of significant positive-charge densities, wholly dependent on pH, has profound implications not only for the ability of the soil to retain cations, but also for the adsorption of anions, two topics that are discussed in the following sections.

7.2 Cation exchange

Exchangeable cations

The general reactions involved in mineral weathering, especially the process of *hydrolysis* (the splitting of water molecules), are described in Section 9.2. A high proportion of the cations released by weathering and organic decomposition are adsorbed by clay and organic colloids, leaving a low concentration in the soil solution to be balanced by mineral anions, and bicarbonate generated from the respiration of soil organisms. Apart from those forming inner-sphere complexes, cations held in the double layer are called *exchangeable* because they can exchange rapidly with cations in the bulk solution. In all but the most acid or alkaline of soils, the major cations are Ca^{2+}, Mg^{2+}, K^+ and Na^+, in typical proportions of 80% Ca^{2+}; 15% Mg^{2+}; 5% ($Na + K$) with variable amounts of NH_4^+, depending on the extent of microbial nitrification (Section 8.3). Trace amounts of other cations such as Cu^{2+}, Mn^{2+} and Zn^{2+} are also adsorbed, although chelation with organic compounds and precipitation play a more prominent part in the retention of these elements (Section 10.5).

Cation exchange capacity and 'base saturation'

Cation exchange capacity (*CEC*) is an expression of the negative charge per unit mass of soil (Section 2.5).

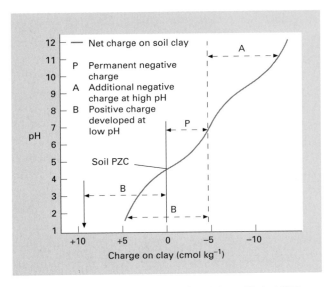

Fig. 7.8 Charge *vs* pH curve for an acid red loam (Oxisol (ST) or Ferrosol (A)) (after Schofield, 1939).

However, *CEC* is measured by the amount of exchangeable cations (or a replacement 'index' cation) held per unit mass of soil, so it is an operationally defined parameter dependent on the conditions of measurements. Several methods have been developed to measure *CEC* which may give different results for the same soil, especially if the soil has appreciable variable charge. Soluble salts can also inflate the results. For these reasons, it is most important to specify the method of measuring *CEC* (Box. 7.3).

In calcareous soils the sum of the cations (Σ Ca, Mg, K, Na) is invariably equal to the *CEC* since any deficit of the cations on the exchanger can be made up by Ca^{2+} ions from the dissolution of $CaCO_3$. In non-calcareous soils, however, Σ Ca, Mg, K, Na is frequently less than the *CEC*, the difference being referred to as the *exchange acidity* (in cmol H^+ kg^{-1} soil). The ratio

$$\frac{\sum \text{Ca, Mg, K, Na}}{CEC} \times 100$$

has been called the *per cent base saturation*. The cations involved are called 'basic' simply to distinguish them from the more acidic cations such as Fe^{3+} and especially Al^{3+}, which hydrolyse to release H^+ ions at comparatively low pH (see below). Base saturation is crudely correlated with soil pH in many mildly acid and neutral soils, ranging from *c.* 20 to > 60% as the pH increases from 5 to > 7.

Acid clays are Al-clays

Exchange acidity is *not* primarily due to H^+ ions. As the soil pH decreases below 5, increasing amounts of Al^{3+} ions can be displaced by leaching with strong salt solutions. Under very acid conditions (pH < 4) the weathering of clay minerals is accelerated, with the release of Al, SiO_2 and smaller amounts of Mg, K, Fe and Mn (Fig. 7.9). While the silicic acid formed is removed by leaching, the Al, Mg, K and Mn are retained initially as exchangeable cations. The hydrated ferric ion $(Fe(H_2O)_6)^{3+}$ is hydrolysed at pH < 3 and therefore readily precipitates, in the pH range 4–5, as ferric hydroxide or as the insoluble carbonate or phosphate if

Box 7.3 Measurement of soil *CEC*.

Soil *CEC* methods fall into two groups, namely:
- exchangeable cations are displaced at a fixed pH by a concentrated salt solution, such as 1 M NH_4Cl at pH 7 or M NH_4OOCH_3 at pH 7. Displacement is effected by mass action. The 'index' cation (NH_4^+) is then displaced by another cation (e.g. Na^+) and the amount of NH_4^+ adsorbed (cmol (+) kg^{-1}) is measured;
- exchangeable cations are displaced by an unbuffered salt solution at no fixed pH and the amount of 'index' cation which has been adsorbed (e.g. Ba^{2+} from 0.1M $BaCl_2$/0.1M NH_4Cl solution), or which remains in solution (Ag^+ from 0.01M silver thiourea solution) is measured. Displacement is effected because of the high affinity of the index cation for the surface.

A pre-wash with aqueous ethanol or glycerol is used to remove soluble salts. The exchangeable cations are measured in the extracts; Al^{3+} and H^+ are usually measured separately by displacement in M KCl.

CEC values determined by methods of the first group are widely reported in soil survey data and are used in some soil classifications. In soils of variable charge, *CEC* at pH 7 will overestimate the effective *CEC* (*ECEC*); that is, the *CEC* of the soil at its natural pH. For such soils, methods of the second group are preferred. Details of all these methods are given in specialist texts, such as Rayment and Higginson (1992) and Page (1982).

the appropriate anions are present. With the exception of illitic clay soils, K^+ is also lost by leaching so that an

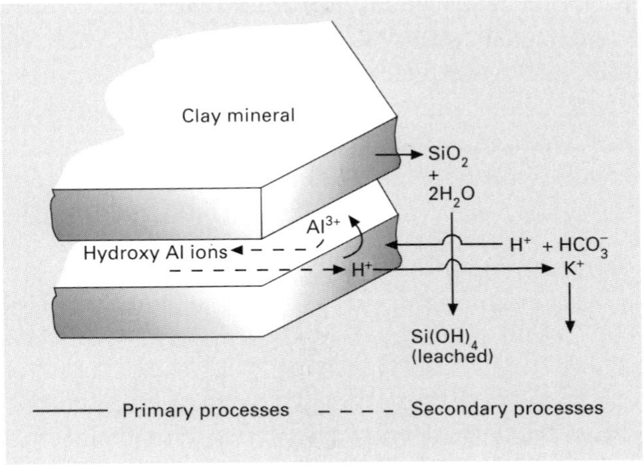

Fig. 7.9 Clay mineral weathering in an acidic environment.

acid clay remains which is dominated by Al^{3+} with some Mg^{2+} (and H^+ ions produced by the hydrolysis of the Al^{3+}).

pH buffering capacity and titratable acidity

High concentrations of exchangeable aluminium greatly increase the soil's capacity to neutralize OH^- ions, which is a measure of the soil's *pH buffering capacity*. The hydrated Al^{3+} ion hydrolyses in water according to the reaction

$$(Al6H_2O)^{3+} + H_2O \leftrightarrow [Al(OH)(H_2O)_5]^{2+} + H_3O^+ \quad (7.5)$$

there being equal concentrations of the di- and trivalent Al species at pH 5. The hydroxyaluminium ions $(AlOH)^{2+}$ show a pronounced tendency to combine through bridging OH groups and build up into polymeric units of six and more Al atoms. The consequent release of H^+ contributes to the buffering capacity until

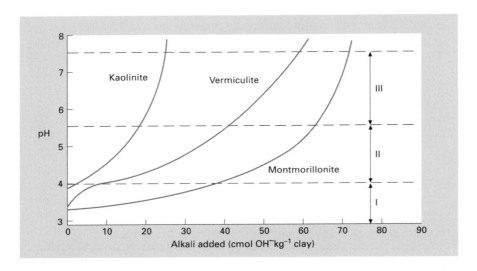

Fig. 7.10 pH-titration curves for kaolinite, vermiculite and montmorillonite in 0.1 M KCl.

eventually neutralization is complete and amorphous $Al(OH)_3$ precipitates on the external surfaces and interlayer spaces of the clay.

Three stages are recognized in the neutralization of acid clays. These are illustrated in Fig. 7.10 where representative curves for kaolinite, vermiculite and montmorillonite titrated in a neutral salt solution are plotted. Provided that sufficient time is allowed for near-equilibrium pH values to be attained, the three clays show markedly different buffering within the ranges:

I   < pH 4   exchangeable H^+

II   pH 4–5.5   exchangeable Al^{3+}

III pH 5.5–7.5   hydroxyaluminium ions.

Some 'weak' acidity due to proton release from hydroxy-Al polymers continues to be neutralized at pH > 7.5. The quantity of acid neutralized when a soil is equilibrated with an 0.1M $BaCl_2$/triethanolamine solution at pH 8.2 is a measure of its *titratable acidity* (cmol H^+ kg^{-1}). In practice, titration of a soil to a specific pH is used to estimate a soil's *lime requirement* (Section 11.3). In soils with much interlayer and surface hydroxy-Al films, the titratable acidity is always greater than the exchange acidity because part of the *CEC* is

blocked by the hydroxy-Al polymers which are not displaced in neutral salt solution.

Al is extensively hydrolysed when adsorbed by organic matter, the average charge per mole of adsorbed Al being +1. The Al^{3+} ions react preferentially with the stronger acid groups, and H^+ ions released by hydrolysis suppress the dissociation of weaker acid groups on the organic matter. The apparent pH of Al-free organic matter is between 4 and 5 but rises to *c.* 6 when Al is added. Thus, the adsorption of Al decreases the acidity of soil organic matter and increases its capacity to adsorb anions such as phosphate (Section 7.3).

Cation exchange reactions

Exchange equations

The technique of using a strong solution of NH_4^+ ions to displace the exchangeable cations and determine the *CEC* is an example of *cation exchange*. Cation exchange is a reversible process in which 1 mole of cation charge (or equivalent)* in solution replaces 1 mole of cation charge on the exchanger, as for example

* See Box 2.3 for an explanation of moles of charge and equivalents.

$$(NH_4^+) + (Ca\text{-exchanger}) \leftrightarrow (NH_4\text{-exchanger})$$
$$+ \tfrac{1}{2}(Ca^{2+}) \qquad (7.6)$$

where (NH_4^+) and (Ca^{2+}) are the molar activities of the ions in solution, and $(Ca\text{-exchanger})$ and $(NH_4\text{-exchanger})$ represent the activities of the ions on the exchanger. Following the law of mass action, the equilibrium constant for this exchange reaction can be written as

$$K_{ex} = \frac{(NH_4\text{-exchanger})}{(Ca\text{-exchanger})} \frac{(Ca^{2+})^{1/2}}{(NH_4^+)} . \qquad (7.7)$$

Much has been written about the validity of cation exchange equations, which all founder on the problem of measuring the *activity* of ions in the adsorbed state. However, if the ion activity ratio on the exchanger is assumed equal to the equivalent fraction there, and concentrations are substituted for activities in solution, Equation 7.7 can be rearranged to give the empirical *Gapon equation*

$$\frac{NH_4\text{-exchanger}}{Ca\text{-exchanger}} = K_{NH_4,Ca} \frac{[NH_4^+]}{[Ca^{2+}]^{1/2}} . \qquad (7.8)$$

$K_{NH_4,Ca}$ is called the Gapon coefficient and square brackets indicate concentrations. This equation has proved of wide applicability to cation exchange is soils.

Provided that the monovalent cation (Na^+ and K^+) amounts to $< 50\%$ of the exchangeable Ca^{2+} and Mg^{2+} combined, reasonably constant values of the Gapon coefficients $K_{Na\text{-}Ca,Mg}$ and $K_{K\text{-}Ca,Mg}$ are obtained for soils of similar clay mineral type from the equations

$$\frac{Na\text{-exchanger}}{Ca,Mg\text{-exchanger}} = K_{Na\text{-}Ca,Mg} \frac{[Na^+]}{[Ca^{2+}+Mg^{2+}]^{1/2}} \qquad (7.9)$$

and

$$\frac{K\text{-exchanger}}{Ca,Mg\text{-exchanger}} = K_{K\text{-}Ca,Mg} \frac{[K^+]}{[Ca^{2+}+Mg^{2+}]^{1/2}} . \qquad (7.10)$$

(Ca^{2+} and Mg^{2+} have sufficiently similar adsorption affinities, relative to the monovalent cations, to be regarded as interchangeable.) An equation similar to Equation 7.9 has been of great value in predicting the

likely 'sodium hazard' of irrigation waters (Section 13.2), and Equation 7.10 has been applied to the study of potassium availability in soil.

Soil K availability

Equation 7.10 may be rewritten as

$$\frac{\text{exchangeable K}}{CEC - \text{exchangeable K}} = K' \frac{K}{\sqrt{Ca + Mg}} \qquad (7.11)$$

where the quantities on the left hand side of the equation are in $cmol\ kg^{-1}$ soil. When exchangeable K is small ($< 10\%$ of the CEC), Equation 7.11 reduces to the simple form

$$\text{exchangeable K} = K'' \frac{K}{\sqrt{Ca + Mg}} \qquad (7.12)$$

where K'' includes the Gapon constant and the CEC of soil. Usually the change in exchangeable K (ΔK) is plotted against the activity ratio (AR) to give a *quantity/intensity* relation (Q/I) for the soil's exchangeable potassium (Fig. 7.11). However, the reactions of K in soil are complex, especially in soils containing much partially expanded clay mineral. As described in Section 2.4, K^+ ions form inner-sphere complexes in the unexpanded interlayer spaces of illite and micaceous clays and are therefore non-exchangeable. (NH_4^+, having a similar ionic potential to K^+, may also be held as a non-exchangeable cation in this way.) However, with weathering and the depletion of K in solution, the clay crystals slowly exfoliate from the edges exposing the interlayer K to exchange by other cations in solution (Fig. 7.12). The exposed interlayer sites at the crystal edges (which have been likened to 'wedge-shaped' sites) have a higher affinity for K^+ than the cleavage faces so that the whole crystal exhibits a changing affinity for K^+ relative to Ca^{2+} and Mg^{2+}, as first the edge sites and then the cleavage face sites are occupied. This is reflected in a gradual change in the slope of the Q/I plot, and hence the soil's buffering capacity for K, in going from low to high AR values.

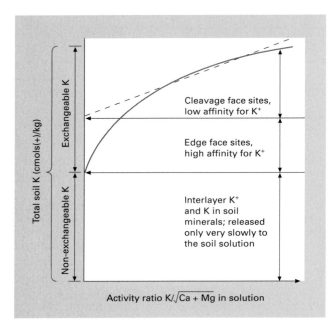

Fig. 7.11 Schematic Q/I relation for K in a micaceous clay soil (after Beckett, 1971).

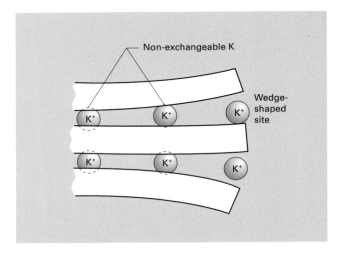

Fig. 7.12 The weathering edge of a micaceous clay crystal.

The ratio law

The Gapon equation indicates that if the proportions of Na^+ and K^+ on the exchanger are unchanged, the ratio Na^+/K^+ in the equilibrium solution remains constant, and similarly, if the proportions of K^+ and Ca^{2+} on the exchanger are constant, the ratio $K^+/\sqrt{Ca^{2+}}$ is also constant. Schofield (1947) was the first to formulate such observations into a general hypothesis called the *ratio law* which states that

'when cations in solution are in equilibrium with a larger number of exchangeable cations, a change in the concentration of the solution will not disturb the equilibrium if the concentrations of all the monovalent ions are changed in the one ratio, those of all the divalent ions in the square of that ratio, and those of all the trivalent ions in the cube of that ratio'.

For 'concentration' one may also read 'activity'.

The constancy of cation activity ratios depends on the effective exclusion of the accompanying anions such as Cl^- from the adsorption sites, since only then will the activities of the exchangeable cations be unaffected by changes in the Cl^- concentration of the bulk solution. Chloride exclusion is determined by the thickness of the double layer (Equation 7.4) so that the ratio law is only valid if:

• there is a preponderance (> 80%) of negative over positive charge on the surface, and

• the total solution concentration is not too high.

As theory predicts, the upper limit of concentration for which the ratio law applies varies with the valency of the cation, being approximately 0.1M for Na and K, 0.01M for Ca and 0.002M for Al. Despite these limitations, the ratio law is very useful in interpreting ion exchange in soils as the following examples illustrate.

Leaching of cations

Percolation of water through the soil leads to losses of solutes by leaching. With percolating rainwater, which is naturally very dilute, all the cations are in low concentration, and exchange occurs between cations of higher valency in solution and adsorbed cations of lower valency so that the activity ratios: $Na : \sqrt{Ca}$, $K : \sqrt{Ca}$ and

Box 7.4 Measurement of soil pH.

Soil pH is usually measured after shaking one part of the soil by weight with 2.5 or 5 parts of distilled water by volume and immersing the tips of a glass and a calomel reference electrode in the supernatant solution. For soils containing predominantly negatively charged clays, dilution of the soil solution by distilled water increases the absolute value of the surface potential and changes the distribution of H^+ ions between the DDL and the bulk solution. The proportion of H^+ ions in the DDL relative to the bulk solution increases so that the measured pH, which is the pH of the bulk solution, is higher than that of the undisturbed soil. This effect is especially noticeable in saline soils. However, if the soil is shaken with a solution containing the most abundant cation at an ionic strength approximating that of the soil solution, the measured pH changes little with dilution. For many soils a solution of 0.01M $CaCl_2$, for which the ionic strength I is given by

$$I = 0.5 \Sigma\, c_i z_i^2 = 0.03 \tag{B7.4.1}$$

is found to satisfy these requirements, where c_i is the molar concentration of ion i (cation or anion) and z_i is the valency. The pH measured in 0.01 M $CaCl_2$ is very close to the natural pH of the soil, as determined in a solution which has been shaken with successive samples of the soil to achieve equilibrium (Table 7.2). Thus, many soils laboratories, particularly in Australia, have adopted pH measured in 0.01M $CaCl_2$ at a soil/liquid ratio of 1 : 5 as the preferred method of pH measurement. pH ($CaCl_2$) is generally 0.6–0.8 units lower than pH(H_2O) at the same soil/liquid ratio.

so on tend to remain constant. Gradually Na followed by K is lost, the exchange surfaces becoming dominated by Ca and Mg and ultimately by Al. The end result is the genesis of an acid soil. Another example of the application of the ratio law is given in Box 7.4.

7.3 Anion adsorption

Adsorption sites and mechanisms

Protonated hydroxyl groups on the edge faces of crystalline clay minerals or on oxide surfaces will adsorb anions from solution according to the reaction.

$$M - OH_2^+ + A^- \leftrightarrow M - OH_2^+ A^-. \tag{7.13}$$

The anion A^- may exchange with other anions in the soil solution, in accordance with the ratio law, that is the ratio of adsorbed $H_2PO_4^-$ to SO_4^{2-} varies with the

Table 7.2 pH measurements on a basaltic red loam. After White, 1969.

Soil/liquid ratio (g ml⁻¹)	pH in H_2O	pH in 0.01M $CaCl_2$	pH in an equilibrium solution*
1 : 2	5.08	4.45	4.45
1 : 5	5.29	4.45	4.45
1 : 10	5.43	4.46	4.45
1 : 50	5.72	4.52	4.45

* 0.01M $CaCl_2$ after successive equilibration with four separate samples of soil.

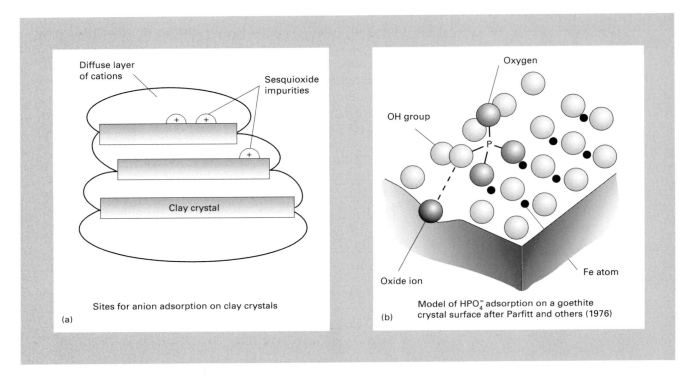

Fig. 7.13 (a) and (b) Physical models of the interrelations between clays, sesquioxides and anions in solution.

activity ratio $H_2PO_4^-/\sqrt{SO_4^{2-}}$ in solution provided that the concentration of cations at the surface is vanishingly small. However, when the positive sites are only isolated spots on a clay-mineral surface, or the soil solution concentration is low enough for the diffuse layer at planar surfaces to envelope the positive sites at the edges, anion adsorption is determined by the characteristics of the double layer at the negatively charged surfaces (Fig. 7.13a). In consequence, the concentration of SO_4^{2-} at the inner surface of the double layer is less than the concentration of $H_2PO_4^-$, and as the soil solution becomes more dilute, the increase in the concentration of SO_4^{2-} in the bulk solution is relatively greater than the increase in $H_2PO_4^-$ concentration. Under these conditions the cation-anion *activity products* $(Ca^{2+}) \times (SO_4^{2-})$ and $(Ca^{2+}) \times (H_2PO_4^-)^2$ measured in

solution should have constant values.

Non-specific adsorption as represented by Equation 7.13, is typical of anions such as Cl^- and NO_3^-. But oxyanions such as $H_2PO_4^-$, HCO_3^-, SO_4^{2-}, $H_3SiO_4^-$ and $B(OH)_4^-$ can be adsorbed by a *ligand exchange* mechanism in which an OH or OH_2^+ group in the mineral surface is displaced by an O of the oxyanion to form an inner-sphere complex at the surface.

Ligand exchange

Examples of two possible ligand exchange reactions between orthophosphate ions and a metal oxide surface are

$$M-OH + H_2PO_4^- \leftrightarrow M-O-\overset{\displaystyle OH}{\underset{\displaystyle OH}{\overset{|}{\underset{|}{P}}}}{=}O + OH^- \qquad (7.14)$$

and

$$M - OH_2^+ + H_2PO_4^- \leftrightarrow M\!-\!O\!-\!\underset{\underset{OH}{|}}{\overset{\overset{OH}{|}}{P}}\!=\!O + H_2O\,. \qquad (7.15)$$

In these reactions, the anion forms an ionic–covalent bond with the cation (M) by displacing either an OH or OH^{2+} group and entering into its coordination shell. The anion is therefore very strongly adsorbed, a condition described as *chemisorbed*. Depending on the oxide and the ambient pH, both types of site M-OH and $M\text{-}OH^{2+}$ occur on the surface, in which case as phosphate is adsorbed, the solution pH can increase *and* the net positive charge on the surface decrease. The latter effect decreases the *PZC* of the surface (Box 7.2).

Experimentally, decreases in surface charge from 1 to 0 have been observed per mole of anionic charge adsorbed onto pure Fe and Al oxides. Studies of phosphate adsorption on goethite using infra-red spectroscopy suggest that phosphate displaces two singly coordinated OH groups of adjacent Fe atoms to form a binuclear HPO_4 bridging complex (Fig. 7.13b). Phosphate can also be adsorbed by ligand exchange on hydroxy-Al polymers that form on clay surfaces between pH 4.5 and 7. Such polymers may constitute the positively charged sites on planar surfaces as in Fig. 7.13a. These compounds are most stable between pH 5 and 6.5 so that P adsorption in acid soils that have a predominance of phyllosilicates in the clay fraction often shows a maximum in this pH range (Section 10.3).

Although ligand exchange is probably the principal mechanism by which phosphate, molybdate, silicate and borate are adsorbed, HCO_3^- and SO_4^{2-} are intermediate in that they can be adsorbed by this mechanism, or by simple electrostatic attraction as in Equation 7.13. Phosphate and other anions non-specifically adsorbed at positive sites on oxide surfaces can be desorbed by decreasing the ion's concentration in the ambient solution, but anions coordinated to the surface can only be desorbed by raising the pH or introducing an anion with a higher affinity for the metal of the oxide.

7.4 Particle interaction and swelling

Attractive and repulsive forces

The interaction between clay-size particles in soil, especially their response to changes in soil water content, depends on the balance between attractive and repulsive forces at their surfaces. The attractive forces between clay particles are of two main kinds:
- *electrostatic*, as between negatively charged planar faces and positively charged edges; or between K^+ ions and the adjacent siloxane surfaces in an illite crystal; or between partially hydrated Ca^{2+} and Mg^{2+} ions and the siloxane surfaces of vermiculite;
- *van der Waals' forces* between individual atoms in two particles. For individual pairs of atoms the force decays with the sixth power of the interatomic distance, but the total force between two particles is the sum of all the interatomic forces which results not only in an appreciable force between macromolecules, but also in a less rapid decay in force strength with distance.

The origin of the attractive forces is such that, apart from the effect of the exchangeable cation, they vary little with changes in the solution phase. On the other hand, the repulsive forces that develop due to the interaction of diffuse layers are markedly dependent on the composition and concentration of the soil solution, as well as on the surface charge density of the clay. As shown in Section 7.1, the thickness of the diffuse layer is controlled by the valency of the counter ion and the concentration of the bulk solution. The model of the double layer presented so far assumes no impediment to the diffuse distribution of ions at the charged surface – the concentration in the midplane between two clay crystals is equal to the concentration in the bulk solution. This is reasonably true of dilute clay suspensions in salt solutions, but in concentrated suspensions and in natural soils the particles are close enough for the diffuse layers to interact, when the balance between forces of attraction and repulsion determines the *net attractive force* between the particles. If the net force is attractive (Fig. 7.14a), the particles remain close together and are described as *flocculated*. It also follows that soil micro-

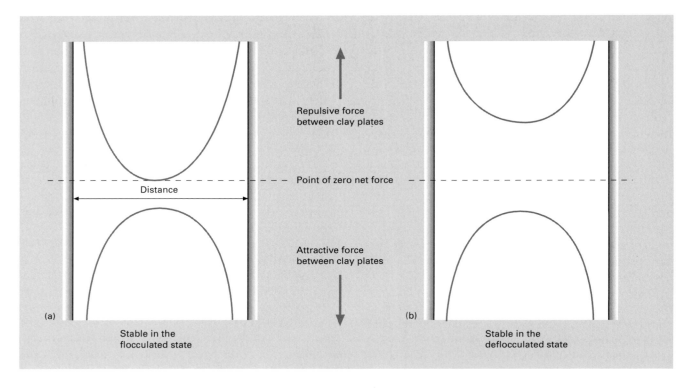

Fig. 7.14 (a) and (b) The interaction of attractive and repulsive forces between clay plates in salt solutions.

aggregates, of which the clay particles form a part, will remain stable. Conversely, if the net force is repulsive (Fig. 7.14b), the particles move farther apart until a new equilibrium position is attained under the constraint of an externally applied force, or the particles move far enough to exist as separate entities in the solution – a state described as *deflocculated*.

The excess concentration of ions at the midplane between two particles where the diffuse layers overlap gives rise to an osmotic or *swelling pressure* between the two plates. Swelling pressures of 10–20 bar may develop for particle separations from < 1 nm upwards, depending on the type of clay mineral, the interlayer cations and the concentration of the bulk solution.

The interpretation of clay interactions relies heavily on an understanding of colloid behaviour in dilute sus-

pensions. Nevertheless, clays in soil are normally in very close contact and the water : solid ratio is much lower than in a suspension, even in wet soil. As a result, additional short-range forces dependent on the surface charge density of the clay, ion–ion correlations and the structure of water at surfaces come into play in natural soils. These have been reviewed by Quirk (1994) and are responsible for some unusually stable microstructures as described below.

Flocculation–deflocculation and swelling behaviour

Face-to-face flocculation

The stacking of single layers of Ca-montmorillonite in roughly parallel alignment with much overlapping, to form *quasi-crystals*, and of thicker crystals of Ca-illite or Ca-vermiculite to form *domains*, is briefly described in

Section 4.4. Effective surface areas for Ca-montmorillonite of $100-150$ m^2 g^{-1} compared with 750 m^2 g^{-1} for the completely dispersed clay suggest that each quasi-crystal consists of 5–8 layers (Fig. 7.15). Anions are totally excluded from the interlayer regions and a diffuse double layer develops only at the external surfaces of the quasi-crystal. The interlayer cations appear to occupy a midplane position and provide a strong attractive force that stabilizes the interlayer spacing, irrespective of the external solution concentration.

Swelling behaviour

Except in atmospheres drier than *c.* 50% RH (Section 6.1), Ca-montmorillonite holds at least two layers of water molecules in the interlayer spaces due to the hydration of the cations. The basal spacing is then 1.5 nm. Additional water is taken up as the soil solution becomes more dilute, but only to the extent of expanding the basal spacing to 1.9 nm which is maintained even in distilled water. This behaviour is typical of *intracrystalline* swelling in Ca-montmorillonite or *intra-domain* swelling in Ca, Mg-illite or vermiculite.

As observed in Section 4.4, however, the quasi-crystals and domains exhibit long-range order over distances up to 5 μm. Diffuse double layers at the outer surfaces of the quasi-crystals (and domains) lead to swelling and the creation of wedge-shaped spaces where water can be held by surface tension. Swelling of this kind, described as *intercrystalline* or domain swelling, is illustrated in Fig. 7.16. It induces particle realignments that release the strains imposed by the crystal packing and bending during the previous drying cycle. Conversely, the high surface tension forces that develop during a drying phase contribute to the net attractive force holding the particles in close contact.

The swelling pattern of montmorillonite is entirely changed when divalent cations are replaced by monovalent cations such as Na$^+$ and Li$^+$ (Table 7.3). For Na-montmorillonite, once the external salt concentration drops below 0.3M NaCl, repulsive forces prevail and the swelling pressure forces the layers apart to distances that can be predicted from diffuse layer thicknesses calculated from DDL theory. Swelling is a precursor to eventual deflocculation which occurs at solution concentrations $\leqslant 0.01$M NaCl.

Edge-to-face flocculation

Kaolinite crystals are larger than those of montmorillonite and the hydrous micas and they flocculate face-to-face in concentrated solutions at high pH. Leaching and a reduction in the concentration of the soil solution cause the diffuse layers to expand, and the increased swelling pressure to disrupt the face-to-face arrangement. If there is a concurrent increase in soil acidity, the edge faces become positively charged and the mutual attraction of double layers at cleavage and edge faces encourages *edge-to-face* flocculation (Fig. 7.17).

Of the two mechanisms, face-to-face flocculation is the norm for montmorillonitic and micaceous clays, provided that di- and trivalent exchangeable cations predominate on the surfaces. In acid soils, this kind of flocculation can be enhanced by the presence of hydroxy-Al polymers on the planar surfaces which act as an 'electrostatic bridge' between the crystals. However, edge-to-face flocculation is more common in kaolinites because of the positive edge charge at pHs < *PZC* for the edge faces (*c.* pH 6). At higher pHs (6–8), flocculation may be aided by positively charged sesquioxide particles (*PZC* 8–9) forming 'bridges' between the negatively charged edge faces. Edge-to-face

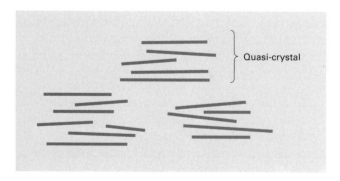

Fig. 7.15 Face-to-face flocculation of Ca-saturated montmorillonite.

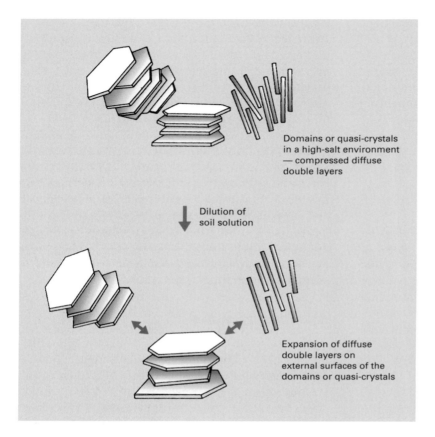

Domains or quasi-crystals in a high-salt environment — compressed diffuse double layers

Dilution of soil solution

Expansion of diffuse double layers on external surfaces of the domains or quasi-crystals

Fig. 7.16 Intercrystalline or domain swelling of a Ca-saturated 2 : 1 clay.

Table 7.3 Intracrystalline swelling of 2 : 1 clay minerals saturated with different cations. After Quirk, 1968.

Clay mineral	Charge per unit cell	Exchangeable cation	Swelling in dilute solution (nm)
Mica	− 2.0	K, Na	None
Vermiculite	− 1.3	Ca, Mg, Na	1.4–1.5
		K	1.2
		Cs	1.2
		Li	≥ 4.0
Montmorillonite	− 0.67	Ca, Mg	1.9
		K	1.5 and ≥ 4.4
		Cs	1.2
		Na, Li	≥ 4, 0

flocculation is insensitive to changes in the solution concentration, but is weakened by the adsorption of anions such as $H_2PO_4^-$ at the edge faces. In the same way, specific adsorption of phosphate and organic anions by sesquioxides may actually reverse the surface charge, rendering the oxide negatively charged and ineffective as a clay bridging agent.

7.5 Clay–organic matter interactions

Much of the organic matter in soil is complexed in some way with crystalline clay minerals and free oxides. Knowledge of these interactions is therefore of great importance to the understanding of virtually every pro-

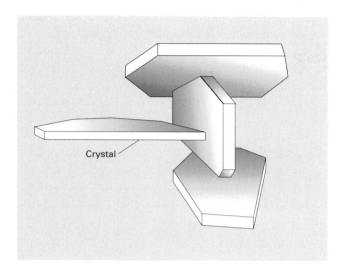

Fig. 7.17 Edge-to-face flocculation of kaolinite at low pH.

cess – physicochemical and biological – that occurs in the soil. Broadly, organic compounds may be divided into cationic, anionic and non-ionic compounds of low (< 1000) to high (> 100 000) molecular weight. Because of the polarity of charge on clays below pH 6, organic cations are adsorbed mainly on the planar faces and anions at the edges, except that small organic anions can form links with exchangeable cations on the cleavage faces. Uncharged organic molecules interact with the surfaces through polar groups and non-specific van der Waals' forces.

Electrostatic interactions

Cations

Many organic cations are formed by protonation of amine groups, as in the alkyl* amines (R-NH_2) and amino acids (RCHNH$_2$COOH). Protonation is enhanced by the greater acidity of water molecules hydrating cations adsorbed on clay surfaces. The higher the ionic potential of the cation M and its polarizing effect

* A chain hydrocarbon group such as CH_3, C_2H_5 . . . R.

on the M-OH$_2$ bond formed with the hydration water molecules, the greater is the tendency for a proton to be expelled from one of these molecules. The resultant protonation of a neutral organic molecule can be represented by

$$M(H_2O)_n^{m+} + R - NH_2 \leftrightarrow$$
$$MOH(H_2O)_{n-1}^{(m-1)+} + R - NH_3^+ \qquad (7.16)$$

On dry clay surfaces the polarizing effect of the cations is concentrated on fewer water molecules so the degree of dissociation is increased and the protonation of adsorbed organic molecules correspondingly enhanced. The displacement of the metallic cations by organic cations such as the alkyl amines decreases the affinity of the clay surface for water, making the soil more water-repellent and diminishing the swelling propensity of the clay.

Organic cation adsorption takes place at the internal and external surfaces of 2 : 1 expanding-lattice clays. The adsorption affinity increases with the chain length of the compound due to van der Waals' forces between the adsorbed molecules, so that adsorption in excess of the CEC may be achieved. Proteins adsorbed in interlayer spaces are less susceptible to microbial attack; conversely, extracellular enzymes, which are functional proteins, often show reduced biochemical activity in the adsorbed state.

Anions

Organic compounds are also adsorbed due to the attraction of —COO$^-$ groups to positively charged sites on clay edge faces and oxide surfaces. In addition, the carboxyl groups may undergo ligand exchange with OH and OH$_2^+$ groups in these surfaces which provides a much stronger form of adsorption than the former.

Organic anions and uncharged polar groups can form inner- and outer-sphere complexes with adsorbed cations on negatively charged organic or clay mineral surfaces. The former complex involves direct coordination of the anion with the metallic cation following the displacement of a hydrating water molecule (the *cation-bridge* shown in Fig. 7.18). The outer-sphere complex is

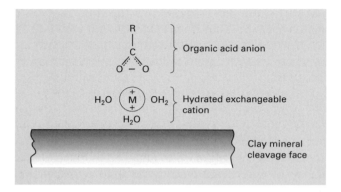

Fig. 7.18 Organic acid molecule forming a 'cation-bridge' (after Greenland, 1965a).

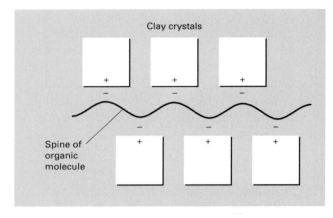

Fig. 7.20 'String of beads' arrangement of organic polyanion and kaolinite crystals.

a weaker bond formed between the anion (or polar molecule) and the metallic cation without the displacement of a water molecule (the *water-bridge* of Fig. 7.19). Adsorption of organic molecules by cation-bridging is more likely when the exchangeable cations are monovalent (because these cations dehydrate more easily), whereas water-bridging is favoured when the cations are di- or trivalent.

Phenolic-type compounds such as tannins, which are active constituents of the litter of many coniferous trees, form coordination complexes with Fe and Al, and in small concentrations are very effective *deflocculants* of clays. The compound must contain at least three phenolic groups, of which two coordinate the metal

atom and the third dissociates to confer a negative charge on the whole complex. Conversely, long chain polyanions are particularly effective *flocculants* of clays under acid conditions, because the clay particle edges become attached to the anionic groups of the molecule in the manner of a 'string of beads' (Fig. 7.20). Such compounds have therefore been applied successfully, if expensively, as soil conditioners to improve soil structure (Section 11.4).

van der Waals' forces and entropy stabilization

Low molecular weight alcohols, sugars and oligosaccharides are not adsorbed in competition with water unless present in high concentration relative to the water. Highly polar compounds can interact with the surface by forming water-bridges, as discussed above, or by direct hydrogen-bonding to another organic compound that is already adsorbed (Fig. 7.21).

High molecular weight alcohols (e.g. polyvinyl alcohol) and polysaccharides are strongly adsorbed at clay surfaces, even in competition with water. The attraction occurs partly through hydrogen-bonding to O and OH groups in the surface and partly through van der Waals' forces. The greatest contribution to the adsorption energy probably comes from the large increase in en-

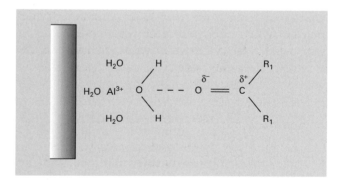

Fig. 7.19 Organic molecule forming a 'water-bridge' with exchangeable cation.

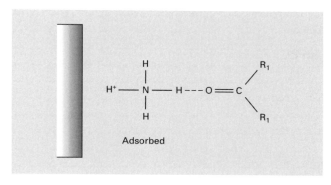

Fig. 7.21 Hydrogen-bonding between two organic compounds at a clay surface.

tropy when the organic macromolecule displaces many water molecules from the surface – the entropy increase stabilizes the molecule in the adsorbed state (Table 4.2).

7.6 Summary

Retention and exchange of cations and anions, soil acidity, water adsorption, and the formation and stabilization of soil aggregates are but a few of the many soil properties that are determined by reactions at mineral and organic surfaces.

Apart from the large specific surface areas of clay-size particles, these surfaces are important because of the charge they bear either:

• a *permanent charge* (usually negative) due to isomorphous substitutions in the crystal lattice, or
• a *pH-dependent (variable) charge* due to the reversible dissociation of protons from OH and OH_2^+ groups (at clay edge faces and oxide surfaces), or COOH and ⬡—OH groups of organic matter.

The preferential attraction of cations and repulsion of anions at planar clay surfaces give rise to an *electrical double layer*, comprising Stern layer cations tightly adsorbed at the surface, plus a diffuse layer in which cations (counterions) are accumulated and anions (co-ions) are in deficit, relative to the bulk solution. The pH at which the diffuse layer charge is zero defines the *point of zero charge (PZC)* of the surface.

Cations in the Stern layer lose their hydration shell and form *inner-sphere* complexes with the surface. The tendency to do this is greatest for monovalent cations whose overall strength of adsorption falls in the order $Cs^+ > Rb^+ > K^+ > Na^+ > Li^+$. Of higher hydration energy, divalent cations form *outer-sphere* complexes in which one layer of water molecules is retained between the cation and the siloxane surface. The overall strength of adsorption falls in the order $Ba^{2+} > Sr^{2+} > Ca^{2+} > Mg^{2+}$. *Specific adsorption* occurs when forces such as dipole–dipole interaction and hydrogen-bonding act on ions at surfaces, in addition to the simple electrostatic force.

Cation exchange capacity (CEC) is an operationally defined parameter which measures the net negative charge on a surface by replacement of the *exchangeable* cations with an 'index' cation. Some methods use a strong salt solution at a fixed pH (e.g. 1 M NH_4Cl at pH 7). Others use an index cation of high affinity for the surface at a lower concentration (e.g. $0.1 M$ $BaCl_2$) at the soil pH. The former methods overestimate *CEC* in acid soils with pH-dependent charge; the latter methods are preferred for such soils because they measure the effective *CEC* (*ECEC*). The difference between the *CEC* and the sum of exchangeable cations Σ Ca, Mg, K, Na defines the *exchange acidity* (Al^{3+} and H^+). Ca^{2+}, Mg^{2+}, K^+ and Na^+ are referred to as 'basic' cations to distinguish them from the 'acidic' cations Al^{3+} and H^+. Acid clays are primarily Al^{3+}-clays. Hydrolysis of the exchangeable Al^{3+} and polymerization of the hydroxy-Al products accounts for much of the *pH buffering capacity* of a soil over the pH range 4–7.5.

Exchange between cations in the double layer and the soil solution may be described by the *Gapon equation*, e.g.

$$\frac{\text{K-exchangeable}}{\text{Ca-exchangeable}} = K_{\text{K-Ca}} \frac{K^+}{\sqrt{Ca^{2+}}}.$$

Changes in solution activities (or concentrations) that occur on leaching soils, or in soils influenced by saline water, may be explained in terms of the *ratio law*, which predicts that when cations in solution are in equilibrium with a larger number of exchangeable cations, the activity ratio $a^+/(a^{n+})^{1/n}$ is a constant over a limited range of salt concentration.

Inorganic and organic anions can be adsorbed at pH-dependent sites when these are positively charged (*non-specific* adsorption). However, oxyanions such as $H_2PO_4^-$, $H_3SiO_4^-$, MoO_4^{2-} and $B(OH)_4^-$ are more likely to be adsorbed by *ligand exchange* in which the O of the anion displaces an OH or OH_2^+ group from the coordination shell of metal cation in the surface. This results in a very strong bond and the anion is said to be *chemisorbed*. Organic anions also interact with cations adsorbed at negatively charged surfaces by forming *cation-bridges* (displacement of a hydrating water molecule) or *water-bridges* (outer-sphere complex formed with the cation). Depending on the sign of the surface charge, organic cations and anions can be important in linking together clays and sesquioxides to form stable microaggregates.

The repulsive force developed as diffuse layers of adjacent clay crystals overlap is manifest as a *swelling pressure*, which varies with water content and can disrupt the microstructure of clay domains. Attractive forces between clay crystals include van der Waals' forces and electrostatic forces associated with divalent cations, particularly Ca^{2+}, located at the midplane between two opposing clay surfaces. Hydroxy-Al polymers can also act as electrostatic 'bridges' between the planar clay surfaces. When such forces exceed the repulsive force, *face-to-face* flocculation of the clay ensues. Alternatively, the crystals may flocculate *edge-to-face* through positive and negative charge interaction, as in the case of kaolinite at pH < 6.

References

Beckett P. H. T. (1971) Potassium potentials – a review. *Potash Review*, Subject 5, pp. 1–41.

Greenland D. J. (1965a) Interaction between clays and organic compounds in soils. 1. Mechanisms of interaction between clays and defined organic compounds. *Soils and Fertilizers* **28**, 415–425.

Page A. L. (1982) (Ed.) *Methods of Soil Analysis*, Part 2. Chemical and Microbiological Properties, 2nd edn. Agronomy Monograph No. 9. American Society of Agronomy/Soil Science Society of America, Madison, Wisconsin.

Parfitt R. L., Russell J. D. & Farmer V. C. (1976) Confirmation of the surface structures of goethite (α-FeOOH) and phosphated geothite by infrared spectroscopy. *Journal of the Chemical Society, Faraday Transactions* **72**, 1082–1087.

Quirk J. P. (1994) Interparticle forces: a basis for the interpretation of soil physical behaviour. *Advances in Agronomy* **53**, 121–183.

Quirk J. P. (1968) Particle interaction and soil swelling. *Israel Journal of Chemistry* **69**, 213–234.

Rayment G. E. & Higginson F. R. (1992) *Australian Laboratory Handbook of Soil and Water Chemical Methods.* Australian Soil and Land Survey Handbook. Inkata Press, Melbourne.

Schofield R. K. (1939) *The electric charges on clay particles.* International Society of Soil Science. British Empire Section, Cambridge, pp. 1–5.

Schofield R. K. (1947) A ratio law governing the equilibrium of cations in the soil solution. *Proceedings of the 11th International Congress of Pure and Applied Chemistry (London)* **3**, 257–261.

Sposito G. (1989) *The Chemistry of Soils.* Oxford University Press, New York.

Sposito G. (1994) *Chemical Equilibria and Kinetics in Soils.* Oxford University Press, New York.

van Olphen H. (1977) *An Introduction to Clay Colloid Chemistry*, 2nd edn. Wiley Interscience, New York.

White R. E. (1969) On the measurement of soil pH. *Journal of the Australian Institute of Agricultural Science* **35**, 3–14.

Further reading

Barrow N. J. (1985) Reactions of anions and cations with variable-charge soils. *Advances in Agronomy* **38**, 183–230.

McBride M. B. (1994) *Environmental Chemistry of Soils.* Oxford University Press, New York.

Sparks D. L. (1995) *Environmental Soil Chemistry.* Academic Press, San Diego.

Thomas G. W. Hargrove W. L. (1984) The chemistry of soil activity, in *Soil Acidity and Liming* (Ed. F. Adams). Agronomy Monograph, No. 12 (2nd edn). American Society of Agronomy, Madison, pp. 3–56.

Wild A. (Ed.) (1988) *Russell's Soil Conditions and Plant Growth*, 11th edn. Longman Scientific & Technical, Harlow.

Chapter 8
Soil Aeration

8.1 Soil respiration

Respiratory quotient and respiration rate

The importance of soil organisms in promoting the turnover of carbon is outlined in Chapter 3. *Aerobic respiration* involves the breakdown or dissimilation of complex C molecules, oxygen being consumed and CO_2, water and energy for cellular growth being released. Under such conditions *the respiratory quotient*, defined as

$$RQ = \frac{\text{volume of } CO_2 \text{ released}}{\text{volume of } O_2 \text{ consumed}} \qquad (8.1)$$

is equal to 1. When respiration is anaerobic, however, the RQ rises to infinity because O_2 is no longer consumed but CO_2 continues to be evolved. The *respiration rate*, R, of the soil organisms is measured as the volume of O_2 consumed (or CO_2 released) per unit soil volume per unit time.

Soil respiration is augmented by the respiration of living plant roots. Further, the respiration of micro-organisms is greatly stimulated by the abundance of carbonaceous material (mucilage, sloughed-off cells and exudes) in the soil immediately around the root.

This zone, which is influenced by the root, is called the *rhizosphere*. The stimulus to soil respiration provided by the presence of a crop is demonstrated in Table 8.1, where respiration rates for cropped and fallow soil in the south of England are compared for the months of January (mean soil temperature 3°C at 30 cm depth) and July (mean soil temperature 17°C). This soil had an unusually deep A horizon so that the respiration rates were about twice those normally recorded in the field.

Measurement of respiration rates

An instrument designed to measure respiration rates is called a *respirometer*, and one designed specifically for field measurements is illustrated in Fig. 8.1. The soil moisture content is regulated by controlled watering and drainage; temperature is recorded and the circulating air is monitored for O_2, the deficit being made good by an electrolytic generator. Carbon dioxide is absorbed in vessels containing soda lime and measured. Such measurements show that the respiration rate depends on:

- soil conditions, such as organic matter content, O_2 supply and moisture,
- cultivation and cropping practices, and
- environmental factors, principally temperature.

The effect of temperature on respiration rate is expressed by the equation

$$R = R_0 Q^{T/10} \qquad (8.2)$$

Table 8.1 Soil respiration as influenced by crop growth and temperature. After Currie, 1970.

| | Respiration rate (R) ($L\ m^{-2}\ d^{-1}$) | | | |
| | January | | July | |
	Fallow	Cropped	Fallow	Cropped
O_2 demand	0.5	1.4	8.1	16.6
CO_2 release	0.6	1.5	8.0	17.4
RQ	1.20	1.07	0.99	1.05

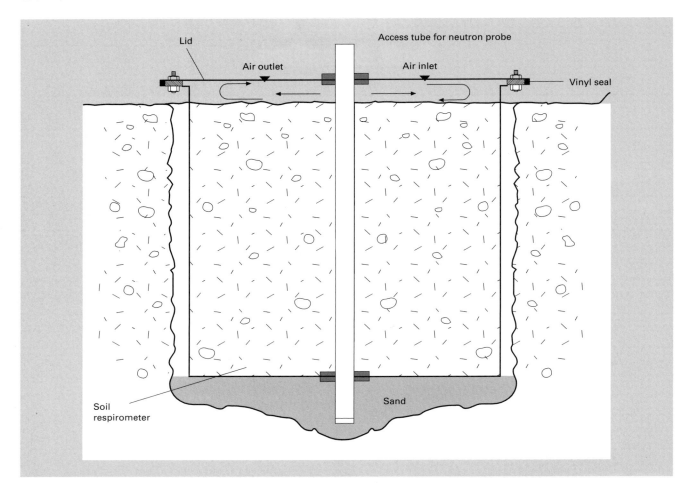

Fig. 8.1 Diagram of a soil respirometer (after Currie, 1975).

where R and R_0 are the respiration rates at temperature T and $0°C$ respectively, and Q is the magnitude of the increase in R for a $10°C$ rise in temperature, called the *Q-10 factor*, which lies between 2 and 3. Temperature change is the cause of large seasonal fluctuations in soil respiration rate in temperate climates, as illustrated in Fig. 8.2.

Cycles or respiratory activity

In temperate regions, the seasonal maximum in R values lags 1–2 months behind that of soil temperature. Superimposed on this seasonal pattern is the effect of weather, because within any month of the year the daily maximum in R may be up to twice the daily minimum, a fluctuation strongly correlated with soil surface temperatures. Finally, there are minor variations in R caused by the diurnal rise and fall of soil temperature. The causes of soil temperature variations are discussed in Section 6.6.

These trends are modified by other factors; for example, at Rothamsted in southern England the respiratory peak is consistently lower during a dry summer than a wet one, and the rate tends to be higher in spring than

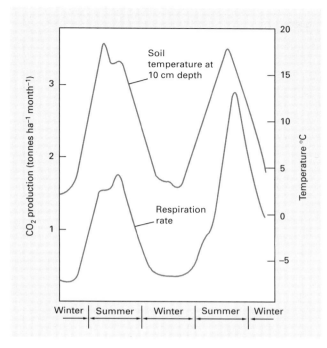

Fig. 8.2 Seasonal trends in soil temperature and respiration rate in southern England.

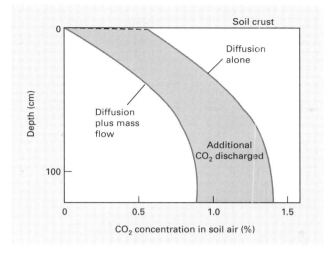

Fig. 8.3 Concentration *vs* depth profiles for CO_2 in the soil air under a surface crust (after Currie, 1970).

in autumn for the same average soil temperatures. The latter effect is attributable to two factors, primarily a greater supply of organic residues at the end of the quiescent winter period than after a summer of active decomposition, together with spring cultivations that break up large aggregates and expose fresh organic matter to attack by micro-organisms.

8.2 Mechanisms of gaseous exchange

In Section 4.5, the average composition of the air in a range of soils was given as 79% N_2, 20% O_2 and *c.* 1% CO_2. The soil air composition is buffered against change by the much larger volume of the Earth's atmosphere. There is also a large reservoir of HCO_3^- in the oceans. Buffering is achieved by the dynamic exchange of gases as O_2 travels from the atmosphere to the soil

and CO_2 moves in the reverse direction, a process called *soil aeration*. Three transport processes are involved in soil aeration.

1 Dissolved O_2 is carried into the soil by percolating rainwater. The contribution is small owing to the low solubility of O_2 in water (0.028 mL mL^{-1} at 25°C and 1 atmosphere pressure).
2 Mass flow of gases due to pressure changes of 1–2 mbar created by wind turbulence over the surface.
3 Diffusion of gas molecules through the soil pore space.

Gaseous diffusion is normally far more important than mass flow in either the liquid or gas phase (1 and 2). This is illustrated by the gas concentration profiles shown in Fig. 8.3, for a soil with a thin surface crust, where diffusion alone reduces the concentration of CO_2 in the soil air at a depth of 1 m to below 1.5%. The effect of mass flow, on the other hand, is merely to reduce the CO_2 concentration by a further 0.5%, virtually all of this decrease occurring across the surface crust where mass flow due to barometric pressure changes is effective.

Gas exchange by diffusion

The rate of gas diffusion or *diffusive flux* J_g is given by the equation

$$J_g = -D \, grad \, C \tag{8.3}$$

where J_g is the volume of gas diffusing across a unit area perpendicular to the direction of movement in unit time, *grad C* is the change in gas concentration C ($mL \, mL^{-1}$ air-space) per unit distance in the direction of diffusion, and D is the gas diffusion coefficient (*cf.* Equation 6.11 of water flow). Values of D for CO_2 and O_2 in water and air are given in Table 8.2, from which it can be seen that for the same *grad C* the diffusive flux is potentially 10 000 times faster through air-filled pores than water-filled pores. Further, since D_{O_2} is slightly larger than D_{CO_2}, when the soil RQ is 1, O_2 diffuses into the soil faster than CO_2 can diffuse out. A pressure difference builds up and mass flow of N_2 (the most abundant gas in the pore space) occurs to eliminate it. Nevertheless, under the most steady-state aerobic conditions it is sufficiently accurate to equate the diffusive flux in O_2 into the soil with that of CO_2 out.

Air-filled porosity and tortuosity

D values for O_2 and CO_2 through the air-filled pores are always less than D values in the outside air because of the restricted volume of the pores and the tortuosity of the diffusion pathway.

These effects are expressed in the equation

$$D_e = \alpha \varepsilon D_0 \tag{8.4}$$

where D_e is the effective diffusion coefficient, and D_0 is the diffusion coefficient of the gas in the bulk air. ε is the air-filled porosity (Section 4.5) and α is an impedance factor, which ranges from 0 to 1 and decreases as the pathway becomes more tortuous.

The value of α varies with ε in a complex way that depends on a soil's structure, texture and water content. Some of these effects are illustrated in Fig. 8.4. The change in α in dry sand (Fig. 8.4a) is limited by the extent to which pore-size distribution within the matrix can be altered by changing the packing density and by the addition of finer particles. However, when the sand is wet (Fig. 8.4b), water films fill the narrowest necks between pores so that the shape as well as the size distribution of the pores is altered. When water is added to dry soil peds so that as θ increases, ε decreases (Fig. 8.4c), α initially increases as the intraped pores fill and the tortuosity and roughness of the gas diffusion path decreases. However, once the larger interped pores begin to fill, α declines rapidly as it does in the wet sand, approaching a minimum value at $\varepsilon \approx 0.1$, a value corresponding to the average proportion of discontinuous or 'dead-end pores' in the soil. It can be concluded that in structured soils, the intraped pores make little contribution to gas diffusion and the major pathway is

Table 8.2 Diffusion coefficients ($cm^2 \, s^{-1}$) of O_2 and CO_2 in air and water.

	Oxygen	Carbon dioxide
Air	2.3×10^{-1}	1.8×10^{-1}
Water	2.6×10^{-5}	2.0×10^{-5}

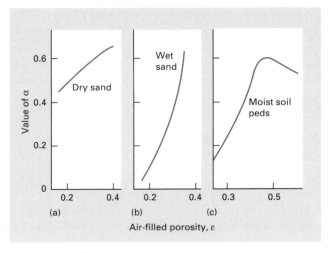

Fig. 8.4 The relation between impedance factor α and air-filled porosity ε for dry and wet sand and soil peds (after Currie, 1970).

through the interped pores, most of which are drained at water contents less than the field capacity. This being so, the value of α will reflect management practices that change the proportion of large pores in the total pores space. Soil structural deterioration through continuous cultivation, for example, may not necessarily result in a marked decrease in total porosity, but it is usually associated with a disproportionate decrease in the macroporosity. Consequently, the value of α, which may be as high as 0.6 in a well-drained, structured soil, may fall to value as low as 0.16 in an over-worked arable soil.

Profiles of O_2 and CO_2 in field soils

In the field, the concentrations of O_2 and CO_2 in the soil pores change in space and time due to variations in the air-filled porosity, tortuosity and respiratory activity. Examples of changes in the O_2 partial pressure in the soil air during summer and winter for soils of contrasting texture are given in Fig. 8.5. A quantitative description of O_2 and CO_2 movement under transient conditions in the field requires a combination of the diffusive flux equation with a continuity equation that includes a term for soil respiration (*cf.* Equation 6.18). For transfer of O_2 or CO_2 in the vertical direction z, we may write

$$J_g = -D_e \frac{\partial C}{\partial z} \tag{8.5}$$

and

$$\varepsilon \frac{\partial C}{\partial t} = -\frac{\partial J_g}{\partial z} \pm R \tag{8.6}$$

where R is the rate of CO_2 production ($+R$) or O_2 consumption ($-R$). C is usually expressed in mL gas mL^{-1} gas and R in mL gas cm^{-3} soil s^{-1}, which accounts for the appearance of ε (volume of air/volume of soil) in Equation 8.6. Combining Equations 8.5 and 8.6 gives

$$\varepsilon \frac{\partial C}{\partial t} = D_e \frac{\partial^2 C}{\partial z^2} \pm R \tag{8.7}$$

Using some simplifying assumptions, Equation 8.7

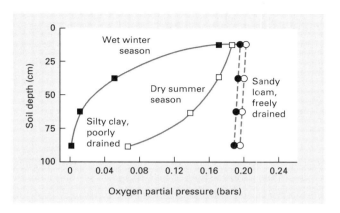

Fig. 8.5 Changes in O_2 partial pressure in contrasting soils with depth and seasons of the year.

can be solved to obtain estimates of the change in O_2 concentration (ΔC) in the air-filled pore space in the upper part of a soil profile which is either freely drained (sandy loam) or poorly drained (silty clay) (Box 8.1). These show good agreement with the actual changes in O_2 concentration (or partial pressure) shown in Fig. 8.5. However, the restriction imposed on gas diffusion when there is a predominance of water-filled pores in a poorly drained soil is discussed below.

Gas diffusion through water-filled pores

The cytochrome oxidase enzyme system, which catalyses the reduction of O_2 to H_2O in plants and microorganisms, has a very high affinity for O_2. This is indicated by a K_M value of 2.5×10^{-8} M – the concentration of dissolved O_2 at which the rate of reaction is 50% of the maximum rate. However, in stirred bacterial suspensions it is found that aerobic respiration can only be sustained at dissolved O_2 concentrations above c. 4×10^{-6} M, indicating that diffusion of O_2 to the site of the enzyme in the cell's mitochondria is the rate-limiting step in respiration. The concentration of O_2 in water in equilibrium with the normal partial pressure of O_2 in the atmosphere at 25°C is

$0.21 \times 0.028 = 0.0059$ mL mL^{-1} water,

Assuming that steady-state conditions are attained (i.e. $\partial C / \partial t = 0$) and that R and D_e do not change substantially with depth, Equation 8.7 can be integrated to give an expression relating the change in O_2 concentration (ΔC) in the gas phase to the depth z over which that change occurs, that is

$$\Delta C = \frac{Rz^2}{2D_e} \qquad \qquad \text{(B8.1.1)}$$

Substitution of appropriate values for R, z and D_e in this equation gives changes in O_2 concentration consistent with field measurements, as illustrated by two contrasting situations.

(a) For a well-drained sandy soil, putting $\varepsilon = 0.4$ and $\alpha = 0.6$ in Equation 8.4 gives $D_e(O_2) = 5.5 \times 10^{-2}$ cm^2 s^{-1}, and taking $R = 2 \times 10^{-7}$ mL cm^{-3} s^{-1} (Table 8.1), gives

$\Delta C = 0.004$ mL mL^{-1} gas phase at a depth of 50 cm

This corresponds to a negligible fall in O_2 partial pressure of 0.004 bar by Dalton's law of partial pressures (Section 4.5).

(b) On the other hand, in a clay soil where poor structure impedes diffusion in the gas phase, D_e could fall to 1.5×10^{-3} cm^2 s^{-1} and R to 1×10^{-7} mL cm^{-3} s^{-1}, so that

$\Delta C = 0.083$ mL mL^{-1} gas phase at 50 cm

and the O_2 partial pressure falls to $(0.21 - 0.08) = 0.13$ bar.

Box 8.1 Calculated depth profiles of soil O_2 concentration.

which is approximately equal to 2.4×10^{-4} M.* Thus, soil micro-organisms continue to respire aerobically provided that the O_2 partial pressure in their immediate surroundings is one-sixtieth or more of that in the atmosphere, that is ≥ 0.004 bar. Although partial pressures above this value may be maintained in the few air spaces remaining in a waterlogged soil, the respiring organisms live in the water-filled pores *within* peds so that O_2 must diffuse through water, a slow process due to the low solubility of oxygen and the low diffusion coefficient for O_2 in water (Table 8.2). In many soils, therefore, anaerobic pockets develop at the centre of peds larger than a certain size when the peds remain wet for any length of time.

*The volume of an ideal gas at 0°C and 1 atmosphere's pressure is 22.4 L mol^{-1}.

Figure 8.6 shows the effect of respiration rate on the minimum size of wet ped in which anaerobic conditions can develop at its centre. This relationship is obtained by integrating Equation 8.7 for appropriate initial and boundary conditions (Box 8.2).

Provided there is some continuity of air-filled pores ($\varepsilon > 0.1$) and respiration rates are low (< 1 mL $\times 10^{-7}$ cm^{-3} s^{-1}), anaerobic conditions are unlikely to occur unless the sites of active respiration are separated from the gas phase by more than 2 cm of water-saturated soil. However, as respiration rates rise, anaerobic conditions are likely in peds of radius of 1 cm or more. This is probably a common occurrence in clay soils in Britain and other humid temperate regions in spring, and predisposes to denitrification, as discussed in Section 8.4. Because the solubility of CO_2 is much greater than that

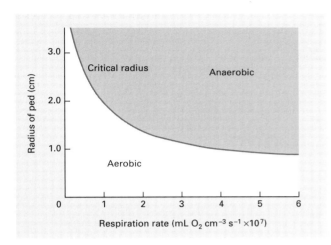

Fig. 8.6 Critical radius for the induction of anaerobic conditions at ped centres (after Greenwood, 1975).

of O_2, its diffusive flux through water is larger with the result that CO_2 concentrations do not rise markedly, even at the centre of wet peds (Box 8.3). Nevertheless, once the pores between peds fill with water, as in a waterlogged soil, the CO_2 partial pressure can rise to 0.1 bar or more and may have detrimental effects on the growth of plants and micro-organisms.

8.3 Effects of poor soil aeration on root and microbial activity

Plant root activity

The picture emerging from Section 8.2 is one of the mosaic of aerobic and anaerobic zones in heavy textured soils during wet periods of the year, as illustrated in Fig. 8.7. The anaerobic zones exert no serious ill effect (other than through the reduction of NO_3^- discussed later) because most of the roots grow through the larger pores (> 100 μm diameter) where O_2 diffusion is rapid. Oxygen can also diffuse for limited distances through plant tissues from regions of high to low O_2 concentration, an attribute best developed in aquatic species and paddy rice (*Oryza sativa*), because of large intercellular air species in which the value of D_{O_2} approaches 1×10^{-1} cm^2 s^{-1}. Figure 8.8 shows a healthy young rice crop growing in a completely puddled and waterlogged soil.

However, when wet weather coincides with temperatures high enough to maintain rapid respiration rates (typically in spring), the whole root zone can become anaerobic. Even if sustained for only a day, this condition can have serious effects on the metabolism and

Box 8.2 Estimating the critical ped size for the onset of anaerobic conditions.

Assuming the peds to be spherical (so z is put equal to r, the ped radius) and that steady-state conditions apply, Equation 8.7 can be integrated to give

$$r = \sqrt{\frac{6\Delta C D_e}{R}}, \tag{B8.2.1}$$

where ΔC is the difference in O_2 concentration between the ped surface and its centre, D_e is the effective diffusion coefficient of O_2 *through the water phase*, and R is the respiration rate inside the ped. It the partial pressure of O_2 in the air space at the ped surface is assumed to be 0.21 bar, while at the centre of the ped it is zero, Equation B8.2.1 can be solved for a range of R values to give the minimum radius of ped, the centre of which will just be anaerobic. When D_e has a value of 1.5×10^{-5} cm^2 s^{-1}, the results shown in Fig. 8.6 are obtained.

Consider the conditions of a wet ped of radius r, just less than the critical radius for the onset of anaerobic conditions. The partial pressures of O_2 and CO_2 in the gas phase outside the ped are 0.21 and 0.0003 bar, respectively. The partial pressure of O_2 at the ped centre is zero. Assuming steady-state conditions, the influx of O_2 can be equated with the efflux of CO_2, so that on substitution in Equation 8.5 we have

$$D_{O_2} \frac{\Delta C'}{r} = D_{CO_2} \frac{\Delta C''}{r}, \tag{B8.3.1}$$

where $\Delta C'$ and $\Delta C''$ are the concentration differences in the water phase for O_2 and CO_2 respectively. Substituting in Equation B8.3.1 for D_{O_2} and D_{CO_2} from Table 8.2, putting $\Delta C' = 0.0059$ mL cm^{-3} and rearranging gives

$$\Delta C'' = \frac{D_{O_2}}{D_{CO_2}} \Delta C' = 0.0077 \text{ mL cm}^{-3}. \tag{B8.3.2}$$

As the solubility of CO_2 at 25°C and 1 atmosphere's pressure is 0.759 mL cm^{-3}, it follows that the CO_2 partial pressure at the ped centre is $c.$ 0.01 bar. An increase in partial pressure of this magnitude would have a negligible effect on plant and microbial metabolism.

growth of sensitive plants. Species intolerant of water-logging experience an acceleration in the rate of glycolysis and the accumulation of endogenously produced ethanol and acetaldehyde. Cell membranes become

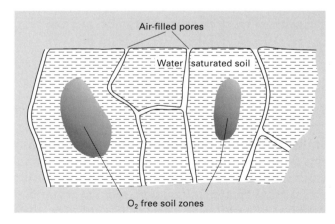

Fig. 8.7 Distribution of water and oxygenated zones in a structured soil (after Greenwood, 1969).

leaky, and ion and water uptake is impaired. The characteristic symptoms ar wilting, yellowing of leaves and the development of adventitious roots at the base of the stem. The production of endogeneous ethylene (C_2H_4) may be stimulated and cause growth abnormalities, such as leaf epinasty – the downward curvature of a leaf axis.

Soil microbial activity

Soil aeration determines the balance between aerobic and anaerobic metabolism which has profound effects on the energy available for microbial growth and the type of end-products. Some general points about aerobic and anaerobic organisms are summarized in Box 8.4.

Aerobic processes: nitrification

Nitrification entails the biological oxidation of inor-

Fig. 8.8 Healthy young rice plants growing in a completely puddled and waterlogged soil.

ganic N forms to nitrate (NO_3^-). Nitrification is discussed here because it is an essential prior step to the loss of N by NO_3^- reduction and the evolution of N_2O and N_2 – a process called *denitrification*.

The principal nitrifying organisms are chemoautotrophic bacteria of the genera *Nitrosomonas* and *Nitrobacter*. These organisms are unique in deriving energy for growth solely from the oxidation of NH_4^+ produced by saprophytic decay organisms, or accruing to the soil from fertilizer and rain. The oxidation occurs in two steps

$$NH_4^+ + \frac{3}{2}O_2 \rightarrow NO_2^- + 2H^+ + H_2O + \text{energy} \qquad (8.8)$$

carried out by the *Nitrosomonas* species, and

$$NO_2^- + \frac{1}{2}O_2 \leftrightarrow NO_3^- + \text{energy} \qquad (8.9)$$

carried out by *Nitrobacter* species. Traces of N_2O gas are also produced during nitrification. These energy-releasing reactions are coupled to the reduction of CO_2 and cell growth. Because CO_2, which is the sole source of C for these organisms, is plentiful, the growth rate and hence bacterial numbers are limited primarily by the supply of oxidizable substrates (NH_4^+ and NO_2^-). For example, the demand for NH_4^+ created by the more numerous soil heterotrophs restricts the rate of NH_4^+ oxidation by the slower growing nitrifiers, and the absence of detectable NO_2^- in well-aerated soil indicates that NO_2^- supply is potentially rate-limiting for the growth of *Nitrobacter*.

Growth of the organisms and rate of nitrification are also influenced by temperature, moisture, pH and O_2 supply as follows:

• the temperature optimum lies between 30 and 35°C and nitrification is very slow at temperatures $< 5°C$;

• the optimum moisture content is *c.* 60% of field capacity. *Nitrosomonas* species are more susceptible to dry conditions than *Nitrobacter*, but sufficient bacteria survive short periods of desiccation for increased nitrification to occur during the flush of decomposition which follows the rewetting of an air-dry soil (Fig. 10.6);

• the growth of *Nitrosomonas* is more sensitive to low

Aerobic organisms
- are generally beneficial to soil and plants,
- the heterotrophs break down complex C compounds to produce much microbial biomass, humus, water and CO_2 (Section 3.2),
- mineralize essential elements such as N, P and S during decomposition,
- specialized groups of autotrophs, such as the nitrifying bacteria and sulphur-oxidizing bacteria, produce NO_3^- and SO_4^{2-},
- some rhizosphere-dwelling species of bacteria may produce growth regulators, including gibberellic acid and indoleacetic acid, which stimulate plant growth.

Anaerobic organisms
- are much less numerous than aerobes, comprising about 10% of the total soil population, but some aerobes are facultative anaerobes (can grow anaerobically when the O_2 supply is limiting),
- generate less metabolic energy during anaerobic respiration, so that cell growth is slower than for aerobic respiration,
- form more varied end-products, including CO_2, H_2O and reduced C compounds such as organic acids, C_2H_5OH, CH_4, C_2H_4 and other hydrocarbon gases,
- use alternative electron acceptors (NO_3^-, MnO_2, $Fe(OH)_3$ and SO_4^{2-}) and produce reduced inorganic forms, some of which are undesirable (Section 8.4).

Box 8.4 Aerobic and anaerobic organisms compared.

pH than *Nitrobacter* which might be expected from reactions 8.8 and 8.9. Growth rates measured in pure culture or in laboratory soil perfusion experiments are at a maximum at pH $\geqslant 7.6$ for *Nitrosomonas*, and pH 6.6–7.6 for *Nitrobacter*. Nitrifying organisms in field soils also produce less NO_3^- per unit time at low pH, and hence grow more slowly. But measurements of the short-term nitrification rate at non-limiting NH_4^+ concentrations suggest that soil nitrifiers adapt to the prevailing pH such that the optimum pH for nitrification is close to the soil pH (Fig. 8.9). This relationship also suggests that the pH optimum for nitrifiers in any soil is unlikely to exceed 6.6.

Anaerobic processes: fermentation and reduction products

The metabolic pathway for carbohydrate oxidation in plants and micro-organisms is the same, in the presence and absence of O_2, as far as the key intermediate, pyruvic acid ($CH_3COCOOH$). This is called glycolysis. Normally, reduced coenzymes in the cell are reoxidized as electrons are passed along a chain of respiratory enzymes to the terminal acceptor O_2. However, in the absence of O_2, as in a waterlogged soil, there are two possible alternatives:
- other organic compounds serve as final electron acceptors and are reduced in turn – the process of *fermentation*, or
- inorganic compounds such as NO_3^-, MnO_2, $Fe(OH)_3$, and SO_4^{2-} are reduced (Section 8.4).

As indicated in Fig. 8.10, several organic acids of low molecular weight – the *volatile fatty acids* (VFAs) – are formed as intermediates in carbohydrate fermentation. The more abundant of these VFAs are acetic and butyric acids, the concentrations of which can sometimes

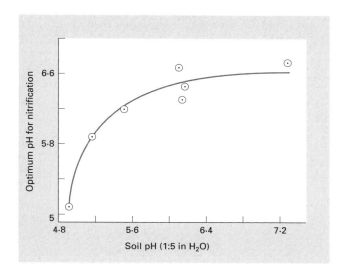

be > 1–10 mM for several weeks in flooded soils. Even in normally aerobic soils, when unfavourable conditions develop (through surface compaction and excess water), crop residues may ferment to produce sufficiently high localized concentrations of soluble phytotoxins, usually acetic acid, for seed germination and seedling establishment to be adversely affected.

In addition to the VFAs, a range of low molecular weight hydrocarbon gases (CH_4, C_2H_4 and C_2H_6) are produced during the first few days of waterlogging. However, as anaerobiosis is prolonged, the acids and hydrocarbon gases are dissimilated to the final end-products, mainly CH_4, CO_2 and some H_2, by the *obligate* anaerobes which multiply slowly in completely waterlogged soil. Thus, although some methane is produced during the initial stages of anaerobiosis, the bulk is produced by the reduction of CO_2 as reducing conditions intensify, according to the reaction

$$CO_2 + 8H^+ + 8e^- \rightarrow CH_4 + 2H_2O . \qquad (8.10)$$

Fig. 8.9 pH optimum for nitrifying organisms in a range of acid to neutral soils. The pH optimum was based on short-term nitrification rates measured under non-limiting NH_4^+ supply (after Bramley and White, 1991).

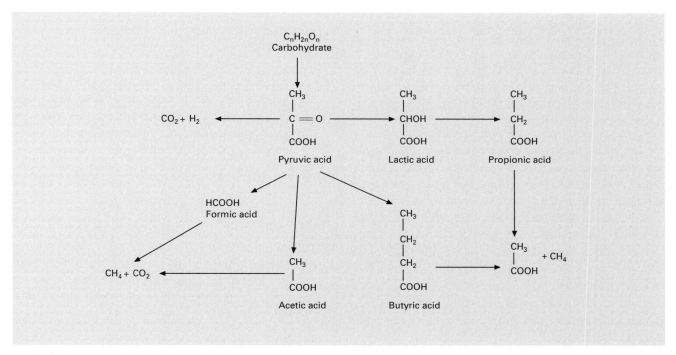

Fig. 8.10 Products of the fermentation of carbohydratres (after Yoshida, 1975).

Methane is a radiatively active gas accounting for *c.* 15% of the Enhanced Greenhouse Effect (Box 3.1). Waterlogged soils and marshes are the major source of CH_4, contributing around 20% of the global flux to the atmosphere annually. Less N is immobilized in a given time under anaerobic conditions because decomposition of the substrate is slower and the growth yield of the micro-organisms is lower (Section 3.3). Consequently, more N appears as NH_4^+, and putrefaction products, such as volatile amines, thiols and mercaptans, may also occur.

Physiological effects of anaerobiosis

Of the range of fermentation products – VFAs, hydrocarbon gases, CO_2 and H_2 – produced in poorly aerated and waterlogged soils, only acetic and butyric acids and ethylene (C_2H_4) are likely to have adverse effects on root activity and plant growth. The biochemistry of C_2H_4 production by soil micro-organisms is not well understood. It appears that an anaerobic/aerobic interface may be necessary for C_2H_4 production in soil. Very low concentrations of ethylene (~ 0.1 ppm in the gas phase) stimulate root elongation, but higher concentrations, which may persist for several weeks in wet, poorly drained soil, or beneath the smeared plough layer in a clay, seriously inhibit root growth (Fig. 8.11). Some aerobic micro-organisms metabolize C_2H_4, so the C_2H_4 concentration in soil reflects a balance between microbial production and diffusive losses and decomposition. This explains why very little C_2H_4 accumulates in well-aerated soils where it is destroyed or lost as rapidly as it is formed.

Methane is also metabolized by aerobic micro-organisms in soil, so even though waterlogged soils (such as paddy rice soils) are important sources of CH_4 that is released to the atmosphere, other well-drained soils can be significant sinks for the gas.

Indices of soil aeration

From the foregoing it can be seen that the aeration status of a soil can be assessed in several ways, as

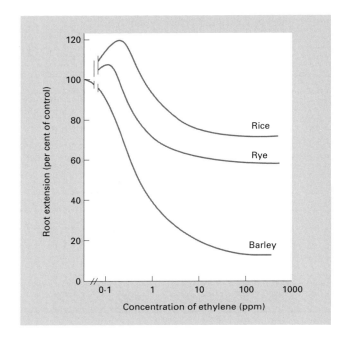

Fig. 8.11 The effect of ethylene on root growth (after Smith and Robertson, 1971).

outlined in Box 8.5. However, once the O_2 concentration in the bulk soil air has fallen below *c.* 0.004% by volume (Section 8.2), and anaerobic respiration becomes significant, the *redox potential* is a useful measure of the intensity of the ensuing reducing conditions.

8.4 Oxidation–reduction reactions in soil

Redox potential

Oxidation–reduction or *redox* reactions are those in which a chemical species (a molecule or ion) goes from a more oxidized to a less oxidized (reduced) state, or vice versa, through the transfer of electrons. One such reversible reaction is

$$H_2 \text{ (gas)} \leftrightarrow 2H^+ + 2e^- \tag{8.11}$$

where e^- stands for a free electron. This reaction may be

Box 8.5 Ways of assessing soil aeration status.

- Measurement *in situ* of the rate of O_2 diffusion through the soil. A popular method relies on the principle that reduction of O_2 at the surface of a platinum (Pt) electrode inserted into the soil, to which a fixed voltage is applied, causes a current flow proportional to the rate of oxygen diffusion to the electrode. The problems encountered in applying this method have been reviewed by McIntyre (1970).
- Alternatively, the concentration of O_2 (or CO_2 and CH_4) in a sample of air extracted from the soil can be measured by gas–liquid chromatography (GLC). Obviously, where aerobic and anaerobic zones exist in close proximity (as in Fig. 8.7), such a measurement is only a spatial average for a volume of soil in which the exact distribution of the O_2 concentration remains undefined.
- The detection of fermentation and putrefaction products by their odour (especially the volatile amines and mercaptans), and where possible, by quantitative analysis using GLC (the VFAs, CH_4 and C_2H_4).
- By noting the visual evidence of chemical changes in the soil profile, such as the solution and reprecipitation of Mn as small black specks of MnO_2, or the appearance of reddish brown mottles in a background of bluish-grey reduced Fe compounds.

coupled with another reversible reaction, for example

$$\tfrac{1}{2}O_2 \text{ (gas)} + 2e^- \leftrightarrow O^{2-}, \tag{8.12}$$

in such a way that electrons flow from the species of lower electron affinity (H_2) to higher electron affinity (O_2). The resultant coupled reaction may be written as

$$\tfrac{1}{2}O_2 \text{ (gas)} + 2H^+ + 2e^- \leftrightarrow H_2O \text{ (liquid)} . \tag{8.13}$$

The general form of the redox reaction is therefore

$$Ox + mH^+ + ne^- \leftrightarrow Red \tag{8.14}$$

for which (Ox) and (Red) are the activities of the oxidized and reduced species, respectively, and m and n are stoichiometric coefficients. If the free electron is assigned an activity (e^-), the expression for the equilibrium constant K_e for reaction 8.14 can be written as

$$K_e = \frac{(\text{Red})}{(\text{Ox})(H^+)^m(e^-)^n}. \tag{8.15}$$

Taking -$\log_{10}$ of the terms in Equation 8.15 and re-

arranging gives

$$pe = \frac{1}{n}[\log K_e + \log(\text{Ox}) - \log(\text{Red}) - mpH], \tag{8.16}$$

where pe is the negative logarithm of the free electron activity. If K_e, (Ox), (Red), pH, m and n are known, pe can be calculated. Large positive values of pe (low electron activity) favour the existence of oxidized species or those that are electron acceptors. However, in heterogeneous soil systems, it is virtually impossible to assign values to all these parameters and variables. Instead, the oxidation–reduction status of the system is measured by the *redox potential*, which is the difference in electrical potential between the two half-reactions of the coupled reaction (8.14) (Box. 8.6).

The redox potential will show least change on the addition or removal of electrons when (Ox) = (Red): this is the condition of maximum *poise* and occurs, as can be seen from Equation B8.6.6, when $E_h = E_7^0$. Equation B8.6.6 also reveals that the change in E_h of

Box 8.6 An expression for the redox potential.

By combining the equation for the free energy change in a reversible reaction

$$\Delta G = \Delta G^0 + RT \ln \frac{\text{(products)}}{\text{(reactants)}} \tag{B8.6.1}$$

with the equation for electrical work done for a given free energy change (ΔG)

$$\Delta G = -nFE, \tag{B8.6.2}$$

where n is the number of electrons transferred through an electrical potential difference E, and F is Faraday's constant (the electron has charge but no mass), we may write

$$E = E^0 - \frac{RT}{nF} \ln \left\{ \frac{\text{(Red)}}{\text{(Ox)(H}^+)^m} \right\}, \tag{B8.6.3}$$

that is

$$E_h = E^0 + 2.3 \frac{RT}{nF} \left\{ \log_{10} \frac{\text{(Ox)}}{\text{(Red)}} - m\text{pH} \right\}. \tag{B8.6.4}$$

In Equation B8.6.4, E_h is the equilibrium redox potential and E_0 is the standard redox potential of the reaction $H^+ + e^- \leftrightarrow \frac{1}{2}H_2(\text{gas})$. E_0 is zero when the pressure of H_2 gas at 25°C is 1 atmosphere and the activity of H^+ ions in solution is one (i.e. pH = 0). It is clear that oxidizing systems will have relatively positive E_h values and reducing systems more negative E_h values. It also follows that for a given ratio of (Ox)/(Red), the lower the pH the higher the E_h value at which reduction occurs.

In practice, it is more realistic to refer the measured E_h to pH 7, rather than pH 0, in which case the standard redox potential is defined by

$$E_7^0 = E^0 - \left(2.3 \frac{RT}{F} \right) \frac{m}{n} \text{pH}. \tag{B8.6.5}$$

Therefore, substituting for E^0 in Equation B8.6.4 gives

$$E_h = E_7^0 + \frac{0.059}{n} \log_{10} \frac{\text{(Ox)}}{\text{(Red)}}. \tag{B8.6.6}$$

To measure the redox potential, a Pt electrode and suitable reference electrode (usually a calomel electrode) are placed in the wet soil and the electrical potential difference recorded. Aerobic soils have E_h values between 0.3 and 0.8 volts (V), but reproducible values are only obtained in anaerobic soils, in which the potential ranges from c. 0.4 to −0.4 V.

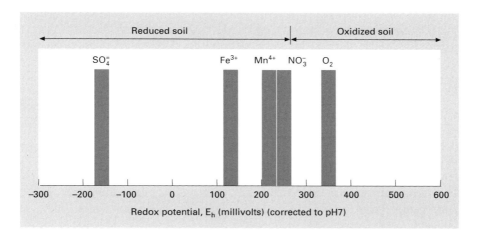

Fig. 8.12 Threshold redox potentials at which the oxidized species shown become unstable (after Patrick and Mahapatra, 1968).

the system as the ratio (Ox)/(Red) changes is inversely related to n, the number of electrons transferred in the redox reaction.

Sequential reductions in soil deprived of oxygen

Once part or all of the soil has become anaerobic, substances other than O_2 are used as terminal acceptors for the free electrons produced during respiration. A redox reaction will proceed spontaneously in the direction of decreasing free energy ($-\Delta G$), but the rate at which it proceeds may be very slow if the activation energy is high or if the redox half-reactions are physically not well coupled. In soil, redox reactions are coupled through various micro-organisms growing facultatively or obligately under anaerobic conditions (Section 8.3). The following sequence of reactions, involving different inorganic compounds as terminal acceptors for the electrons produced during respiration, may occur:

$$2NO_3^-\,(\text{soln}) + 12H^+(\text{soln}) + 10e^- \leftrightarrow N_2(\text{gas})$$
$$+ 6H_2O\,(\text{liquid}) \qquad (8.17)$$

$$MnO_2(\text{solid}) + 4H^+(\text{soln}) + 2e^- \leftrightarrow Mn^{2+}(\text{soln})$$
$$+ 2H_2O\,(\text{liquid}) \qquad (8.18)$$

$$Fe(OH)_3(\text{solid}) + 3H^+(\text{soln}) + e^- \leftrightarrow Fe^{2+}(\text{soln})$$
$$+ 3H_2O\,(\text{liquid}) \qquad (8.19)$$

$$SO_4^{2-}\,(\text{soln}) + 10H^+(\text{soln}) + 8e^- \leftrightarrow H_2S(\text{gas})$$
$$+ 4H_2O\,(\text{liquid}). \qquad (8.20)$$

The approximate E_h value at which each oxidized form becomes unstable is shown in Fig. 8.12. Each reaction poises the soil in a particular E_h range until most of the oxidized form has been consumed and then the E_h drops to a lower value as the reducing power intensifies. Note that those reactions such as (8.17) in which a gas of low water-solubility is produced are essentially irreversible.

Denitrification

The first reaction (8.17) summarizes the process of *denitrification* that is carried out by facultative anaerobic bacteria, predominantly *Pseudomonas* and *Bacillus* species, once the partial pressure of O_2 has fallen to a low level (< 0.004 bar). The complete pathway for NO_3^- reduction is

$$\underbrace{NO_3^- \leftrightarrow NO_2^-}_{\text{aerobic}} \underbrace{\rightarrow NO\,(\text{gas}) \rightarrow N_2O\,(\text{gas}) \rightarrow N_2\,(\text{gas})}_{\text{anaerobic}}.$$

Only trace amounts of nitric oxide (NO) are produced, and the main products are nitrous oxide (N_2O) and dinitrogen (N_2), with different suites of micro-organisms being involved in NO_2^- reduction and N_2O reduction. Laboratory incubations under anaerobic conditions indicate a temperature optimum of *c*. 40°C for denitrification, with the rate being very slow at temperatures < 10°C. Temperature and pH affect the ratio of $N_2O:N_2$ evolved during denitrification in closed systems. At low temperatures and pH < 5, the ratio of $N_2O:N_2$ is 1 or greater, but at temperatures of 25°C or greater and pH > 6, most of the N_2O is reduced to N_2. In the field, however, both gases can escape from the site of formation by diffusion and some N_2O, which is about as soluble as CO_2, is lost in the drainage water from very wet soils. The very variable ratio of $N_2O:N_2$ produced during denitrification makes difficult the measurement of field losses of N by this pathway (Section 10.2).

Denitrification losses can be serious in agricultural soils, especially those exhibiting a mosaic of aerobic and anaerobic zones (Fig. 8.7), and in which the supply of C substrates and NO_3^- as a terminal electron acceptor is not limiting. An extreme situation can arise in paddy rice soils which are completely waterlogged during the early period of crop growth (Fig. 8.8). The sequences of processes occurring in the oxidized and reduced soil layers is illustrated in Fig. 8.13. The NO_3^- produced aerobically diffuses or is leached into the anaerobic zone where it is denitrified. Accordingly, N fertilizer in such systems is best applied in an NH_4^+ or urea form and placed directly in the anoxic zone below the soil surface.

More intensive reducing conditions

Reactions 8.18 and 8.19 are characteristic of *gleying*, a pedogenic process discussed in Section 9.2. The former is less effective in poising the soil than either NO_3^- or Fe^{3+} reduction because MnO_2 is very insoluble and relatively few micro-organisms use it as a terminal electron acceptor. Nevertheless, in very acid soils, waterlogging may lead to high levels of exchangeable Mn^{2+}, which cause manganese toxicity in susceptible plants.

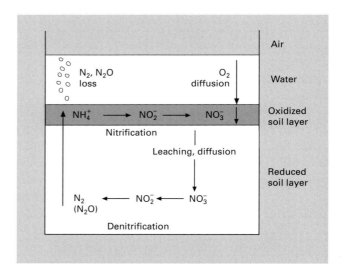

Fig. 8.13 Nitrification and denitrification in the oxidized and reduced zones of a flooded soil (after Patrick and Mahapatra, 1968).

Anaerobiosis and Fe^{3+} reduction generally increase the phosphate concentration in the soil solution, especially in acid soils where phosphate is specifically adsorbed on iron oxide particles and coatings on clay surfaces (Section 7.3), or precipitated as very insoluble Fe phosphates. The effect is complicated, however, by the inevitable rise in pH on prolonged waterlogging ($3H^+$ ions are consumed for each electron transferred in Reaction 8.19). At pH > 6 ferrous compounds become less soluble and a metastable iron oxyhydroxide containing both ferric and ferrous iron precipitates – ferrosoferric hydroxide $Fe_3(OH)_8$ – according to the reaction

$$2Fe(OH)_3 + Fe^{2+} + 2OH^- \leftrightarrow Fe_3(OH)_8 . \qquad (8.21)$$

At the raised CO_2 partial pressures, $FeCO_3$ can also precipitate and the pH eventually stabilizes between pH 6.5 and 7. Freshly precipitated $Fe_3(OH)_8$ presents a highly active surface for the readsorption of phosphate and certainly all the dissolved P is readsorbed when aerobic conditions return and ferrihydrite precipitates (Section 2.4).

At very low E_h values (Fig. 8.12), SO_4^{2-} is reduced due to the activity of obligate anaerobes of the genus *Desulphovibrio*. As the concentration of weakly dissociated H_2S builds up in the soil solution, ferrous sulphide (FeS) precipitates and slowly reverts to *iron pyrites* (FeS$_2$), according to the reaction

$$FeS(solid) + S(solid) \leftrightarrow FeS_2(solid) . \qquad (8.22)$$

Iron pyrites is a mineral characteristic of marine sedimentary rocks and of sediments currently deposited in river estuaries and mud flats.

8.5 Summary

During respiration by soil organisms and plant roots, complex C molecules are dissimilated to provide energy for cellular growth. When conditions are *aerobic*, 1 mol of CO_2 is released for each mole of O_2 consumed and the *respiratory quotient* (*RQ*) is 1. Respiration continues aerobically, at a rate determined by soil moisture, temperature and substrate availability, provided that O_2 from the atmosphere moves into the soil fast enough to satisfy the organisms' requirements, and CO_2 can move out. This dynamic exchange of gases is called *soil aeration*.

Diffusion of gases through the air-filled pore space is the most important mechanism of soil aeration. The rate of diffusion or diffusive flux J_g is given by

$$J_g = - D_e \ grad \ C$$

where D_e is the *effective gas diffusion coefficient*, which takes account of the effect of differences in air-filled porosity ε and the tortuosity α of the diffusion pathway. Given the possibility that CO_2 and O_2 fluxes vary with time t and depth z, the flux equation must be combined with a continuity equation to calculate O_2 and CO_2 concentration profiles in soil; that is

$$\varepsilon \frac{\partial C}{\partial t} = D_e \frac{\partial^2 C}{\partial z^2} \pm R$$

where R, the soil respiration rate, is positive when measured as CO_2 release and negative when measured as O_2 uptake.

Sandy soils with ε values > 0.1 maintain O_2 partial pressures close to the atmospheric level of 0.21 bar. Similarly, well structured clay soils at field capacity can maintain O_2 partial pressures close to 0.2 bar in their macropores, but sites of active microbial respiration within peds are likely to become *anaerobic* if separated from air-filled pores by more than 1 cm of water-saturated soil in spring, or > 2 cm in winter. The reason for this is the fact that the D values of O_2 and CO_2 in water are 10 000 less than in air, and gas exchange through water-filled pores is very slow. This is especially true of O_2 which has a much lower water solubility than CO_2. Even well-drained clay soils can exhibit a mosaic of aerobic and anaerobic zones which is particularly conducive to *denitrification*, if the temperature is favourable and C and NO_3^- supplies adequate. Denitrification is carried out by facultative anaerobes: the switch from aerobic to anaerobic respiration occurs at O_2 partial pressures $\leqslant 0.004$ bar.

Once the O_2 supply is exhausted, which may happen within 1–2 days of a soil becoming completely waterlogged, bacteria that can respire either aerobically or anaerobically continue to grow and obligate anaerobes begin to multiply. Most plants are intolerant of waterlogging, but some such as marsh plants and paddy rice survive because of O_2 diffusion through their large intercellular air spaces. During anaerobic respiration, carbohydrates are incompletely oxidized or fermented to *volatile fatty acids* (VFAs) – which can attain localized concentrations high enough to be toxic to seedlings – and low molecular weight hydrocarbon gases, including ethylene (C_2H_4). As anaerobiosis continues, these VFAs are further dissimilated, the end-products being CO_2, CH_4 and H_2.

In the absence of O_2, other substances serve as electron acceptors in bacterial respiration, and reduction of NO_3^-, MnO_2, Fe^{3+} compounds and SO_4^{2-} occurs, in that sequence, as reducing conditions intensify. The reducing power of the soil system is measured by the *redox potential* E_h, which for the coupled reversible reaction

$$Ox + mH^+ + ne^- \leftrightarrow Red$$

is defined by the equation

$$E_{\mathrm{h}} = E^{0} + 2.3\,\frac{RT}{nF}\left\{\log_{10}\frac{(\mathrm{Ox})}{(\mathrm{Red})} - m\mathrm{pH}\right\}.$$

References

Bramley R. G. V. & White R. E. (1991) The variability of nitrifying activity in field soils. *Plant and Soil* **126**, 203–208.

Currie J. A. (1970) Movement of gases in soil respiration, in *Sorption and Transport Processes in Soils*. Society of Chemical Industry Monograph **37**, pp. 152–169.

Currie J. A. (1975) Soil respiration, in *Soil Physical Conditions and Crop Production*. MAFF Bulletin **29**, pp. 461–468.

Greenwood D. J. (1969) Effect of oxygen distribution in soil on plant growth, in *Root Growth* (Ed. W. J. Whittington). Butterworths, London, pp. 202–221.

Greenwood D. J. (1975) Measurement of soil aeration, in *Soil Physical Conditions and Crop Production*. MAFF Bulletin **29**, pp. 261–272.

McIntyre D. S. (1970) The platinum electrode method for soil aeration measurement. *Advances in Agronomy* **22**, 235–283.

Patrick W. H. & Mahapatra I. C. (1968) Transformation and availability to rice of nitrogen and phosphorus in waterlogged soils. *Advances in Agronomy* **20**, 323–359.

Smith K. A. & Robertson P. D. (1971) Effect of ethylene on root extension of cereals. *Nature New Biology* **234**, 148–149.

Yoshida T. (1975) Microbial metabolism of flooded soils, in *Soil Biochemistry*, Vol. 3. (Eds E. A. Paul and A. D. McLaren). Marcel Dekker, New York, pp. 83–122.

Further reading

Bartlett R. J. & Bruce B. R. (1993) Redox chemistry of soils. *Advances in Agronomy* **50**, 152–208.

Glinski J. & Stepniewski W. (1985) *Soil Aeration and its Role for Plants*. CRC Press, Boca Raton, Florida.

Rowell D. L. (1988) Flooded and poorly drained soils, in *Russell's Soil Conditions and Plant Growth* 11th edn. (Ed. A. Wild). Longman Scientific and Technical, Harlow, pp. 899–926.

Tiedje J. M., Sextone A. J., Parkin T. B., Revsbech N. P. & Shelton D. R. (1984) Anaerobic processes in soil. *Plant and Soil* **76**, 197–212.

Chapter 9
Processes in Profile Development

9.1 The soil profile

Soil horizons

In Chapter 1, the course of profile development from bare rock to a *brown forest soil* (Inceptisol (ST) or Tenosol (A)) is described in outline. Except in peaty soils, the downward penetration of the weathering front greatly exceeds the upward accumulation of litter on the surface, and the mature profile may have the vertical sequence previously illustrated in Fig. 1.4, that is

Litter layer, L
A horizon (mainly eluvial)
B horizon (mainly illuvial)
C horizon (weathering parent material).

Predictably, profile development can be much more complex than this simple model suggests. A more comprehensive description requires an enlarged glossary of descriptive terms – the *horizon notation* – intended to convey briefly the maximum information about the soil to an observer. Unfortunately, as with much of the descriptive terminology in soil science (Box 1.1), systems of horizon notation differ from country to country. The system adopted by the Soil Survey of England and Wales (Avery, 1980) is consistent with that proposed for international use (International Society of Soil Science, 1967). The system currently used in Australia (McDonald *et al.*, 1990) shows some variations which are summarized in Box 9.1. In all systems, there are two main kinds of horizon, *organic* and *mineral*, which are distinguished by their organic matter content.

Organic horizons

An organic horizon must have:
• greater than 30% organic matter (18% C) when the mineral fraction has 50% or more clay, or
• greater than 20% organic matter (12% C) when the mineral fraction has no clay, or
• more than a proportionate amount of organic C (between 12 and 18%) if the clay content is intermediate.

Organic horizons that are derived from a superficial *litter layer L*, and are rarely saturated with water for more than 30 consecutive days, are designated as:
• an *F horizon* when the organic matter is only partly decomposed, and
• an *H horizon* when the organic matter is low in minerals and well decomposed.

An organic horizon that remains saturated with water for at least 30 consecutive days most years is called a *peaty* or *O horizon*. Wetness favours an increasing thickness of the O horizon, which qualifies as *peat* when there is:
• more than 40 cm of O horizon material in the upper 80 cm of the profile, excluding litter L and living moss, or
• more than 30 cm of O horizon material resting directly on bedrock R, or very stony material.

Mineral horizons

A *mineral* horizon is low in organic matter and overlies consolidated or unconsolidated rock. It may be designated A, E, B, C or G, as indicated in Table 9.1, depending on its position in the profile, its mode of

formation, colour,* structure, degree of weathering and other properties that can be recognized in the field. Three kinds of *complex* horizon are also recognized:
• a horizon with distinctive features of two horizons, such as a strongly gleyed B horizon, is designated B/G,
• a relatively homogeneous horizon with subordinate features of two contiguous horizons, for instance A and B, is designated AB,
• a transitional horizon which may comprise discrete parts (more than 10% by volume) of two contiguous horizons is called A & B, or E & B for example.

Specific attributes of each major horizon are indicated by lower-case suffixes attached to the horizon symbol (Table 9.1). Again, there is not complete consistency between national systems so the source books should be consulted for details.

Soil layers

Unless the record of rock lithology or stratigraphy suggests the contrary, it is assumed that the parent material originally extended unchanged to the soil surface. As indicated in Section 5.2, this assumption is often invalid for soils on very old land surfaces subjected to many cycles of weathering, erosion and deposition that have resulted in the layering of parent materials. An example of layering is the *K cycle* of soil formation on weathered and resorted materials in southeast Australia (Fig. 9.1).

Layers of different lithology are identified by numbers prefixed to the horizon symbols, the uppermost layer being 1 (prefix usually omitted) and those below labelled sequentially 2, 3, etc. Where an old soil has formed on a lower layer and subsequently been buried, its horizons are prefixed by the letter *b* (or suffixed in the Australian system) (Fig. 5.7). *Buried horizons* can be found in very old cleared areas where humans have shifted topsoil for farming or construction purposes. Similarly, the course of soil formation near old settlements has frequently been altered by the deliberate deposition of organic residues, such as seaweed and

* Standard colour descriptions are obtained from Munsell colour charts.

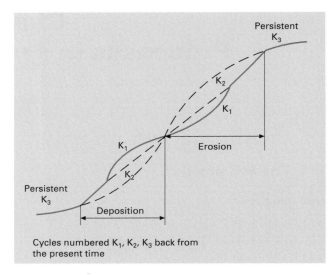

Fig. 9.1 K cycles of soil formation on a hill slope (after Butler, 1959).

waste. Soils of this kind are called man-made or plaggen soils.

9.2 Pedogenic processes

In this section, the more important processes operative in profile development (pedogenesis) are discussed. The magnitude of their effect is governed by the interaction of factors defining the initial state of soil formation (parent material, relief), the environment (climate, organisms) and the extent of reaction (time), as discussed in Chapter 5.

Congruent and incongruent dissolution

Hydrolysis

Rock minerals become unstable when exposed to a changed environment. Chemical reactions occur that may be relatively subtle (e.g. oxidation of Fe^{2+} to Fe^{3+}) or drastic, such as the complete dissolution of the mineral in water. Minerals such as *halite* (NaCl) or *gypsum* ($CaSO_4.2H_2O$) dissolve to release their constituent ele-

Table 9.1 Soil horizon notation. After Avery, 1980.

Horizon	Brief description
O – organic	
Of	Composed mainly of fibrous peat
Om	Composed mainly of semi-fibrous peat
Oh	Humified peat; uncultivated
Op	Humified peat; mixed by cultivation
A – at or near the surface, with well-mixed humified organic matter (by natural processes or cultivation)	
Ah	Uncultivated with >1% organic matter
Ap	Mixed by cultivation; normally an abrupt lower boundary
Ahg or Apg	Partially gleyed due to intermittent waterlogging; rusty mottles along root channels
E – eluvial, subsurface horizon of low organic content, having lost material to lower horizons	
Ea	Lacking mottles, bleached particles due to the removal of organic and sesquioxidic coatings
Eb	Brownish colour due to uniformly disseminated iron oxides: often overlies a Bt horizon
Eg	Partial gleying with mottles; often overlies a relatively impermeable Bg or Btg horizon, and sometimes a thin iron-pan
B – subsurface horizon without rock structure, showing characteristic structure, texture and colour due to illuviation of material from above and/or weathering of the parent material	
Bf	Thin iron-pan (< 1 cm); often enriched with organic matter and aluminium
Bg	Dominant gley colours on ped faces, rusty mottles within peds; blocky or prismatic structure
Bh	Translocated organic matter (> 0.6%) with some Fe and Al in coatings on sand grains and small peds
Bs	Enriched with sesquioxides, usually by illuviation; orange to red colour
Bt	Accumulation of translocated clay as shown by clay coatings on ped faces
Bw	Shows alterations to parent material by leaching, weathering and structural reorganization under well-aerated conditions
C – unconsolidated or weakly consolidated mineral material that retains rock structure	
Ck	Contains > 1% secondary $CaCO_3$, as concretions or coatings (also Ak and Bk)
Cg	Dominantly grey to green to blue colours due to waterlogging, typically in sands or recent alluvium
Cgy	As for Cg but contains secondary $CaSO_4$ as gypsum crystals
Cm	Continuously cemented, other than by a thin iron-pan
Cr	Weakly consolidated but dense enough to prevent root ingress except along cracks
Cu	Unconsolidated and lacking evidence of gleying, cementation, etc.
Cx	With fragipan properties, as in Pleistocence deposits; dense but uncemented; firm when dry but brittle when moist
CG – intensely gleyed, bluey-grey to green colours that fade on exposure to air and predominate throughout a structureless soil matrix	

Box 9.1 The Australian system of soil horizon notation.

O horizons – Dominated by organic materials in varying stages of decomposition that have accumulated on the mineral soil surface; subdivided into O1 and O2, depending on the degree of decomposition and humification (*cf.* F and H horizons).

P horizons – Organic horizons that have accumulated under water or conditions of excessive wetness, commonly called *peat* (no thickness criteria).

A horizons – Surface mineral horizons, subdivided into:
  A1 – mineral horizon at or near the surface with some accumulation of humified organic matter;
  A2 – mineral horizon having, either alone or in combination, less organic matter, sesquioxides or clay than contiguous horizons;
  A3 – transitional between A and B, dominated by the properties of an overlying A1 or A2.

B horizons – Subsurface mineral horizons characterized by one or more of the following: an accumulation of silicate clay, iron, aluminium or organic matter; a structure and/or consistence different from that of A horizons above or any other horizon immediately below, and stronger colours.
  B1 – transitional between A and B, dominated by the properties of an underlying B2.
  B2 – dominant feature is one or more of the following:
  • an illuvial, residual or other accumulation of silicate clay or iron, aluminium or humus;
  • maximum expression of pedological organization as evidenced by a different structure and/or consistence and/or stronger colours than horizons above and below.
  B3 – Transitional between B and C, dominated by properties of an overlying B2.

C horizons – Below the A–B soil consisting of consolidated or unconsolidated material, usually partly weathered but little affected by pedogenesis.

ments in the same molar ratio as they occur in the solid, that is

$$NaCl(solid) \leftrightarrow Na^+(liquid) + Cl^-(liquid) \quad (9.1)$$

and

$$CaSO_4.2H_2O(solid) \leftrightarrow Ca^{2+}(liquid) + SO_4^{2-}(liquid) + 2H_2O. \quad (9.2)$$

This is called *congruent* dissolution, as opposed to incongruent dissolution, where the molar ratio of elements in solution differs from that in the solid. Because of this inequality, a compound different from the original mineral gradually forms in the solid phase. Chemical weathering of the primary silicate minerals is a good example of *incongruent* dissolution by surface hydrolysis. Water molecules at the mineral surface dissociate into H^+ and OH^- and the small, mobile H^+ ions (actually H_3O^+) penetrate the crystal lattice, creating a charge imbalance which causes cations such as Ca^{2+}, Mg^{2+}, K^+ and Na^+ to diffuse out. The potash feldspar-orthoclase hydrolyses to produce a weak acid (silicic), a strong base (KOH) and leaves a residue of the clay mineral *illite*.

$$3KAl^{IV}Si_3O_8 + 14H_2O \leftrightarrow K(AlSi_3)^{IV}Al_2^{VI}O_{10}(OH)_2 + 6Si(OH)_4 + 2KOH. \qquad (9.3)$$

Hydrolysis of the calcic feldspar *anorthite*, on the other hand, leaves a residue of vermiculite-type clay, which is a weak acid, and the strong base $Ca(OH)_2$

$$3CaAl_2^{IV}Si_2O_8 + 6H_2O \leftrightarrow 2H^+2[(AlSi_3)^{IV}Al_2^{VI}O_{10}(OH)_2] + 3Ca(OH)_2O. \qquad (9.4)$$

Equations (9.3) and (9.4) are specific examples of the general reaction

$$[M^+silicate] + H_2O \leftrightarrow H^+[silicate\ secondary\ mineral]^- + M^+OH^- \qquad (9.5)$$

in which M^+ is an alkali or alkaline earth cation, or sometimes a transition element (Fe, Mn). Owing to the formation of the strong base MOH, nearly all the primary silicates have an alkaline reaction – the *abrasion pH* – when placed in CO_2-free distilled water (pH $\cong$ 6).

However, Reaction 9.5 proceeds more rapidly to the right when more H^+ ions are present to neutralize the OH^- ions of the strong base. The main sources of this extra acidity in natural, well-drained soils are:
• CO_2 dissolved in the soil water, at partial pressures up to at least 0.01 bar (Section 8.2), and
• carboxyl and phenolic groups of humified organic matter (Section 3.4).

The hydrolysis, neutralization and cation exchange reactions involving a weathering primary silicate, organic matter and the clay mineral residues are illustrated in Fig. 9.2. The reactions leading to the formation of acid clays in soil are outlined in Box 9.2 (see also Section 7.2).

Oxidation

The weathering of iron-bearing rock minerals in water containing dissolved O_2 and CO_2 is another example of incongruent dissolution. Hydrolysis of the mineral releases the iron, present mainly as Fe^{2+}, which then oxidizes to the Fe^{3+} form and precipitates as an insoluble oxyhydroxide – usually either ferrihydrite ($Fe_2O_3.2FeOOH.nH_2O$), which is a necessary precursor of hematite (α-Fe_2O_3), or the stable mineral goethite (α-FeOOH) (Section 2.4). The precipitate may form a coating over the mineral surface which slows down the

Box 9.2 The formation of acid clays.

As alkali and alkaline earth cations Na^+, K^+, Ca^{2+} and Mg^{2+} are leached from the soil, clay mineral surfaces and organic acid groups gradually become saturated with H^+ ions. However, H-clays are unstable at pH < 4 and disintegrate slowly from the crystal edges inwards to release Al and SiO_2 (Fig. 7.9). The silica goes into solution as $Si(OH)_4$ and the Al is adsorbed as Al^{3+} by both clay minerals and humified organic matter. The subsequent slow hydrolysis of the Al^{3+} produces more H^+ ions which are neutralized by further lattice decomposition, so the resultant pH is higher than that of a pure H-clay or organic acid. Ultimately, clay minerals are completely destroyed to yield gibbsite, or goethite/hematite mixtures in soils high in ferromagnesian minerals. The change in mineralogy and pH with depth in a highly weathered soil on basalt (Table 9.2) illustrates this sequence of events.

Table 9.2 Mineralogical changes in weathering basalt. After Black and Waring, 1976.

	Depth cm	pH (in water)	Kaolinite (%)	Other clay fraction minerals
Soil surface				
	10–20	6.5	15–25	H, Q, G
Decreasing	60–90	6.2	30–40	H, Q, G
intensity	120–150	6.1	40–50	H, G
of	180–210	6.0	45–55	H
weathering	270–300	5.7	55–65	H
	570–600	4.9	65–75	H

H, hematite; Q, quartz; G, gibbsite.

subsequent rate of hydrolysis. Accordingly, in an anaerobic environment where ferric oxide precipitation is inhibited, the rate of weathering of iron-bearing minerals is expected to be faster.

Note that the oxidation of Fe^{2+} to Fe^{3+} according to

$$2Fe^{2+} + 5H_2O + \tfrac{1}{2}O_2 \leftrightarrow 2Fe(OH)_3 + 4H^+ \qquad (9.6)$$

is an acidifying reaction. The localized production of H^+ ions will generally accelerate the rate of mineral weathering, as discussed under 'Hydrolysis'. Some rock minerals alter directly from one form to another by *in situ* oxidation of Fe^{2+} in the lattice; the weathering of biotite mica to vermiculite is one example.

Leaching

Differential mobilities

The mobility of an element during weathering reflects its solubility in water and the effect of pH on that solubility. Relative mobilities, represented by the Polynov series in Table 9.3, have been established from a comparison of the composition of river waters with that of igneous rocks in the catchments from which they drain. The low mobility of Al and Fe can be explained in terms of the formation and strong adsorption of hydroxy-Al and hydroxy-Fe ions at pH > 4 and 3, respectively, and the precipitation of the insoluble oxyhydroxides. The least mobile element is titanium, which forms the oxide TiO_2,

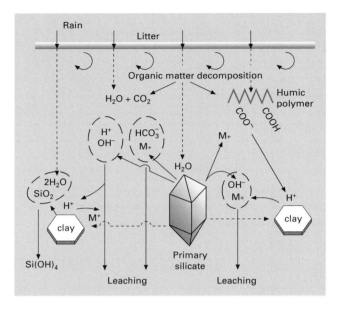

Fig. 9.2 Scheme for silicate mineral hydrolysis in the presence of CO_2-charged rainwater and humified organic residues.

insoluble at pH > 2.5. It is therefore used as a reference material to estimate the relative gains or losses of other elements in the profile (Section 2.4).

Percolation depth

Leaching of an element depends not only on its mobility, but also on the rate of water percolation through the

Table 9.3 The relative mobilities of rock constituents. After Loughnan, 1969.

Constituent elements or compounds	Relative mobility*
Al_2O_3	0.02
Fe_2O_3	0.04
SiO_2	0.20
K	1.25
Mg	1.30
Na	2.40
Ca	3.00
SO_4	57

* Expressed relative to Cl taken as 100.

soil. In arid areas, even the most soluble constituents, mainly NaCl and to a lesser extent, the chlorides, sulphates and bicarbonates of Ca and Mg, tend to be retained and give rise to *saline soils* (Section 9.6). As the climate becomes more humid, losses of salts and silica increase and the soils are more highly leached, except in the low-lying parts of the landscape where the drainage waters and salts accumulate and modify the course of soil formation (Section 9.4). The effect of leaching is well illustrated by the steadily increasing depth to the carbonate layer in soils formed on calcareous loess in the USA, along a line drawn from semi-arid Colorado to humid Missouri (Fig. 9.3).

Cheluviation

The leaching of the immobile Al^{3+} and Fe^{3+} cations can be enhanced by the formation of soluble organometal complexes or *chelates* (Section 3.4) – hence the term *cheluviation*. Aluminium forms an electrostatic-type complex with the carboxyl groups of humified organic matter in which the average charge per mole of complexed Al is *c.* +1, that is

$$R\text{-COO-}[Al(OH)_2(H_2O)_4]^+$$

where *R* is the stem of the organic molecule. Ferric iron, on the other hand, is likely to be reduced to ferrous iron at the expense of plant-derived reducing agents, with the concurrent formation of a soluble ferrous–organic complex

$$\text{Organic reductant } (e^-) + Fe(OH)_3 + 3H^+ \leftrightarrow Fe^{2+}\text{-} \\ \text{organic ligand} + 3H_2O. \qquad (9.7)$$

The organic reductant supplies the electrons that drive this reaction to the right. The stability of the Fe^{2+} organic ligand complex is favoured by anaerobic conditions (low redox potential E_h – see Section 8.4); but because H^+ ions are also involved on the left-hand side of Equation 9.7, the complexes will form at higher E_h values (more oxidizing conditions), the lower the local pH. Thus, the most active complexing agents are organic acids, particularly those of the fulvic acid (FA) fraction leached from litter, and strongly reducing poly-

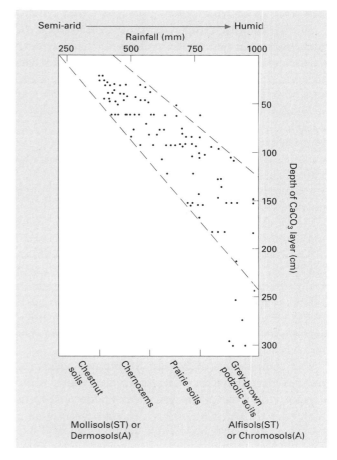

Fig. 9.3 Depth of $CaCO_3$ accumulation with increasing rainfall (after Jenny, 1941).

phenols that are leached from the canopies of conifers and heath plants and their freshly fallen litter.

Polyphenols not involved in cheluviation are oxidatively polymerized and contribute to the characteristic F and H horizons of mor humus found under coniferous forest and heath vegetation (Section 3.3).

Podzolization

Podzolization involves the cheluviation of Al and Fe from insoluble oxides in the upper soil zone and their deposition, with organic matter, deeper in the profile.

There is also evidence that during podzolization Al can move in colloidal form as part of short-range order hydroxy-Al silicates (proto-imogolite – Section 2.4). The removal of iron oxide and organic coatings renders the sand grains *bleached* in appearance, which is a feature of an Ea horizon. As the ferrous–organic complex is leached more deeply it becomes less stable through a combination of factors:

- microbial decomposition of the organic ligands,
- complexing of more metal ions, which increases the ratio of cation to ligand in the complex,
- a rise in pH, which favours the oxidation of Fe^{2+} at a given E_h, and
- the flocculating effect of higher concentrations of salts in the soil solution near the weathering parent material.

The iron precipitates as ferrihydrite, giving a uniform orange-red colour to the B horizon. The precipitated hydroxide adsorbs fresh ferrous-organic complexes, in which the iron subsequently oxidizes, and also acts as a coarse filter for humus and Al-organic complexes eluviated from the A horizon. Hence, the Bs horizon that forms is usually overlain by a Bh horizon.

The combination of Bs and Bh horizons generally satisfies the criteria for a *spodic horizon*, which is the diagnostic horizon of the soil order Spodosol (ST) or Podosol (A). The end-product of podzolization is a *podzol* in the Great Soil Group terminology of Table 5.4, the exact form of which is determined by the interplay of vegetation, drainage and parent material. Podzolization is favoured by permeable, siliceous parent material, a *P/E* ratio > 1 for most of the year and slow rates of organic decomposition because of the type of litter or the acidity of the soil, or both. Two types of podzol profiles are shown in Fig. 9.4.

Lessivage

Duchaufour (1977) used the term *lessivage* to describe the mechanical eluviation of clay particles from the A horizon without chemical alteration; alternatively, the term clay translocation can be used. The clay is washed progressively downwards in the percolating water to be deposited in oriented films (argillans) on ped faces and

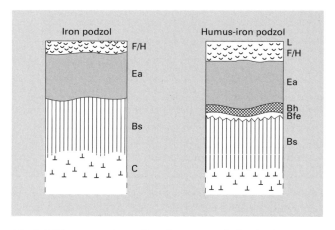

Fig. 9.4 Diagrammatic profiles of an *iron podzol* and a *humus-iron podzol* (Spodosols (ST) or Podosols (A)).

pore walls, gradually forming an horizon of clay accumulation (Bt) or an *argillic horizon*. Water flow through macropores (old root and worm channels and cracks between peds) favours lessivage when the physicochemical conditions are favourable (see below).

Lessivage is unlikely to occur in calcareous soils, or neutral soils that retain a high proportion of exchangeable Ca^{2+} ions, because the clay remains flocculated. Under acid conditions, if exchangeable Al^{3+} and hydroxy-Al ions are predominant on the clay surfaces, the clay should also remain flocculated (Section 7.4). However, when adsorbed Al and the Fe of oxyhydroxide films stabilizing microaggregates are complexed by soluble organic compounds, the clay particles are more likely to deflocculate and be translocated. Similarly, when Na^+ rises to more than 10–15% of the exchangeable cations, clay deflocculation and translocation are likely, giving rise to *solonized soils* (Section 9.6).

The formation of a lessived soil (Duchaufour's *sol lessivé*) predisposes the soil to further pedogenic processes, the course of which depends on drainage in the profile. When the soil is freely drained, with increasing lessivage and acidification in the upper profile, podzolization ensues and the soil may pass through successive stages in the maturity sequence

lessived brown soil → *acid lessived soil* → *podzolized lessived soil* → *podzol.*

The first three stages correspond to the podzolic or podzolized soils of the Great Soil Group terminology (Table 5.4) (see also Box 9.3). Essentially, the difference between a podzol and lessived soil is that clay is destroyed rather than translocated in the podzol, and the intensity of cheluviation of Fe and to a lesser extent Al is greater in the podzol. The latter effect is illustrated by the free oxide ratios in Table 9.4. In the lessived soil the ratios SiO_2/Al_2O_3 and Al_2O_3/Fe_2O_3 remain reasonably constant down the profile; but in the podzol these ratios are high in the Ea horizon and much lower in the Bh and Bs horizons.

Where the drainage of the lessived soil is impeded, waterlogging and consequent gleying processes may supervene at variable depths.

Gleying

Collectively, soils that are influenced by waterlogging are called *hydromorphic soils* (Section 9.5). The most important feature of such soils is *gleying* (Section 8.4). True gley soils have uniformly blue-grey to blue-green colours and are depleted of iron due to its solution and removal as Fe^{2+} ions. The reduction of iron is primarily biological and requires both organic matter and

Table 9.4 Free oxide ratios in the clay fractions of a typical podzol and lessived soil.* After Muir, 1961.

Podzol (Spodosol (ST) or Podosol (A))			Lessived soil (Alfisol (ST) or Chromosol (A))		
Horizon	$\dfrac{SiO_2}{Al_2O_3}$	$\dfrac{Al_2O_3}{Fe_2O_3}$	Horizon	$\dfrac{SiO_2}{Al_2O_3}$	$\dfrac{Al_2O_3}{Fe_2O_3}$
Ea	3.22	5.72	Eb	3.15	3.98
Bh Bs	1.85	0.84	Bt1	3.05	3.66
Bs	1.83	2.40	Bt2	3.70	3.12
C	2.12	3.71	C	3.21	3.10

* Grey-brown podzolic soil.

micro-organisms capable of respiring anaerobically. However, most of the iron exists as Fe^{2+}–organic complexes in solution or as a mixed precipitate of ferric and ferrous hydroxides, $Fe_3(OH)_8$, which is responsible for the characteristic gley colour (see Reaction 8.21). Mottling is a secondary effect in gleys resulting from the reoxidation of iron in better aerated zones, especially around plant roots and in the larger pores. The orange-red mottles have a much higher Fe content than the surrounding blue-grey matrix. Thus, a C/G or B/G horizon frequently has a coarse prismatic structure in which the ped interiors have a uniform gley colour and the exterior faces, pores and root channels have rusty mottling. This is typical of a *groundwater gley* formed when the watertable resides more or less permanently in the subsoil.

By contrast, in soils with impeded drainage in the upper part of the profile (e.g. many duplex soils), gleying is confined to the A, E or top of the B horizon and is subject to seasonal weather conditions. When a soil moisture deficit develops (Section 6.4), Fe^{2+} is oxidized and rusty mottles develop against the blue-grey colours of the ped faces, darkened in the A horizon by the presence of organic matter. This is typical of the *pseudogleys*. In colder and wetter climates, the A horizon can remain wet for most of the year even though the mineral soil profile is well drained, allowing a thick O horizon or even peat to develop, and blue-grey gley colours to persist in the mineral soil. This is typical of a *stagnogley*. Pseudogleys and stagnogleys fall into the category of *surface water gleys*. Profile features of groundwater and surface water gleys are illustrated in Fig. 9.5.

Illustrative examples of these pedogenic processes in action in temperate and tropical soils are given in the following sections.

9.3 Freely drained soils of humid temperate regions

The soils over substantial areas of Europe and North America have formed on comparatively uniform parent materials of glacial or periglacial origin for a similar

Box 9.3 Soils with an A–B texture contrast.

Lessivage generally leads to a soil in which the clay content increases markedly from the A to the B horizon. Often the A–B boundary is sharp as in the Australian *red brown earths* (Plate 9.1). Such soils are widespread in Australia, where the generic term 'duplex soil' is used, and in other parts of the world. In the Australian Soil Classification (Isbell, 1996), they fall into one of three orders.

- *Chromosols* – soils of strong texture contrast and pH (1 : 5 in water) > 5.5 in the B horizon.
- *Kurosols* – soils of strong texture contrast and pH (1 : 5 in water) < 5.5 in the B horizon.
- *Sodosols* – soils of strong texture contrast and exchangeable Na^+ content of the fine earth $\geq$ 6%.

The distinction between Chromosols and Kurosols broadly reflects the effect of parent material, climate and the stage reached in the lessivage sequence. In the case of Sodosols, the physicochemistry of the clay environment is overriding.

In Soil Taxonomy (Soil Survey Staff, 1992), lessived soils generally fall into the orders Alfisol, or Ultisol if the base saturation in the subsoil is < 35%.

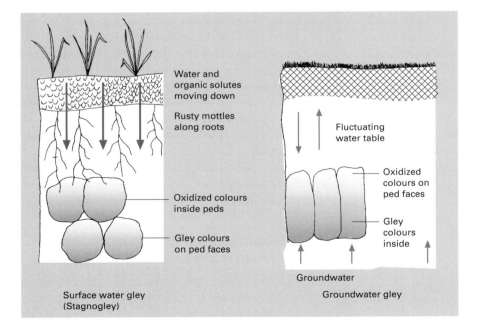

Surface water gley
(Stagnogley)

Groundwater gley

Fig. 9.5 Profile features of groundwater gley and surface water gley soils (after Crompton, 1952).

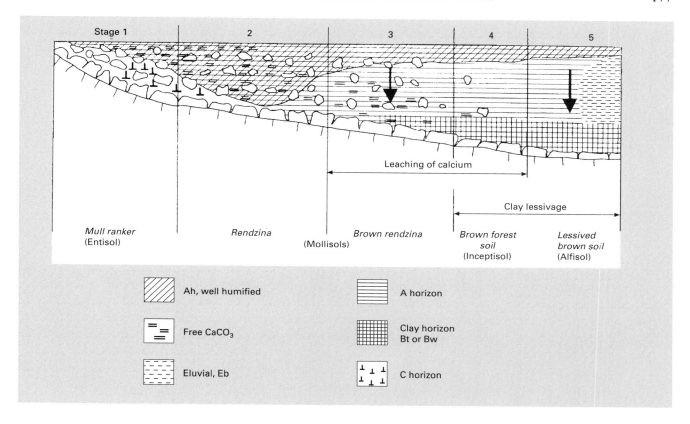

Fig. 9.6 Soil maturity sequence on calcareous clayey parent material under a humid temperature climate (after Duchaufour, 1982).

period of time. On well-drained sites of slight to moderate relief, two soil sequences are common: one on calcareous, the other on siliceous, parent materials.*

Soils on calcareous, clayey parent material

The general sequence of soil formation with the passage of time, and increased leaching, is shown in Fig. 9.6. The plant succession advances from grassland to deciduous forest. In eastern Canada under deciduous forest,

* Since the examples given are from humid temperate regions in the northern hemisphere, no attempt is made in this section to allocate the soils to Australian soil orders.

for example, the soil may change from a *brown forest soil* (Inceptisol) to *grey-brown podzolic soil* (Alfisol); or to a *grey wooded soil* (Alfisol) under cooler conditions where conifers flourish. The major difference is that the *grey-brown podzolic soil* has a mull-like Ah horizon and the *grey wooded soil* a mor humus layer. Both soil types degenerate, with further leaching, into *brown podzolic soils* (Ultisols), in which the whole profile becomes acidic and podzolization more obvious. The latter corresponds to the intermediate stages (4 and 5) of Duchaufour's maturity sequence on siliceous parent rock (Fig. 9.7).

Soils on permeable, siliceous parent material

The early stages of this soil sequence, illustrated in Fig. 9.7, feature lessivage under a deciduous forest

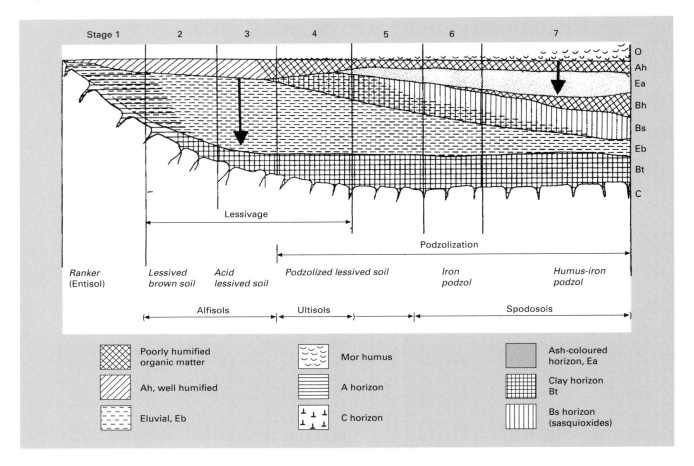

Fig. 9.7 Soil maturity sequence on permeable, siliceous parent material under a humid temperate climate (after Duchaufour, 1982).

cover, leading to the formation of a *lessived brown soil* (Alfisol) and *acid lessived soil* (Alfisol). With further time and leaching, the podzolizing and lessivage influence strengthens and a *podzolized lessived soil* forms (Ultisol). Finally, an *iron podzol* or *humus-iron podzol* (Spodosol) forms in the surface of the deep permeable Eb horizon of the lessived soils. Mackney (1961) has described the terminal stages (5–7) of this sequence under oakwoods on heterogeneous acid sands and gravels of the Bunter Beds in the English Midlands. On less

acid parent material, however, continued degradation of the *podzolized lessived soil* (Alfisol) to the *humus-iron podzol* (Spodosol) is normally associated with a change in the vegetation from climax deciduous forest to coniferous forest or heathland.

Soils on mixed siliceous and calcareous parent material

A sequence similar to that shown in Fig. 9.7 occurs under beechwoods in the Chiltern Hills of England, where the original parent material consisted of some siliceous glacial drift mixed with soliflucted material from the Chalk that has, however, become acidic after

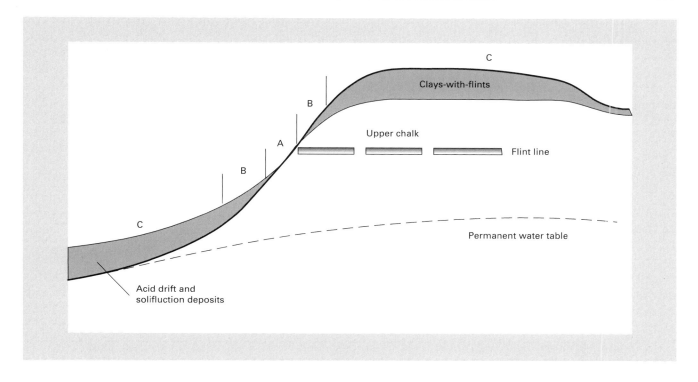

Fig. 9.8 Parent material variations with relief in an old Chalk landscape of the Chiltern Hills (after Avery, 1958).

prolonged weathering and leaching (Avery, 1958). As shown in Fig. 9.8, the soils on the steepest slopes A, where erosion is active, are dominated by the Chalk rock and a typical *rendzina* (Mollisol) is formed (*cf.* stage 2, Fig. 9.6). On the intermediate slopes B, mantled by thin drift and solifluction deposits, immature *brown forest soils* (Inceptisols) are found. On the deeper drift and acid clay-with-flints of the plateaux and gentle slopes C, *lessived brown soils* and *acid lessived soils* (Alfisols) occur. Some gleying occurs in the Bt horizons of the lessived soils in the valleys, whereas on the plateaux, mor humus accumulates and occasionally *micropodzols* (Spodosols) are found in the Eb horizons of the lessived soils.

9.4 Soils of the tropics and subtropics

The tropics are delineated by the Tropics of Cancer and Capricorn (latitude 23° north and south, respectively). The subtropics extend into the low to mid 30° latitudes, depending on the influence of the sea or a large land mass on the regional climate. Some broad generalizations can be made which account for differences in pedogenesis in these lands compared to temperate regions. See the following examples.

• Tropical and subtropical lands are generally hot so that chemical weathering and biochemical reactions, such as organic matter decomposition, proceed more rapidly than in temperate climes (Section 5.3). Tropical soils have an 'iso' temperature regime, that is, the difference between mean summer and winter soil temperatures is 5°C or less.

• Fluctuations in soil moisture are more extreme, except for the humid equatorial regions where rainfall is generally high and well distributed throughout the year.
• Many land surfaces in the tropics and subtropics are pre-Pleistocene (> 2 million y BP) and some date back to the early Tertiary (40–60 million y BP). There has been little recent folding apart from orogenic belts along the Andes in South America and in south and southeast Asia, but the ancient stable shields and tablelands have been gently upwarped or downwarped to form continental swells and basins on a grand scale. Partial alteration of crystalline rocks (igneous and metamorphic) can extend to depths of 50–100 m and give rise to deep *saprolite* (rotten rock) profiles. Thus, many of the soils are very old and deep, having developed on these deeply weathered materials, and the influence of surface organic matter on profile development is much less than in the temperate regions.
• There has been no rejuvenation of parent materials

through deposits from Pleistocene ice sheets. However, despite the stability of the old tablelands and basins, there have been periods of active erosion and reworking of old weathered regoliths on slopes and in valleys. This has produced extensive areas of complex colluvial deposits on slopes and alluvial deposits in valley bottoms (Fig. 9.9). The former are often underlain by a *stone line* or stone layer which separates the reworked, unconsolidated material – the pedisediment – from the underlying saprolite. The soil formed on the pedisediment may bear little relationship to the underlying saprolite.
• As indicated in Chapter 5, age and relief and important factors influencing pedogenesis in the tropics and subtropics, although trends associated with differences in parent material are also discernible.

Duchaufour (1982) considers there are three basic kinds of weathering in hot climates, which lead to distinctive soil types. In the humid tropics all three may occur together, the change from one to another being

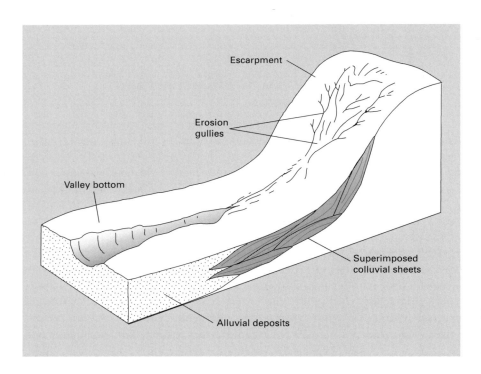

Fig. 9.9 Unconsolidated colluvial and alluvial deposits in old deeply weathered material in a subtropical landscape (after Thomas, 1994).

determined by local parent material and relief, so that it is possible to treat the processes as stages in the same weathering sequence

fersiallitization → ferrugination → ferrallitization.

Fersiallitization

During this process, primary minerals weather to 2 : 1 clays partly inherited and partly neoformed (Section 2.3); Fe^{2+} is liberated, but is rapidly oxidized and precipitated as ferric oxyhydroxides. The formation of *hematite*, in particular, as a coating on clays gives many of these soils a distinctive red colour – the process of *rubification*. Such soils are readily formed in humid Mediterranean-type climates on a wide range of parent materials. Lessivage leading to the development of a red Bt horizon is a characteristic feature, probably aided by the rapid movement of water down cracks as the soil reverts at the end of the dry summers. These soils are generally of high base saturation (> 35%) and therefore are classed as Alfisols (ST). In the Australian system, they may be classed as Dermosols (lacking strong texture contrast) or Chromosols (with strong texture contrast).

In the humid tropics, fersiallitization of relatively young basic rocks may produce *eutrophic brown soils* (Inceptisols (ST) or Tenosols (A)), which, if freely drained, go on to form ferruginous soils (see below). On poorly drained sites, however, such soils develop into mature *vertic eutrophic brown soils* (Vertisols (ST) or Vertosols (A)) which have > 35% clay (predominantly 2 : 1 types) and a pronounced tendency to shrink and swell with changing moisture content.

Ferrugination

Weathering is more complete (only resistant primary minerals like orthoclase, muscovite and quartz remain), and 1 : 1 clays are synthesized. The iron oxides formed may or may not be rubified, and 2 : 1 clays are translocated to form a Bt horizon. Base saturation is variable, depending on the climate, for example:

• under the marked wet and dry seasonality of the savanna regions, base saturation is usually > 35% and *tropical ferruginous soils* (Alfisols (ST) or Chromosols (A)) form;

• in humid climates, however, ferrugination leads to the formation of deep (> 3 m), more highly leached *ferrisols* (Ultisols (ST) or Chromosols and Kurosols (A)). In the humid tropics and on old landscapes, ferrisols may persist at higher altitudes (where they are humus-rich and very acidic) and on the steeper slopes; otherwise pedogenesis proceeds to the next stage of ferrallitization (see below).

Ferrallitization

For a range of parent materials, on reasonably level but well drained sites, soil formation in the humid tropics eventually proceeds to the end member of the series

tropical fersiallitic soil → tropical ferruginous soil → ferrisol → ferrallitic soil.

Virtually all the primary minerals have been decomposed leaving quartz, kaolinite (neoformed), and iron and aluminium oxides. Cation exchange capacities < 16 cmol charge (+) kg^{-1} clay and the absence of a Bt horizon place most *ferrallitic soils* in the order Oxisol (ST) or Ferrosol (A). They are generally acidic (pH 4–5) and of low chemical fertility, but have good physical properties.

Parent material and drainage exert their influence on the degree of ferrallitization in the following way. On acidic parent materials with imperfect drainage, sufficient silica is retained to allow the neoformation of kaolinite in the lower profile, which can be up to 50 m deep overall (*ferrallitic soil with kaolinite*). However, basic rocks weathering on well drained sites produce shallower soils containing much less kaolinite and an abundance of gibbsite and goethite/hematite (Table 9.2). The latter soils are true *ferrallites*, also known as *krasnozems* in Australia.

Laterite

The accumulation of a thick zone of sesquioxide nod-

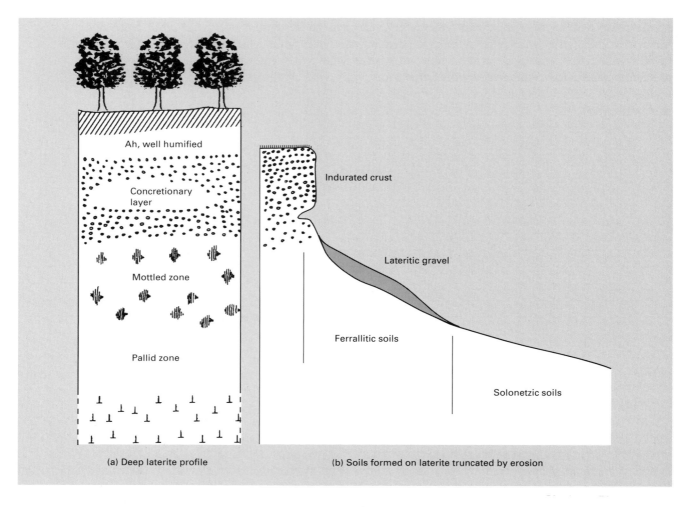

Fig. 9.10 (a) A deep *laterite* (Oxisol (ST) or Ferrosol (A)) profile. (b) Soil catena on a truncated *laterite* (Oxisol (ST) or Ferrosol (A)).

ules is characteristic of ferrallitic soils. While the natural vegetation remains intact, the soil is very stable and the sesquioxide nodules usually soft. However, if the vegetative cover is disturbed or completely removed, the organic-rich A horizon of the old ferrallitic soil can be stripped by erosion to expose the sesquioxide zone which becomes indurated by dehydration at high temperatures. The result is a geological structure called

laterite. The profile of a typical *ferrallitic soil with kaolinite* that degenerates to a deep laterite is illustrated in Fig. 9.10. It comprises:

1 an uppermost *concretionary zone* of iron oxides and gibbsite, which may be massive if cemented, but loosely nodular or pisolitic* if uncemented,

2 an intermediate *mottled zone* – orange to red iron oxide deposits in a pale-grey matrix of kaolinite, and

3 lowermost, a *pallid* zone of pale-grey to white kaolinite and bleached quartz grains, sometimes weakly

* Resembling small dried peas in size and shape.

Plate 9.1 *Red brown earth* showing sharp texture contrast between the A and B horizons.

(*Facing page 182*)

cemented by secondary silica, and usually increasing in salt concentration towards the weathering zone.

Ancient laterites that were much eroded and dissected during the Pleistocene often survive as prominent escarpments and crusts in the landscapes of tropical Africa, Australia and southeast Asia. Truncation by erosion gives rise to a new weathering surface (Fig. 9.10b) on which distinctive contemporary soils have formed: impoverished ferrallitic soils on the upper two laterite zones, and solonetz-type soils (Section 9.6) in the pallid zone where salts have accumulated. Whereas zones 1 and 2 of the laterite profile represent the residuum of prolonged intense weathering, the pallid zone appears to form due to the reduction and solution of iron under anaerobic conditions in a zone of groundwater influence. The reduced iron moves in solution both laterally and vertically, the vertical movement following seasonal, or longer-term, fluctuations in the height of the watertable. Reoxidation and precipitation of insoluble $Fe(OH)_3$ may in some cases account for the development of the overlying mottled zone.

Fluctuating aerobic and anaerobic conditions associated with variations in regional drainage also produce other distinctive pedological features in tropical and subtropical landscapes. For example, Fe^{2+}-complexed iron (Section 9.2) going into solution in a zone of fluctuating watertable can move vertically and laterally in the percolating drainage until it is oxidized and precipitated at a point in the landscape where better aeration supervenes. This better aeration may occur during the 'dry season' of a seasonal subtropical climate, or due to a change in parent material. The result is the formation a layer of ironstone or *plinthite* typical of a *groundwater laterite* (Oxisol (ST) or Ferrosol (A)) (Fig. 9.11). The plinthite becomes hardened and impenetrable when exposed at the soil surface, or when the regional watertable falls due to long-term climate change and the soil profile becomes drier.

9.5 Hydromorphic soils

Soils showing some hydromorphic features (Section 9.2) are found in catenas in many parts of the world:

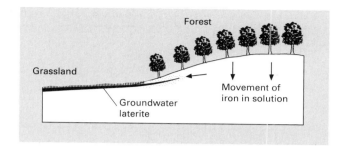

Fig. 9.11 *Groundwater laterite* formation in a humid tropical environment.

the *vertic eutrophic brown soils* and *groundwater laterites* are two examples already cited. *Acid sulphate soils* are distinctive hydromorphic soils formed in tidal swamps and on the sediments from brackish lakes. Hydromorphism is especially prevalent in humid temperate regions, such as northwest Europe, where groundwater gleys and surface water gleys can be found on most parent materials. Two soil sequences found in the north and west of England and Wales serve to illustrate the point.

1 At altitudes > 300 m, where the annual rainfall is > 1250 mm, the high *P/E* ratio (Section 5.3) and low temperatures favour blanket peat formation (see below), even on well-drained parent material. Proceeding downslope, a typical catena consists of

blanket peat (Histosol)* ↔ *thin iron-pan soil* ↔ *brown podzolic soil* (Ultisol) or *podzol* (Spodosol).

This is illustrated in Fig. 9.12. Gley colours predominate under the peat and to the base of the wet soil zone as the peat attenuates to a thin O horizon downslope. It is roughly in this midslope area that the distinctive iron-pan – shiny black above and rusty-brown beneath – is found, as reduced iron in solution moves down into the better aerated lower horizons and is oxidized. This is an example of a *surface water gley* (stagnogley type).

2 At lower altitudes under an annual rainfall of 750–1000 mm, a catenary sequence similar to that of

* See footnote for Section 9.3.

Fig. 9.13 forms on slopes of gentle relief, often over-lying relatively impermeable glacial till. The sequence consists of

brown podzolic soil ↔ *peaty podzol* ↔ *peaty gley soil.*
(Ultisol)   (Spodosol)   (Histosol)

The better drained upslope soils show rusty mottling along root channels in an otherwise weakly gleyed Ea or Eb horizon underlying a thin Ah. But as the watertable rises downslope, the whole profile becomes gleyed, apart from weak rusty mottling near the surface, and a *peaty gley soil* (Histosol) forms. This is an example of a *groundwater gley.*

Peat formation

A prerequisite of peat formation is a very slow rate of organic decomposition due to a lack of oxygen under continuously wet conditions. This condition is achieved by various combinations of relief or climate (especially *P/E* ratio), together with the permeability and base status of the underlying rocks. The two main types of peat are described below.

Basin or topogenous peats

Irrespective of climate, basin peats form where drainage water collects. If the water is neutral to alkaline, even though plant decomposition rates are substantial, the greater growth of sedges, grasses and trees produces an accumulation of well humified remains as *fen peat*. The ash content of this peat is high (10–50%) due to the accession of colloidal mineral particles in the drainage waters. Larger areas of fen peat are found, for example, in Florida, the Fens of England and in Botswana, the latter two being areas of relatively low rainfall.

Blanket bog (upland or climatic peat)

A *P/E* ratio > 1 for all or part of the year and cool temperatures predispose to the formation of *blanket peats*, described as ombrogenous deposits, over much of Ireland, the west of Scotland and the Pennines in Eng-

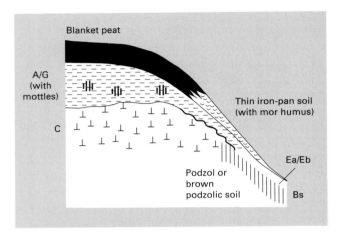

Fig. 9.12 Catenary sequence on cool wet uplands in the west of England and Wales.

land, especially on acidic rocks (Fig. 9.12). The vegetation of hardy grasses, ericas and *Sphagnum* moss provides an acid litter, and because the only source of water is rain which is low in bases, the rate of plant decomposition is extremely slow and the humification imperfect. The ash content of the peat is usually < 5%.

The formation of basin and fen peats began earlier in the Holocene period (post-Pleistocene) than did the blanket peats, many of which originated during the warm, wet Atlantic phase lasting from *c.* 7500 BP to 4000 BP. Where present-day climate is favourable (*P/E* ratio > 1), basin and fen peats are often transformed to raised bog (Fig. 9.14). The build-up of sediment and peat in a wet hollow eventually raises the living vegetation above the level of the groundwater so that continued plant growth depends solely on rainwater. Slowly, acid-tolerant and hardy species such as heather, cottonwood and *Sphagnum* invade and displace the species typical of the fen peat.

Acid sulphate soils

In tidal swamps and brackish sediments, in the presence of abundant organic matter, SO_4^{2-} from the tidewater and ferric oxides from the sediments are reduced

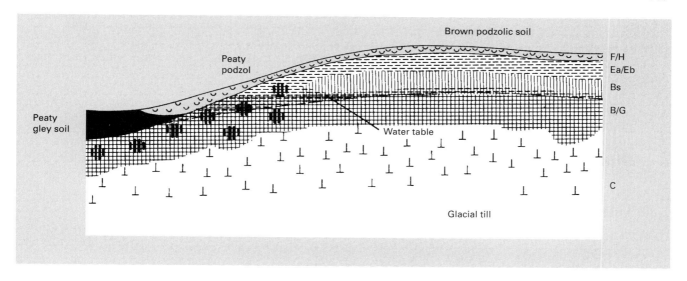

Fig. 9.13 Catenary sequence on cool, imperfectly drained lowlands in the west of England and Wales.

according to Reactions (8.20) and (8.21). The end-products are Fe^{2+} ions (solution), H_2S (gas) and insoluble FeS_2 (Reaction 8.22). While these 'sulphidic' sediments remain submerged, or at least saturated through the influence of a high watertable, they are relatively harmless; but if exposed through falling water levels the FeS_2 will oxidize and an acid sulphate soil forms (Box 9.4). Worldwide, there are some 12–14 million hectares of coastal plains and tidal swamps where the surface soil is already very acid, or will become so if drained. An area at least as large of sulphidic sediments, covered during the Holocene period by peat and alluvial deposits, is also at risk of acidification if cleared and drained.

9.6 Salt-affected soils

Sources of salt

'Salt' is a generic term covering all the soluble salts in soil, mainly comprising the chlorides and sulphates of Na, K, Ca and Mg, and the bicarbonates and carbon-

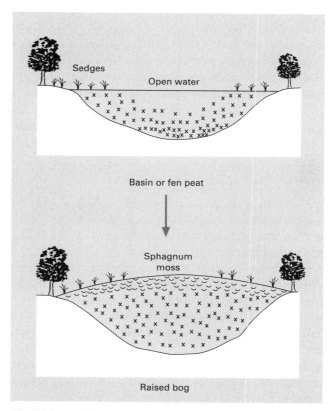

Fig. 9.14 Transformation of a basin or fen peat to a raised bog.

Box 9.4 Genesis of acid sulphate soils.

A schematic representation of the build-up of sulphidic sediments in a tidal swamp and their coverage with recent peat and alluvium, is shown in Fig. 9.15. Such sediments are widespread in West Africa, southeast Asia, northern South America and northern Australia. Dent and Pons (1995) describe three stages in the development of an acid sulphate soil on these sediments.

- *Unripe sulphidic clay* – under tidal water or peat and/or alluvium, up to 15% FeS_2, smell of H_2S, high soluble salts, pH 5–7.
- *Raw acid sulphate soil* – drained but subject to periodic flooding, lower soluble salts, pH 3–5 in the upper oxidized zone to 5–7 in the subsoil, rusty mottles along roots and yellow mottles of jarosite ($KFe_3(SO_4)_2(OH)_6$). The reactions occurring in the presence of O_2, namely

$$FeS_2 + 7/2O_2 + H_2O \leftrightarrow Fe^{2+} + 2SO_4^{2-} + 2H^+ \qquad (B9.4.1)$$

$$Fe^{2+} + 1/4O_2 + 5/2H_2O \leftrightarrow Fe(OH)_3 + 2H^+, \qquad (B9.4.2)$$

are highly acidifying and cause the pH to drop as low as 2. This is lethal to most plant and animal life.

- *Ripe acid sulphate soils* – drained; gleyed at depth with H_2S and FeS_2 present, pH 5–7 and high salts; overlain by a layer with jarosite and goethite mottles, pH 3–4, coarse prismatic structure; overlain by a layer with many dark brown mottles, prismatic to angular blocky structure, pH 3–4; topped by an Ah horizon of reddish brown clay, low in salts, pH *c.* 4.

Acid sulphate soils require large amounts of lime to counteract the acidity. Watertables should be kept as high as possible to minimize FeS_2 oxidation and for the pH to rise through reduction reactions (Section 8.4).

ates of Na. Salts derived from past cycles of geologic weathering have accumulated as salts in oceans, lakes and groundwater, and act as a source of salt in soil in the following ways.

- *Inheritance* – sediments laid down under water naturally entrain salts (*connate salts*) which are inherited by the soils that ultimately form on these sediments.
- *Weathering* – during soil formation, salts continue to be released by weathering of residual primary minerals and secondary silicates.
- *Atmospheric deposition* – wind turbulence over the sea creates aerosols of salt crystals, which may subsequently be deposited on land as *dry deposition*; or they may act as condensation nuclei for raindrops which

subsequently fall on land. Salt derived from the sea in this way is called *cyclic salt* and can amount to 100–200 kg ha^{-1} y^{-1} for distances up to 50–150 km from the sea. These are significant inputs for very old soils containing few weatherable minerals, such as those found in the western part of Australia.

- *Human activity* – changes in hydrology, through large scale clearing of deep-rooted vegetation or the use of excess irrigation water, can result in saline groundwater rising too close to the soil surface. Examples of this are discussed in Chapter 13.

A soil is *saline* if it has a salt concentration exceeding about 2500 ppm, corresponding to an electrical conductivity of the saturation extract > 4 dS m^{-1} (Section

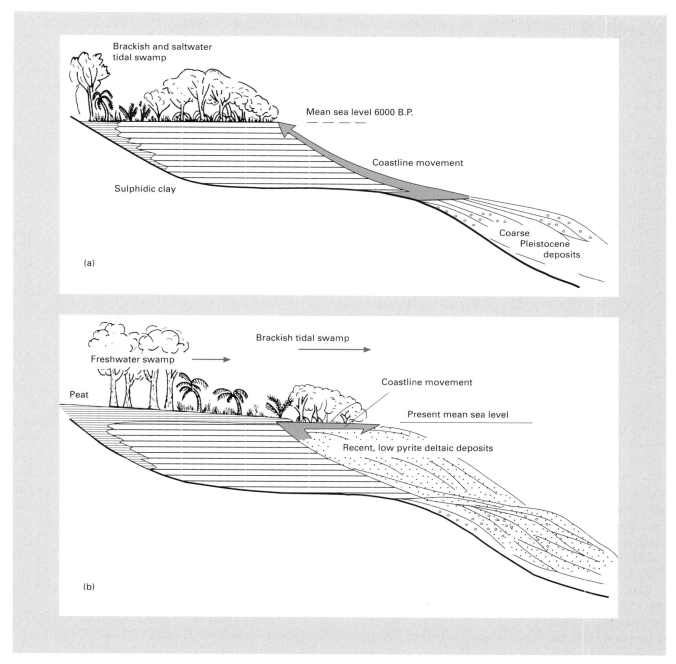

Fig. 9.15 (a) Schematic cross-section of a tropical delta showing the older sulphidic sediments accumulated under a steadily rising sea level. (b) Schematic cross-section of a tropical delta before forest clearance showing accumulation of newer sediment of low pyrite content and peat during a period of a relatively stable sea level (after Dent and Pons, 1995).

13.3). Naturally occurring saline soils, common in arid regions, are called *white alkali soils* or *solonchaks* (Aridisols (ST)).* Typically, they have a uniform dark-brown Ah horizon containing superficial salt efflorescences as well as interflorescences within the pores. The salts are brought up by capillary rise of the saline groundwater. Below the Ah horizon the soil is structureless and frequently gleyed. Although there is exchange of Na^+ for Ca^{2+} and Mg^{2+} on the clay surfaces, the clay remains flocculated and the structure stable provided that a high salt concentration is maintained. Frequently, sufficient secondary $CaCO_3$ precipitates out at depth to form an Ak horizon and the soil pH ranges from 7 to *c.* 8, depending on the partial pressure of CO_2 in the soil air.

Solonchak–solonetz–solod soil sequence

Climatic change with rainfall increasing, or a lowering of the regional groundwater, can have serious consequences for a saline soil (*solonchak*), especially if the exchangeable Na^+ content exceeds 10–15% of the CEC (Section 13.3). Dilution of the soil solution as the salts are slowly leached causes the diffuse double layers at clay surfaces to expand and increased swelling pressure to be exerted within clay domains (Section 7.4). The soil pH also rises (Box 9.5).

Owing to swelling pressure and dispersion of organic matter at high pH, the Na-clay becomes unstable and deflocculates. It is illuviated into the lower profile where it is flocculated by the higher salt concentration there. The Bt horizon that develops is very poorly drained, often with signs of gleying, and has massive columnar peds capped with organic coatings (Fig. 4.5). The overall process of clay deflocculation and illuviation is called *solonization*, and the soil formed is a *solonetz* (Alfisol (ST) or Sodosol (A)).

Continued leaching of the *solonetz* leads to the complete removal of soluble salts, and most of the basic exchangeable cations, from the A horizon which becomes acid, and the soil becomes a *solodized solonetz*

---

* Not well accommodated in the Australian Soil Classification; could be in several orders.

(Alfisol (ST) or Sodosol (A)). The H-clay decomposes to release Al^{3+}, Mg^{2+} and SiO_2 (Box 9.2). Thus, with the advance of salt leaching and acidification down the profile into the Bt horizon, the H,Na-clay becomes an Al,Mg-clay. The massiveness of that horizon decreases and permeability improves, facilitating further leaching. This, the final step in the degradation of a *solonchak*, is called *solodization*, and the resultant soil with its sandy, bleached E horizon and disintegrating Bt horizon is called a *solod* or *soloth* (sodic Alfisol (ST) or Kurosol (A)). The typical profiles of this maturity sequence are illustrated in Fig. 9.16.

9.7 Summary

A *soil profile* develops as the weathering front penetrates more deeply into the parent material and organic residues accumulate at the surface. The principal pedogenic processes operating are mineral dissolution (congruent and incongruent), hydrolysis, oxidation–reduction, leaching, cheluviation, clay decomposition, lessivage (clay translocation), gleying and salinization, variations in the intensity of which with depth give rise to one or more *soil horizons*.

Superimposed lithological discontinuities are recognized as *soil layers*, numbered from the top down.

Plant residues on the surface form a *litter layer L*, which may give rise to a superficial *organic horizon* comprising F and H subdivisions, depending on the degree of decomposition and humification. An organic horizon that remains wet for more than 30 consecutive days in most years is called a *peaty* or *O horizon*. An O horizon > 30 cm thick on bedrock R, or > 40 cm thick in other situations is called *peat*. *Mineral horizons* are designated A, B and C from the surface down, with C being weathering parent material. This notation is expanded by using suffixes to denote particular pedological features (e.g. iron oxide accumulation) or processes (e.g. lessivage, gleying). Clay translocation which produces profiles with a marked A–B texture contrast ('*duplex soils*') is predisposed by the cheluviation of sesquioxide cementing agents, or the deflocculation of clay particles due to high exchangeable Na^+ and low salt

Box 9.5 pH change on leaching a
solonchak (saline soil).

The rise in pH of a solonchak soil as it is leached by rain water can be due to a combination of effects.

- While Ca^{2+} and Mg^{2+} remain the dominant exchangeable cations, the pH will be controlled by the activity ratio $(H^+)/\sqrt{(Ca^2 + Mg^2)}$ (see the ratio law, Section 7.2)) as the soil solution becomes more dilute. Thus, if the soil solution concentration decreases 100-fold from 10^{-2} to 10^{-4} M, for the activity ratio to remain constant, H^+ ion activity in solution must decrease 10-fold; that is, the pH would rise by 1 unit from, say, pH 7 to 8.
- With an increase in the amount of exchangeable Na^+, as the soil is leached with a more and more dilute solution, the weakly adsorbed Na^+ ions are displaced by H^+ ions produced by the hydrolysis of water at the clay surface, according to the reaction

$$Na^+\text{-clay} + HOH \leftrightarrow H^+\text{-clay} + NaOH. \qquad (B9.5.1)$$

The pH rises because NaOH is a strong base. The pH may rise as high as 10–10.5, the actual value being dependent on reversible hydrolytic equilibria involving NaOH, CO_2, $NaHCO_3$ and Na_2CO_3, according to the reactions

$$NaOH + CO_2 \leftrightarrow NaHCO_3 \qquad (B9.5.2)$$

$$NaOH + NaHCO_3 \leftrightarrow Na_2CO_3 + H_2O. \qquad (B9.5.3)$$

The soil is now at the *black alkali stage* because the high pH causes the organic matter to disperse, further weakening the cohesion of the soil aggregates.

concentrations. The resulting 'podzolic' and 'solodic' soils are widespread in Australia, falling into three orders of the Australian Soil Classification – the Chromosols, Kurosols and Sodosols.

Pedogenesis in temperate climates is often dominated by the influence of surface organic matter, and the soluble complexing agents released from it. Many of the parent materials are also young, having been deposited when the Pleistocene ice sheets retreated. With the passage of time, a soil passes through a *maturity sequence*, the stages of which differ from calcareous to siliceous parent materials, and are also influenced by vegetation and drainage. Examples are given in Fig. 9.6, 9.7 and 9.8. Furthermore, subtle changes in the influence of parent material, climate and drainage can produce a

comparable range of profile variations in soils of *the same age*, as illustrated in Fig. 9.12 and 9.13.

In tropical and subtropical regions, temperatures are higher and more constant than in temperate regions. Chemical weathering and organic matter decomposition proceed more rapidly, and extremes of moisture are common. Compared with recently glaciated and periglacial areas of the northern hemisphere, much of the tropical and subtropical land surface is very old and flat so that soils have been forming on the same parent material for millions of years. However, on slopes and in valley bottoms in these old landscapes, soils have formed on deep colluvial and alluvial deposits of reworked material overlying deeply weathered rock (*saprolite*). The junction between these frequently dis-

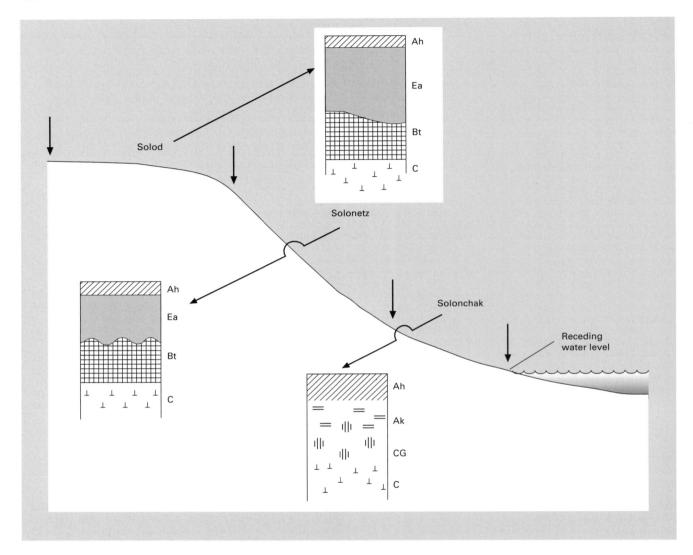

Fig. 9.16 *Solonchak–solonetz–solod* maturity sequence resulting from a falling watertable (after Kubiena, 1953).

parate materials is usually marked by a *stone line*. The main trends in pedogenesis can be differentiated by the extent of primary mineral decomposition, the leaching of bases and silica, and the neoformation of clay-fraction minerals (kaolinite and the sesquioxides).

In the humid tropics, the most severe weathering environment, the three processes of

fersiallitization ↔ ferrugination ↔ ferrallitization

can occur side by side, the change from one to another being determined by local parent material and relief, or in time as a sequence. The end-stage – ferrallitization – produces characteristic *laterite* formations, the soils on which (if undisturbed) are called Oxisols (ST) or Ferrosols (A).

Hydromorphic processes (including gleying) can affect soils under any climate. In cold wet climates, *surface-water gleys (pseudogleys* and *stagnogleys)* occur on many parent materials, even ones that are well drained. Continuously wet and cold climates produce *peaty gleys* and *blanket peats* (Histosols (ST) or Organosols (A)). A watertable within the soil profile for all or part of a year gives rise to *groundwater gleys*, a particularly nasty variant of which are the *acid sulphate soils* found in tidal marshes and forming on drained sediments in river deltas.

Salinization due to *inherited salts, cyclic salt* or the rise of saline groundwater produces *solonchak* soils (Aridisols (ST)) which, on being leached, undergo solonization (*solonetz* phase – Alfisols (ST) or Sodosols (A)) and finally solodization (*solod* or *soloth* phase – Sodic Alfisols (ST) or Kurosols (A)).

References

Avery B. W. (1958) A sequence of beechwood soils on the Chiltern Hills, England. *Journal of Soil Science* **9**, 210–224.

Avery B. W. (1980) *Soil Classification for England and Wales.* Soil Survey Technical Monograph No. 14. Lawes Agricultural Trust, Harpenden.

Black A. S. & Waring S. A. (1976) Nitrate leaching and adsorption in a krasnozem from Redland Bay, Qld. II. Soil factors influencing adsorption. *Australian Journal of Soil Research* **14**, 181–188.

Butler B. E. (1959) *Periodic Phenomena in Landscapes as a Basis for soil studies.* CSIRO Soil Publication No. 14.

Crompton E. (1952) Some morphological features associated with poor soil drainage. *Journal of Soil Science* **3**, 277–289.

Dent D. L. & Pons L. J. (1995) A world perspective on acid sulphate soils. *Geoderma* **67**, 263–276.

Duchaufour P. (1977) *Pedologie. I. Pedogenese et Classification.* Masson, Paris.

Duchaufour P. (1982) *Pedology Pedogenesis and Classification* (Translated by T. R. Paton). Allen & Unwin, London.

International Society of Soil Science (1967) *Proposal for a uniform system of soil horizon designations.* Bulletin of the International Society of Soil Science No. 31, 4–7.

Isbell R. F. (1966) *The Australian Soil Classification.* Australian Soil and Land Survey Handbook. CSIRO Publishing, Melbourne.

Jenny H. (1941) *Factors of Soil Formation.* McGraw-Hill, New York.

Kubiena W. L. (1953) *The Soils of Europe.* Murby, London.

Loughnan F. C. (1969) *Chemical Weathering of the Silicate Minerals.* Elsevier, New York.

McDonald R. C., Isbell R. F., Speight J. G., Walker J. & Hopkins M. S. (1990) *Australian Soil and Land Survey Field Handbook*, 2nd edn. Inkata Press, Melbourne.

Mackney D. (1961) A podzol development sequence in oakwoods and heath in central England. *Journal of Soil Science* **12**, 23–40.

Muir A. (1961) The podzol and podzolic soils. *Advances in Agronomy* **13**, 1–56.

Soil Survey Staff (1992) *Keys to Soil Taxonomy*, 5th edn. Soil Management Support Services Technical Monograph No. 19, Pocahontas Press, Blacksburg.

Thomas M. F. (1994) The Quaternary legacy in the tropics: a fundamental property of the land resource, in *Soil Science and Sustainable Land Management in the Tropics* (Eds J. K. Syers and D. L. Rimmer). CAB International, Wallingford, pp. 73–87.

Further reading

Anderson H. A., Berrow M. L., Farmer V. C., Hepburn A., Russell J. D. & Walker A. D. (1982) A reassessment of podzol formation processes. *Journal of Soil Science* **33**, 125–136.

Avery B. W. (1990) *Soils of the British Isles.* CAB International, Wallingford.

Duchaufour P. (1995) *Pedologie: Sol, Vegetation, Environment*, 4th edn. Masson Editeur, Paris.

Lal R. & Sanchez P. A. (Eds) (1992) *Myths and Science of Soils of the Tropics.* Soil Science Society of America Special Publication No. 29. Soil Science Society of America, Madison, Wisconsin.

Newbould P. J. (1958) Peat bogs. *New Biology* **26**, 89–104.

Thomas M. F. (1994) *Geomorphology in the Tropics: a Study of Weathering and Denudation in Low Latitudes.* Wiley, Chichester.

Chapter 10
Nutrient Cycling

10.1 Nutrients for plant growth

The essential elements

There are 16 elements without which green plants cannot grow normally and reproduce. On the basis of their concentration in plants, these essential elements are subdivided into:

- the *macronutrients* C, H, O, N, P, S, Ca, Mg, K and Cl which occur at concentrations > 1000 mg kg^{-1} (plant dry matter basis), and
- the *micronutrients* Fe, Mn, Zn, Cu, B and Mo which are generally < 100 mg kg^{-1}.

Carbon, H and O are only of passing interest, because they are supplied as CO_2 and H_2O, which are abundant in the atmosphere and hydrosphere; likewise Cl, which is abundant and very mobile as the Cl$^-$ ion. Of the others, with the exception of N which comprises 79% of the atmosphere, the major source is weathering minerals in the soil and parent material. *The nutrient supplying power of a soil is a measure of its fertility*, the full expression of which depends on diverse properties of the plant, its environment, and on soil management (Part 3).

Plants absorb other elements, some of which are beneficial though not essential, such as Si (for sugar cane), Na (for sugar beet and marigolds) and Al (for many ferns). Cobalt is essential for symbiotic N_2 'fixation' in legumes and blue-green algae (Section 10.2), while Cr, Se and I, although of no benefit to plants, are vital for the health and normal reproduction of animals. Vanadium can substitute for Mo to some extent in the N_2 fixation process of bacteria. Nickel may also be essential for plants.

Nutrient cycling

The flow of nutrients in the biosphere is continuous between three main compartments:
- an *inorganic store*, chiefly the soil,
- a *biomass store*, comprising living organisms above and below ground, and
- an inanimate *organic store*, on and in the soil, formed by the residues and excreta of living organisms.

The general nutrient cycle is illustrated in Fig. 10.1. The relative size of the nutrient stores differs for the various elements, as does the importance of the pathways for inputs and outputs and the partitioning of an element between 'soil' and 'non-soil'. Transformations within each store often involve complex physicochemical and biochemical processes.

There are also differences in the way the elements are distributed within the soil. Elements that are released at depth by mineral weathering are absorbed by plant roots and transported to the shoots, to be deposited finally on the soil surface in litter and animal excreta. Accessions from the atmosphere add to the surface store. Nitrogen always accumulates in the organic-rich A horizon, the content declining gradually with depth; P is similar, but the decline with depth is more abrupt because of the immobility of phosphate ions in the soil (Fig. 10.2). Sulphur, like N, accumulates in the surface of temperate soils, but not necessarily in tropical soils, due to the mineralization of organic-S and the leaching of SO_4^{2-}. Many of the latter soils show a bimodal S distribution, with organic S in the surface and retention of sulphate ions in the subsoil at sites of positive charge (Fig. 10.2b). The distribution of most cations – Fe^{3+}, Ca^{2+}, Mg^{2+}, Mn^{2+}, Cu^{2+} and Zn^{2+} for example –

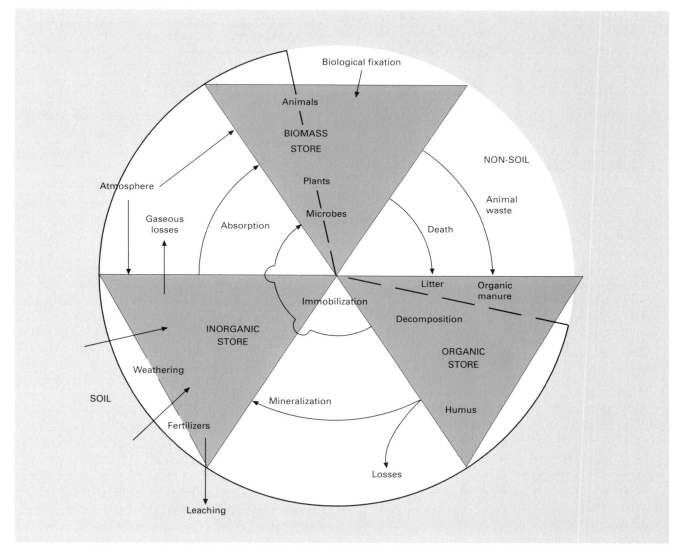

Fig. 10.1 Fundamentals of a nutrient cycle.

usually correlates with clay accumulation. But the 'heavy metals' Cu, Zn, Fe and Mn are also associated with soil organic matter.

The inorganic store

For elements such as P, K, Ca and Mg, plant roots tap only a small fraction of the soil's inorganic store during a single growth season. This fraction is withdrawn from an *available pool*, which is made up of:

• ions in the *soil solution*, and

• *exchangeable ions* adsorbed by clay minerals and organic matter.

Less readily available ions are held in sparingly soluble compounds like gypsum, calcite and inorganic phosphate compounds, or held in non-exchangeable forms

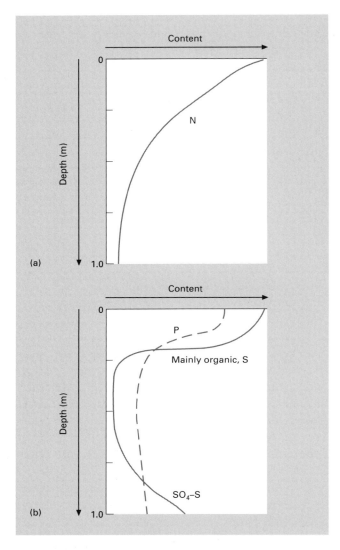

Fig. 10.2 (a) and (b) Distribution of N, P and S in a soil profile.

Table 10.1 The distribution of selected elements between the plant biomass and the soil for different vegetation types in West Africa (kg ha^{-1}). After Nye and Greenland, 1960.

Plant biomass (including roots and litter)				Soil (0–30 cm)			
N	K	Ca	Mg	N	K	Ca	Mg
Mature forest (40 y old)							
2040	910	2670	350	4590	650	2580	370
Imperata cylindrica grassland							
46	105	13	24	1790	190	2910	380

• the application of organic manures, organic and inorganic fertilizers (Section 11.1 and Chapter 12).

The biomass store

The storage of nutrients in the plant biomass depends on the mass of vegetation produced. As Table 10.1 shows, the store of K, Ca and Mg (less so for N) in mature forest under high rainfall in the tropics is comparable to that in the top 30 cm of soil; but the store of these elements in relatively unproductive grassland is much less than in the topsoil. K is a notable exception because it cycles rapidly through the soil–plant system. All biomass is important in nutrient cycling, irrespective of the size of the store, because the remnants of living organisms and their excreta are the substrate on which soil micro-organisms feed.

Although small (0.5–2 t C ha^{-1}), the soil microbial biomass has been identified as the 'eye of the needle' through which all the C returned to the soil must eventually pass. Virtually the same is true of N – 98% or more of the soil N is in organic combination, and the metabolism of this by soil micro-organisms is intimately associated with that of C.

The organic store

The cycle shown in Fig. 10.1 is completed by the decomposition of litter and animal excreta by the soil mesofauna and micro-organisms, a topic discussed in Chapter 3. Specific aspects of mineralization, as they

in clay lattices (K$^+$ and NH$_4^+$).

Inputs to the available pool occur by:
• weathering of soil and rock minerals (Section 5.2 and 9.2),
• precipitation and dry deposition (Section 10.2),
• mineralization of organic matter (Section 3.1 *et seq.*), and

affect the availability of individual nutrients to plants, are discussed in the following sections.

10.2 The pathway of nitrogen

Mineralization and immobilization

Depending on the soil and environmental conditions, the quantity of soil N in the root zone (c. 1 m deep) ranges from 1 to 10 t ha^{-1}, much of which occurs in the top 15–20 cm. Fertile agricultural soils in Britain, for example, have an average N content of 4000 kg ha^{-1} to 15 cm depth (Anon., 1983). Organic N is transformed during microbial decomposition of organic matter according to the reaction

$$\text{Organic N} \rightarrow NH_4^+ + OH^-. \qquad (10.1)$$
(proteins, nucleic acids)

This step is referred to as *ammonification* and it is obviously an alkalizing reaction. Subject to any limitations imposed by temperature, moisture, pH and aeration (Section 8.3), NH_4^+ is oxidized to NO_3^- according to the reaction

$$NH_4^+ + 2O_2 \rightarrow NO_3^- + H_2O + 2H^+, \qquad (10.2)$$

which is called *nitrification*. Note that two H$^+$ ions are produced per mole of NH_4^+ oxidized so that the net effect of Reactions 10.1 and 10.2 is to produce 1 mol H$^+$ per mol of NO_3^- formed during *mineralization*. Mineralization of organic N to NO_3^- is an *acidifying process* which leads to accelerated acidification in the soils prone to NO_3^- leaching (Section 11.3).

NH$_4^+$ and NO_3^- comprise the pool of *mineral N* on which plants feed. Ammonium is held as an exchangeable cation that is readily displaced into the soil solution and, with NO_3^-, moves to the roots in the water absorbed by the transpiring plant – a process of *mass flow*. Both ions can also move from regions of higher to lower concentration by *diffusion*. Thus, all the mineral N (except for non-exchangeable NH_4^+ in micaceous clay lattices) is available to the plant.

The balance between mineralization and immobilization during decomposition is dependent on the C : N

ratio of the substrate (Section 3.3), but the net mineralization rate can be described by the equation

$$\frac{dN}{dt} = -kN \qquad (10.3)$$

where N is the amount of organic N per unit soil volume, and k is a decay coefficient varying between 0.01 and 0.06 y^{-1}. During an interval of 1 year, the amount of N mineralized (N_{min}) is given by kN, the value of which is highly variable, ranging between 10 and 300 kg ha^{-1} (Box 10.1). Whether this is enough to meet the needs of plants depends on their uptake of N, as discussed below.

Plant uptake

In natural ecosystems, plant growth rates are generally slow and the annual uptake of N is relatively small; Cole *et al.* (1967) quote figures of 25–78 kg N ha^{-1} y^{-1} for a coniferous forest in North America. One of the highest figures is 248 kg N ha^{-1} y^{-1} estimated by Nye and Greenland (1960) for mature secondary rainforest in Ghana. Cultivated crops are much more demanding. Nitrogen uptakes range from 80 to 450 kg N ha^{-1} (Table 10.2), so that the mineralizing capacity of the soil is often insufficient to maintain optimum growth. In this case, the magnitude of the gains and losses in the N cycle, shown in Fig. 10.3, are of great significance.

Gains of soil N

Atmospheric inputs

Apart from N released as N$_2$ and N$_2$O during denitrification (Section 8.4), N enters the atmosphere as the oxides NO and NO_2, which are formed during the combustion of fossil fuels (the primary source), biomass burning and by lightning discharges. It is also released as NH$_3$ gas by volatilization from manures, rotting vegetation and soil, especially calcareous soil following the application of ammonium fertilizers (Section 12.2). Dust blown into the air may carry both NH_4^+ and NO_3^- ions, as well as organic N.

The decay coefficient k in Equation 10.3 is an *averaged or lumped* parameter. There are large differences in net mineralization rate between fresh residues (from crops or pastures), animal excreta (dung and urine) and resident soil organic matter (SOM). For example, fresh residues of C : N ratio < 25 have an effective k ≈ 0.3 y^{-1} in the year after their addition, compared to a k value of 0.01–0.03 y^{-1} for SOM. Thus, N_{min} depends on:
• the amount of residues and organic manures returned to the soil, relative to the SOM,
• the C : N ratio of the residues and manure, their placement in the soil and their distribution (which affects their accessibility to microorganisms),
• soil management (soil pH and whether the soil is cultivated or not, drained or not), and
• environmental factors (especially temperature and effective rainfall).

The amount of N in crop residues in the UK range from as low as 20 kg ha^{-1} (linseed) to 150 kg ha^{-1} (oilseed rape), excluding any N returned from roots and rhizosphere deposition (Section 3.1) which is very difficult to measure. Nitrogen returned in residues under highly productive grassland can be even larger (300–400 kg ha^{-1} y^{-1}), and contributes to the large 'flushes' of net mineralization observed if the grassland is ploughed out. For example, in the UK N_{min} in the layer after ploughing a pasture ranges from 60 (1-year-old sward, no N applied) to 300 kg ha^{-1} (old sward, 300 kg N ha^{-1} y^{-1} input). In southeastern Australia, mineral N contents of 100–200 kg N ha^{-1} to 1.2 m depth have been measured at the end of summer under ryegrass–subterranean clover pastures. Both are annual species which die back during the hot, dry summers and add to the pool of plant residues, dung and urine that is mineralized.

Box 10.1 Variation in net mineralization in soils.

Table 10.2 N contents of selected agricultural crops.

Crop	N content (kg ha^{-1})
Spring malting barley, 4.5 t ha^{-1}	80
Wheat, 6 t ha^{-1} (grain and straw)	120
Potatoes, 30 t ha^{-1} (fresh weight)	200
Grass, 10 t ha^{-1} (dry matter)	250
Winter oilseed rape, 3.2 t ha^{-1}	250
Maize, 13 t ha^{-1} (grain and stover)	360
Sugar cane, 120 t ha^{-1} (fresh weight)	450

Nitrogen in the atmosphere can return to the soil–plant system in several forms. Oxides of N (NO$_x$) react with hydroxyl free radicals to form nitric acid, which contributes to '*acid rain*' (about half of the total rainfall acidity in the UK). Some NH$_3$ is absorbed through the stomata in plant leaves while the remainder is dissolved in the rain or forms salts, such as (NH$_4$)$_2$SO$_4$ (Equation 10.11), which comes down in the rain or is deposited on soil and plant surfaces as *dry deposition*. The total N input from the atmosphere ranges from 5 to

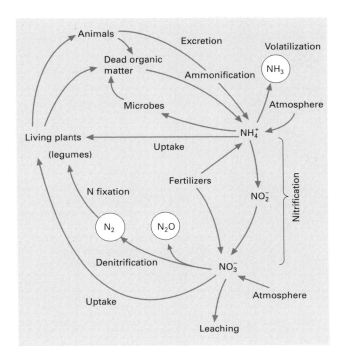

Fig. 10.3 N cycling and transformations in the soil-plant ecosystem.

60 kg ha^{-1} y^{-1}, depending on the extent of air pollution. Exceptionally, total deposition may reach 100 kg N ha^{-1} y^{-1} in forests close to intensive livestock activities. Within the total it is difficult to identify precisely the relative contributions of dry deposition and rainfall, but a ratio of 2 to 3 : 1 is reasonable. The average atmospheric N contribution in northwest Europe (excluding biological N$_2$ fixation) is between 10 and 20 kg ha^{-1} y^{-1}, which is small compared with other inputs to the N cycle for agricultural soils.

Fertilizers and manures

Under natural vegetation, plant litter and organic manures provide the bulk of the nitrogen returning to the soil. In agriculture, however, where much of the crop is removed from the land, the input of N in chemical fertilizers (Chapter 12) or through the growth of le-

gumes is necessary to boost the soil N supplies for subsequent crops.

Biological N fixation

A small minority of micro-organisms is exceptional in being able to reduce molecular N$_2$ to NH$_3$ and incorporate it into amino acids for protein synthesis. These micro-organisms live either independently in the soil (free-living heterotrophs or photosynthetic organisms), or in symbiotic association with plants. The process of N$_2$ reduction, according to the reaction

$$N_2 + 8H^+ + 6e + n\text{Mg.ATP} \rightarrow 2NH_4^+ + n\text{Mg.ADP} + n\text{PO}_4^{3-}, \qquad (10.4)$$

is called *nitrogen fixation*, and enables the organisms to grow independently of mineral N.

The annual turnover of biologically fixed N$_2$ in the biosphere is *c.*140 Mt which is small compared to the total of *c.*15 Gt N in terrestrial vegetation and *c.*1500 Gt N in SOM. Nevertheless, in natural ecosystems, and pastures based on legumes such as clover, the input of biologically fixed N is a most important component of the N cycle. An example of N inputs and losses for a white clover–ryegrass pasture in New Zealand is given in Fig. 10.4. The biochemistry of N$_2$ fixation is summarized in Box 10.2.

Free-living N$_2$ fixers

Regulation of N$_2$ fixation in aerobic organisms is complicated because the reduced nitrogenase must be protected from O$_2$, yet O$_2$ is required for the cell's respiration and oxidative phosphorylation to produce ATP. It is not surprising, therefore, that the ability to fix N$_2$ is not widespread among non-photosynthetic, aerobic organisms. The more important of the free-living or non-symbiotic N$_2$-fixing organisms are listed in Table 10.3.

The distribution and activity of N$_2$-fixing organisms in soil is determined by factors such as aeration, pH, supply of organic substrates, mineral N concentration and the availability of micronutrients, especially Cu,

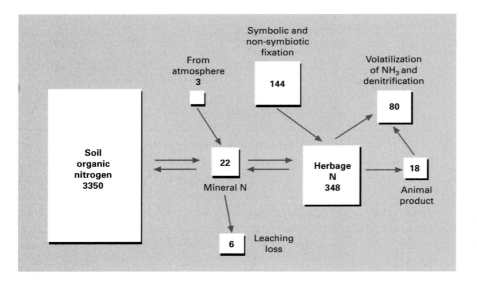

Fig. 10.4 N inputs, outputs and the main soil N components for a soil under a young clover-grass pasture in New Zealand. The quantity of N is given in kg ha^{-1} for soil components or as kg ha^{-1} y^{-1} for fluxes (after White and Sharpley, 1996).

Box 10.2 The biochemistry of N_2 fixation in free-living organisms and symbiotic associations.

The splitting of the dinitrogen molecule (N≡N) and the reduction of each part to NH_3 are catalysed by the nitrogenase enzyme, which consists of two proteins – an Fe protein and a Mo – Fe protein, roughly in the ratio of 2 : 1. The prerequisites of Reaction 10.4 are:
• a reductant of low redox potential (usually ferredoxin),
• an ATP-generating system, and
• a low partial pressure of O_2 at the site of nitrogenase activity.

On average, the amount of N fixed in symbiotic associations, especially involving legumes, is an order of magnitude greater than that fixed by free-living organisms. This reflects the large amount of energy available from carbohydrate metabolism in a higher plant, and the favourable environment for N_2 reduction created within a root nodule (Fig. 10.5). Within the nodule, the site of fixation is the bacteroid surface to which electrons are transported by a reduced co-enzyme and donated to N≡N bound to nitrogenase. Synthesis of the nitrogenase is induced in the bacteria under the favourable conditions within the nodule where the O_2 partial pressure is regulated on a microscale. This is achieved through the pigment *leghaemoglobin* (giving a characteristic pinkish-red colour to nodular tissue that is fixing N_2) which, because of its very high affinity for O_2, is able to transfer O_2 to the respiring bacteroids while keeping the free O_2 concentration in the membrane-bound sacs very low.

Table 10.3 The major free-living N_2-fixing organisms.

Bacteria	Aerobes	*Azotobacter, Beijerinckia* spp.
	Microaerobic	*Azospirillum brasilense* *Thiobacillus ferrooxidans*
	Facultative anaerobes	*Klebsiella* spp., *Bacillus* spp.,
	Obligate anaerobes	*Clostridium pasteuranium, Rhodospirillum, Chlorobium* and *Desulphovibrio* spp.
Cyanobacteria (blue-green algae)	Photosynthetic aerobes	*Nostoc, Anabaena, Cloecapsa* spp.

Mo and Co. *Azotobacter* and the blue-green algae are restricted to neutral and calcareous soils, but spore-forming anaerobes such as *Clostridium* are tolerant of a wide range of pH. At high concentrations of mineral N, the N_2 fixers succumb to competition from more aggressive heterotrophic species. Thus, the rhizosphere of plant roots with its abundance of exudates of high C : N ratio is a favoured habitat for free-living N_2-fixing organisms. In recent years, many of these organisms and associated nitrogenase activity have been identified in the rhizosphere of a variety of crops, including wheat, maize, pasture grasses and sugar cane. This has been called *associative* N_2 fixation.

Nitrogenase activity in soil is detected by the *acetylene reduction technique*. When nitrogen is excluded from the nitrogenase site by an excess of acetylene, this gas is reduced to ethylene according to the reaction

$$C_2H_2 + 2H^+ + 2e^- \rightarrow C_2H_4 \qquad (10.5)$$

and the ethylene detected by gas chromatography. Although the identification of functional nitrogenase activity does not necessarily mean that N_2 is being fixed, an estimate of N fixed can be made by assuming a ratio of 1 mol N_2 fixed for 3 mol of C_2H_2 reduced.

Because of the high energy consumption of N_2 fixation (*c.* 16 mol ATP per mol N_2 reduced), photosynthetic organisms tend to make the greatest contribution to non-symbiotic N_2 fixation: for example, up to 50 kg N ha^{-1} y^{-1} due to blue-green algae and photosynthetic bacteria in paddy rice culture. Estimates for other farming systems lie in the range 0–10 kg ha^{-1} y^{-1}. There is some evidence that contributions from free-living organisms may be higher in temperate deciduous woodlands (20–30 kg N ha^{-1} y^{-1}), and exceptionally up to 100 kg ha^{-1} y^{-1} in some tropical forests.

Symbiotic N_2 fixation

Symbiosis denotes the cohabitation of two unrelated organisms that benefit mutually from the close association. In the case of N_2 fixation, the invasion of roots of the host plant by a micro-organism (the endophyte) culminates in the formation of a nodule in which carbohydrate is supplied to the endophyte and amino acids formed from the reduced N are made available to the host.

There are three main symbiotic associations:
• nodulating species of the family Leguminosae, in which the endophyte is a bacterium of the genera *Rhizobium* and *Bradyrhizobium*,*
• non-legumes, including the genera *Alnus* (alder), *Myrica* (bog-myrtle), *Elaeagnus* and *Casuarina*, in which the endophyte is usually an actinomycete (*Frankia*), and
• lichens, which are an association of a fungus and blue-green alga, and the aquatic fern *Azolla* which lives in association with the blue-green alga *Nostoc*.

In terms of their numbers and widespread distribution, the legumes are by far the most important symbiotic N_2 fixers. The *Rhizobium* and *Bradyrhizobium* bacteria, when not infecting a legume root, live as heterotrophs in the soil. The success of the root infection, nodule initiation and subsequent N_2 fixation is determined by one or more factors which operate at any one of several points in the complex sequence of events, shown diagrammatically in Fig. 10.5. These factors include:
• low pH and low Ca which inhibit root-hair infection and nodule initiation;

*The genus *Bradyrhizobium* was created to include slow growing rhizobial strains isolated from soil which produce alkali, as opposed to acid, when grown in pure culture.

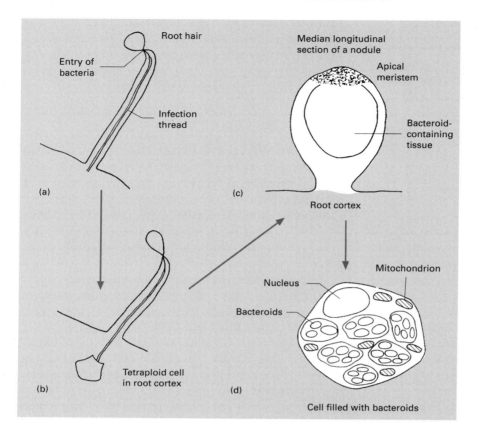

Fig. 10.5 Sequence in nodule formation on a legume root. (a) Bacteria invade a root hair through an invagination of the cell wall – the infection thread. (b) Tetraploid cell penetrated – this cell and adjacent cells induced to divide and differentiate to form a nodule. (c) Bacteria released, divide once or twice and enlarge to form bacteroids which respire but cannot reproduce. (d) Groups of bacteroids enclosed within a membrane and N_2 fixation begins (after Nutman, 1965).

• competition between effective and non-effective bacterial strains at the time of root infection. A non-effective strain produces many nodules which do not fix N_2, yet their presence inhibits nodulation by an effective strain;

• because of the large energy demand of fixation, nodulation is suppressed when mineral N, especially NH_4^+, is readily available;

• inadequate photosynthate supply and moisture stress reduce a nodulated root's fixation capacity;

• nodulated legumes need more P, Mo and Cu than non-legumes, and have a unique requirement for Co;

• nodulation and N_2 fixation are generally stimulated by simultaneous infection of the roots by vesicular–arbuscular mycorrhizas (Section 10.3), due mainly to the enhanced supply of P to the plant.

Because of these environmental and nutritional constraints, and inherent differences in host-bacterial strain performance, the quantity of N fixed is highly variable; but estimates for several temperate and tropical legumes are given in Table 10.4. The legume N is made available by the sloughing-off of nodules, through the excreta of grazing animals returned to the soil, or by the decomposition of legume residues in the soil.

Losses of soil N

Crop removal

Much of the crop N is removed in the harvested material, except under grazing conditions when approximately 85% of the N is returned in animal excreta. The

Table 10.4 Estimation of N_2 fixation by legumes (kg N ha^{-1} y^{-1}).

Temperate species		Tropical and subtropical species	
Clovers (*Trifolium* spp.)	55–600	Grazed grass–legume pastures:	
Lucerne (*Medicago sativa*)	55–400	*Stylosanthes* spp.	10–30
Soyabeans (*Glycine max*)	90–200	*Macroptilium atropurpureum*	44–129
Broad beans (*Vicia faba*)	200	Grain and forage legumes	
Peas (*Pisum* spp.)	50–100	Beans (*Phaseolus vulgaris*)	64
		Pigeon pea (*Cajanus cajan*)	97–152
Median value	*c.* 200	Median value	*c.* 100

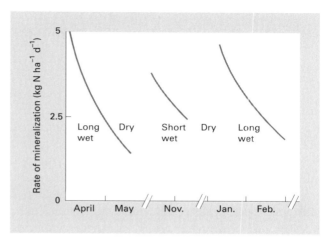

Fig. 10.6 Seasonal changes in mineral N in an East African soil (after Birch, 1958).

fate of cereal crop residues is important, because the straw from a 4 t ha^{-1} wheat crop contains 20–25 kg N. The straw is baled and used as animal bedding, eventually forming farmyard manure, or burnt, when most of the N is volatilized, or occasionally ploughed back into the soil.

Leaching

Nitrogen in the NO_3^- form is very vulnerable to leaching. Nitrate is concentrated mainly in the surface 20–25 cm of soil where it is produced as an end-product of mineralization of organic N, or from fertilizers. Nitrate concentrations fluctuate markedly due to the interaction of temperature, pH, moisture, aeration and species effects on nitrification. The following effects should be noted.

Temperature. In temperate regions, low winter temperatures ($< 5°C$) retard nitrification, which then attains a peak in spring and early summer when NO_3^- concentrations of 40–60 mg N kg^{-1} may be found in surface soils under bare fallow. Higher values are recorded in warmer, drier soils because of capillary rise and minimal leaching.

Soil moisture. In warm climates of little seasonal temperature variation, soil mineral N fluctuates with the cycle of wet and dry seasons, as has been observed in East Africa. Following partial sterilization of the soil during the long dry season, at the onset of the rains the surviving organisms multiply rapidly and produce a flush of mineralization (Fig. 10.6). High NO_3^- concentrations are the result, but these gradually decline as the readily available organic N is consumed and nitrate is removed by plant uptake and leaching.

In temperate climates, during the cold and wet winters, plant N uptake is small and much of the soil NO_3^- is leached. This effect has been observed in well drained soils in northwest Europe, New Zealand and southern Australia (rainfall > 600 mm y^{-1}). However, wetness in a poorly drained soil predisposes to denitrification (when the air-filled porosity is $< 15\%$ at the field capacity (FC)). Thus, there can be an inverse relationship between the proportion of soil NO_3^- lost by leaching and that lost by denitrification, depending on the drainage status and temperature of the soil (Fig. 10.7).

Natural grassland and forest communities. The mineral N found in grassland and forest soils is predominantly NH_4^+, which suggests that nitrification is suppressed,

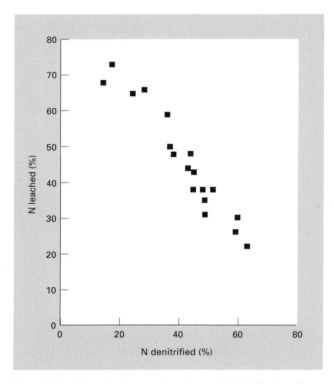

Fig. 10.7 Relationship between leaching and denitrification for grazed grass and white clover-based pastures in a humid temperate environment (after Garrett *et al.*, 1992).

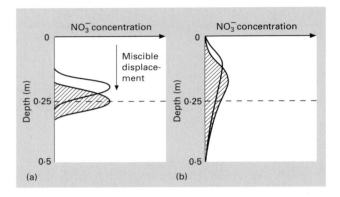

Fig. 10.8 Effects of soil structure on depth of leaching of NO_3^- originally concentrated at the surface, following 100 mm of rain. (a) Structureless clay loam – the symmetrical 'bulge' in NO_3^- concentration reflects the effects of hydrodynamic dispersion (Section 6.4). (b) Well-structured clay loam showing the effect of 'preferential flow' in redistributing NO_3^- between 0 and 50 cm depth.

possibly due to the secretion of inhibitory compounds by the roots of certain species. Normally the input of N in rain roughly balances the loss by leaching in natural plant communities; but clear-felling a forest accelerates the rate of nitrification and, in the absence of plants to absorb the NO_3^-, leads to large N losses in the drainage (Section 10.4).

Soil physical properties. Soil texture and structure influence NO_3^- leaching. Sandy soils, which generally have high hydraulic conductivities and low water-holding capacities, lose NO_3^- rapidly by leaching when rainfall exceeds evaporation. For example, in the soil where $\theta = 0.2$ m^3 m^{-3} at *FC*, 100 mm of rain can displace NO_3^- downwards to a depth of

$100/0.2 = 500$ mm (see Equation 6.16).

Conversely, in a clay loam for which θ at $FC = 0.4$ m^3 m^{-3}, 100 mm of rain is expected to displace nitrate to only 250 mm by 'miscible displacement' (Fig. 10.8a). But if the soil is well structured, some of the nitrate will move to depths > 250 mm due to fast flow down large continuous interped pores ('preferential flow'), while the remainder moves more slowly through the fine pores of the soil matrix (Section 6.3). The net result is that nitrate originally concentrated near the surface becomes distributed through a considerable depth of soil (Fig. 10.8b).

Gaseous losses

Volatilization. Animal excreta containing urea or uric acid (in the case of poultry) are susceptible to volatile losses of N. For example, heavily fertilized grassland (receiving up to 400 kg N ha^{-1} y^{-1}) that is intensively grazed can lose much more N (up to 50% of that applied) by volatilization of NH_3 than comparable grassland which is cut for hay only. Catalysed by the enzyme urease, which is widespread in soil, urea is hydrolysed to ammonium carbonate which then hydrolyses to ammonium hydroxide according to the reactions

$$(NH_2)_2CO + 2H_2O \rightarrow (NH_4)_2CO_3 \qquad (10.6)$$

$$(NH_4)_2CO_3 + 2H_2O \leftrightarrow 2NH_4OH + H_2CO_3. \qquad (10.7)$$

The alkaline NH_4OH solution is unstable, particularly at high temperatures, and releases NH_3 gas according to

$$NH_4OH \leftrightarrow NH_4^+ + OH^- \leftrightarrow NH_3 + H^+. \qquad (10.8)$$

Note that volatilization of NH_3 – whether it be formed from NH_4^+ ions produced by ammonification (Equation 10.1) or hydrolysis of urea in a urine patch – is a *pH neutral* reaction, in contrast to nitrification which is potentially acidifying. Loss of NH_3 also occurs from ammonium and urea fertilizers under certain conditions (Section 12.2).

Denitrification. There are two possible mechanisms of nitrogen loss by denitrification.
• *Chemo-denitrification*, when it occurs, is mainly a result of NO_2^- accumulation following heavy applications of a:nmonium or urea fertilizers. It is therefore discussed in Section 12.2.
• *Biological denitrification* is far more important than chemo-denitrification. It occurs in anaerobic soil culminating in the release of N_2O and N_2 (Section 8.4). Quantification of the N loss is very difficult, because although N_2O concentrations in the soil air can be measured accurately by gas chromatography, N appearing as N_2 gas can only be distinguished from background N_2 gas if the original N substrate is labelled with ^{15}N. Short-term rates of denitrification can be measured in the field by injecting sufficient C_2H_2 into the soil air (10% by volume) to block the reduction of N_2O to N_2, and then determining the flux of N_2O emitted from a limited area of surface. Denitrification rates are highly variable both temporally and spatially and may reach $0.1–0.2$ kg N ha^{-1} day^{-1} for short periods, depending on organic substrate availability, temperature, pH and the distribution of anoxic zones in the soil. As shown in Fig. 10.7, the proportion of available NO_3^- lost by denitrification can range from 15 to 65%. Deep leaching

of NO_3^- to a poorly aerated subsoil during winter does not necessarily predispose to denitrification if the temperature is low ($< 10°C$), and there is insufficient organic substrate for rapid microbial growth.

10.3 Phosphorus and sulphur

Plant requirements

Crop analyses indicate that comparable amounts of P and S are removed in harvested products each year, except for the brassicas (kale, mustard and rape) which have a higher requirement for S than P (Table 10.5). Nevertheless, the two elements differ markedly in the balance of inputs and outputs by the various pathways in the soil–plant cycle, and to a lesser extent in the transformations they undergo in the soil.

The soil reserve

Atmospheric input

Phosphorus concentrations in rain are very low and more P is deposited by dry deposition (mainly dust), particularly during dry weather. The average rate of deposition in the UK is 0.3 kg P ha^{-1} y^{-1}. Amounts of S deposited from the atmosphere are much higher, ranging from < 5 kg ha^{-1} y^{-1} to c. 70 kg ha^{-1} y^{-1}. The major source is SO_2 from the burning of fossil fuels, plus small contributions near the coast in the form of methyl sulphides and H_2S released from marine sediments. Several steps are possible in the deposition of atmospheric S on soil and vegetation.

Table 10.5 Amounts of P and S in crop products (kg ha^{-1}).

Crop	P	S
Barley, 4 t ha^{-1} (grain and straw)	12	15
Potatoes, 30 t ha^{-1}	25	20
Grass, 10 t ha^{-1} (dry matter)	30	15
Maize, 13 t ha^{-1} (grain and stover)	50	50
Kale, 50 t ha^{-1} (fresh weight)	25	100

- Direct absorption of SO_2 through the stomata of leaves.
- Oxidation of SO_2 to SO_3 according to the reaction

$$SO_2 + \frac{1}{2}O_2 \leftrightarrow SO_3. \tag{10.9}$$

Reaction 10.9 is normally slow, but it is accelerated in the presence of O_3 or H_2O_2.
- Sulphuric acid also forms according to the reaction

$$SO_3 + H_2O \leftrightarrow H_2SO_4 \tag{10.10}$$

and is brought down in the rain (hence the term *acid rain*) or neutralized by reaction with NH_3 gas in the atmosphere according to the reaction

$$2NH_3 + H_2SO_4 \leftrightarrow (NH_4)_2SO_4. \tag{10.11}$$

- Some ammonium sulphate is dissolved in rain but most is deposited directly on vegetation, which together with SO_2 absorption by plants, comprises *dry deposition*. In industrial areas where total rates of deposition can be as high as 70 kg S ha^{-1} y^{-1}, 80–90% of the accession is dry deposition. In most areas where rates are much lower (20–30 kg S ha^{-1} y^{-1}), the ratio of wet to dry deposition is approximately 1 : 1. Trends in S emissions and deposition are outlined in Box 10.3.

Forms of sulphur

Soil S is derived originally from sulphidic minerals in rocks that are oxidized to sulphate on weathering. Clay and shales of marine origin are frequently high in sulphides, and recently formed soils on marine muds may have SO_4-S contents in the profile up to 50 000 kg ha^{-1}. Normally, soil S is predominantly in an organic form and ranges between 200 and 2000 kg ha^{-1}.

Organic S is released by microbial decomposition, depending on the C : S ratio of the residues, which lies between 50 and 150 for well-humified organic matter. Net mineralization of organic S is low on average, with k values between 0.01 and 0.03 y^{-1}. Sulphur appears to be stabilized in humus in much the same way as N, and indeed the N : S ratio (between 6 and 10) is much less variable than the C : S ratio. The sulphate released is

In response to concern over the adverse effect of acid deposition on natural ecosystems, particularly in Europe and North America, governments have required industry to reduce S emissions, especially those from fossil fuels. In the UK, for example, total S emissions have fallen by 50% between 1970 and 1993, and the contribution of S to the total acid deposition has fallen from two-thirds to one-half. The distribution of atmospheric inputs of S over the UK in 1992 is shown in Fig. 10.9. As a result of much lower atmospheric S inputs over the last 20–25 years, the area of soils deficient in S for crop and pasture growth has increased. Because of further planned reductions in S emissions, atmospheric inputs are forecast to fall to 5–10 kg S ha^{-1} y^{-1} over the whole UK, so the area of S-deficient soils is likely to continue to expand. This has implications for the use of S fertilizer (Section 12.4).

The reduction in S deposition from the atmosphere should also slow down the rate of acidification of soil and natural waters in industrialized countries (Section 11.3). A worrying trend, however, is that NO_x emissions have risen by about 20% over the same period, and these now account for half the acid deposition in the UK.

Box 10.3 Changes in S emissions and depositions in industrialized countries and their implications.

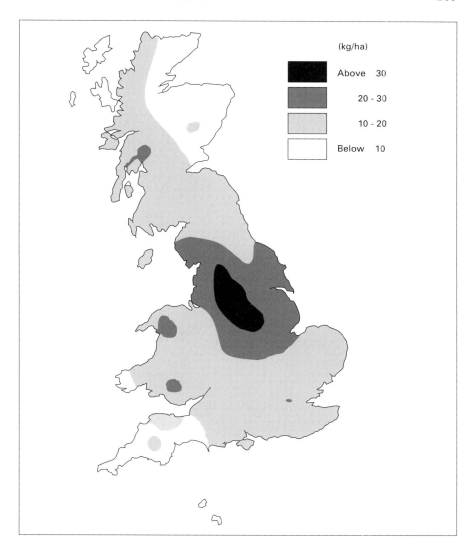

Fig. 10.9 Total deposition of S (non-marine derived) in kg ha^{-1} for Great Britain in 1992 (after McGrath *et al.*, 1996).

only weakly adsorbed at positively charged sites, in competition with phosphate and organic anions, and so it is prone to leaching. SO_4-S adsorbed on surfaces and in solution comprises the *labile pool* of S that is available to plants (see below).

Forms of phosphorus

Soil P contents range from 500 to 2500 kg ha^{-1}, of which 15–70% may exist in the strongly adsorbed or insoluble inorganic forms; the remainder occurs as organic P which is the major source of P for the soil micro-organisms and mesofauna. The phosphate of nucleic acids and nucleotides is rapidly mineralized, but when the C : P ratio rises above *c.* 100, P is immobilized by the micro-organisms, especially bacteria, which have a relatively high P requirement (1.5–2.5% P by dry weight compared to 0.05–0.5% for plants). Bacte-

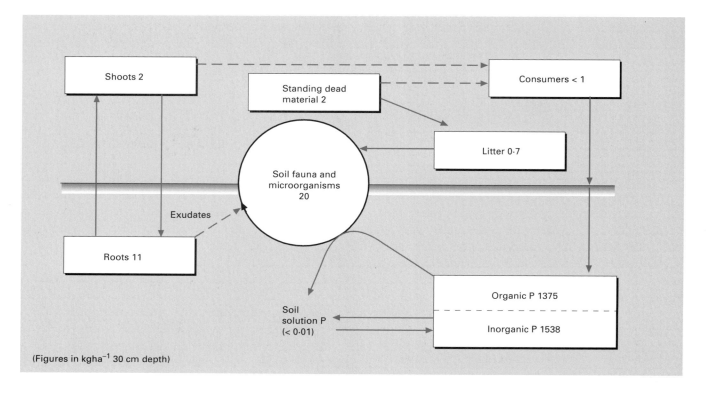

Fig. 10.10 P cycles in a native prairie grassland in Western Canada (after Halm *et al.* 1972).

rial phosphate residues comprise mainly the insoluble Ca, Fe and Al salts of inositol hexaphosphate, which are called *phytates*. Thus, in closed ecosystems with insignificant P inputs, the soil organisms are highly competitive with higher plants for P, as illustrated by the P cycle within a natural grassland (Fig. 10.10). In this case, the soil biomass (mesofauna and microorganisms) is by far the largest pool of P in the cycle apart from the inert inorganic and organic pools in the solid phase. Note also the very low amount of P in the soil solution.

Not only do micro-organisms mineralize organic P, but some groups secrete organic acids, such as α-ketogluconic acid, which attack insoluble Ca phosphates and release the phosphate. Some species of *Aspergillus, Arthrobacter, Pseudomonas* and *Achromo-*

bacter, abundant in the rhizosphere of plants, have this ability. However, most of the P is absorbed by the micro-organisms themselves in the intensely competitive environment of the rhizosphere.

Phosphate availability

Orthophosphate ($H_2PO_4^-$ and HPO_4^{2-}) that is released by mineralization is rapidly adsorbed by soil particles where its availability steadily declines with time, a process called *phosphate 'fixation'*. In terms of its availability to plants, phosphate in soil is often divided into two broad components:

• P on surfaces that can be readily desorbed plus phosphate ions in solution– the *labile P*, and

• P occluded within sesquioxide coatings on surfaces, and held in insoluble compounds or organic matter – the *non-labile P*.

Transformations between the different forms of P in

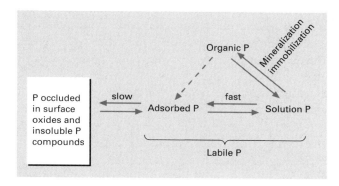

Fig. 10.11 Phosphate transformations in soil.

soil are shown in Fig. 10.11. More information on labile and non-labile P in soil is given in Box 10.4.

Effect of pH change

Soil pH, clay and sesquioxide content and exchangeable
Al^{3+} all influence P availability. In highly weathered
soils where the effects of the Fe and Al oxyhydroxides
are predominant, or in well-fertilized acid soils where
compounds of composition $FePO_4$ and $AlPO_4$ exist and
dissolve incongruently to release more P than Fe or Al,
P in the soil solution increases with pH rise (Fig. 10.12).
By contrast, in P-fertilized alkaline and calcareous soils,
insoluble Ca phosphates such as hydroxyapatite and
octacalcium phosphate (OCP) comprise the bulk of the
non-labile P, and the solubility of these compounds
increases as the pH falls. Consequently, the availability
of P in such soils is greatest between pH 6 and 7
(Fig. 10.12). Note in this figure that P concentration
will increase as the value of the negative log (P activity)
decreases.

However, acid soils of very low P status that retain a
surplus of clay minerals relative to sesquioxides show a
contrary trend in P availability with pH change. Between pH 4.5 and 7, exchangeable Al^{3+} hydrolyses to
form hydroxyaluminium ions which sorb P because of
their residual positive charge. These ions also polymerize readily so that a hydroxyaluminium complex with
$H_2PO_4^-$ ions occluded in its structure forms on the sur-

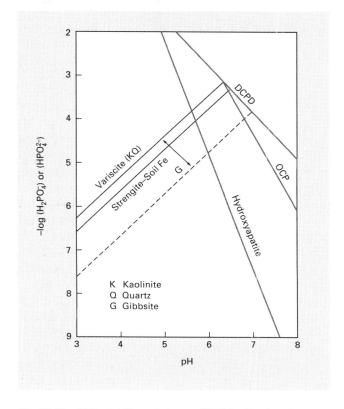

Fig. 10.12 pH-P solubility relations at 25°C for Ca phosphates
(DCPD, OCP and hydroxyapatite) and Fe and Al phosphates
(strengite and variscite, respectively). The position of these lines
depends on the presence of other compounds ($Fe(OH)_3$, kaolinite,
gibbsite and calcite), which also affect the activities of Ca^{2+}, Fe^{3+}
and Al^{3+} ions in the soil solution (after Lindsay, 1979).

face of the clay mineral, resulting in a *minimum* in P
concentration over the pH range 5 – 6.5. Above pH 7,
Al begins to dissolve as the aluminate anion and P is
released into the soil solution.

Mycorrhizas

Because of low P concentrations in soil solutions and
competition from micro-organisms in the rhizosphere,
many plants have evolved mechanisms to enhance the
absorption of P when it is in short supply. One of the
most important of these is the *mycorrhiza*, a symbiotic

Box 10.4 Forms of labile and non-labile inorganic P in soil.

Labile P should be immediately available to plants growing in the soil. Soil P which can be desorbed in a salt solution, such as 0.01M $CaCl_2$, which approximates the ionic strength of the soil solution of many agricultural soils, is classed as labile. Other estimates of labile P are made by extracting soil in a solution of 0.5M $NaHCO_3$ buffered at pH 8.5 for 30 min, or shaking a soil with an anion-exchange resin. By raising the pH and supplying a competitive anion (HCO_3^-), the bicarbonate extraction desorbs some of the surface 'ligand- exchange' P (Section 7.3). The resin, on the other hand, can supply not only a source of competitive anion (e.g. HCO_3^-) but also a large sink for the desorbed phosphate ions. Thus, estimates of labile P vary according to the bonding mechanism of the sorbed P and the desorption/extraction method used (White, 1980).

Non-labile P is not immediately available to plants but may become so if the labile pool is severely depleted, or there is a significant change in the soil's physicochemical environment. Phosphate sorbed by hydroxy-Al polymers on clay surfaces, and that sorbed by ligand exchange on discrete sesquioxide particles, can become *occluded* in these materials. The mechanism is unclear, but may involved solid-state diffusion and/or highly localized dissolution and reprecipitation of the oxyhydroxides due to changing physicochemical conditions in the soil. The hydroxy-Al compounds are most stable in the pH range 4.5–7.0; the Fe oxyhydroxides can dissolve and reprecipitate under fluctuating redox conditions (Section 8.4).

Phosphate also forms insoluble salts with Fe, Al and Ca. The crystalline minerals *strengite* ($FePO_4.2H_2O$) and *variscite* ($AlPO_4.2H_2O$) are only stable below *c.* pH 1.4 and 3.1, respectively, and are unlikely to exist in soil. However, in less acid soils, slightly more soluble compounds with an Al:P or Fe:P mole ratio $\simeq$ 1 (amorphous $AlPO_4$ and $FePO_4$) can form, especially in the vicinity of a dissolving granule of superphosphate (Section 12.3). Above pH 6.5, phosphate forms insoluble salts with Ca, such as *octacalcium phosphate* $Ca_4H(PO_4)_3.5H_2O$, which reverts to the more stable *hydroxyapatite* $Ca_{10}(PO_4)_6(OH)_2$, and phosphate may also be present as an impurity in calcite crystals. In the latter case, the release of phosphate depends on calcite dissolution, which is primarily controlled by the pressure of CO_2 in the soil air (Section 11.3).

association of fungus and root. There are two types of mycorrhiza.

Ectomycorrhizas. These are primarily associated with woody perennials: members of the Pinaceae (pines), Fagaceae (oak, beech), Betulaceae (birch, alder), *Eucalyptus* and poplar. many fungal genera have been identified in ectomycorrhizas, mainly members of the Basidomycetes and Ascomycetes. The fungal mycelium forms a sheath some 0.03 mm thick around the root

cylinder and also grows through the intercellular spaces in the root cortex, forming a distinctive network of hyphae called a *Hartig net*. The external hyphae afford a small increase in the soil volume exploited by the root; but the major benefit derives from:
• the fungus's ability to store P in its tissues, to be released to the host in times of P deficiency, and
• the greater longevity of infected roots which continue to absorb P long after non-mycorrhizal roots have ceased active uptake. In return, the fungus obtains carbon substrates and other growth requirements from the host.

Endomycorrhizas. In this form, the fungus does not form a sheath but grows in the root cortex, forming two characteristic structures – the *vesicles*, which are temporary storage bodies in the intercellular spaces, and the *arbuscules*, which are branched structures growing intracellularly that release solutes into the cell (Fig. 10.13). Hence the common name *vesicular–arbuscular mycor-*

rhiza (VAM). The endophyte is a Phycomycete of the family Endogonaceae. The fungal spores are present in most soils, but only germinate when a suitable host root is near; if the young mycelium does not penetrate a host root, it dies. However, once established in the root (and the host–fungal relationship is not very strain specific), the hyphae ramify widely outside the host tissue.

VAMs are found in nearly all the angiosperms and in many conifers, ferns and lower plants. The incidence of infection is especially high in impoverished soils, where the mycorrhizal state confers a distinct advantage in P uptake, because the external hyphae can add up to 50% to the length of absorbing root and provide a pathway for the rapid transport of phosphate across the 'depletion zone' around the root (Box 10.5). The benefit is most notable in plants such as onion and citrus which have no or very few root hairs. VAMs also enhance plant uptake of Cu and Zn from the soil. They are of considerable importance in cycling of P in natural ecosystems, and for the establishment of pioneer species in

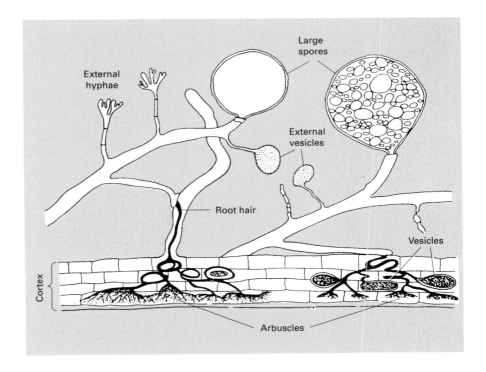

Fig. 10.13 Diagram of a vesicular-arbuscular mycorrhiza (after Nicholson, 1967).

Box 10.5 Nutrient depletion zones around absorbing plant roots.

A depletion zone develops around a plant root when the rate of intake of a nutrient exceeds the rate at which the labile pool of that nutrient can be replenished at the root surface. An example is given in Fig. 10.14. For most nutrients, replenishment from the non-labile pool is too slow to meet the demand of a rapidly absorbing root, so movement of nutrients from the bulk soil into the depletion zone must occur. Ions will move in solution by *mass flow* and *diffusion* (Section 10.2), and *potentially* by surface diffusion in the solid phase. However, it has been found that, although a high proportion of the labile form of a nutrient ion, such as $H_2PO_4^-$, is adsorbed, the contribution of surface diffusion to ion transport near roots is negligible compared to the flux in solution (Nye and Tinker, 1977). For ions such as Ca^{2+}, Mg^{2+} and SO_4^{2-}, the concentration in the soil solution is normally high enough for mass flow alone to meet the root demand, and no depletion zone develops (in some cases a surplus of the ion occurs at the root surface). However, for ions such as $H_2PO_4^-$, K^+, Zn^{2+} and Cu^{2+} which are strongly adsorbed and at low concentration in solution, mass flow is insufficient to meet root demand and movement to the root is primarily by diffusion. The *radial diffusive flux* J_s (Box 6.7) is given by

$$J_s = -D_e \frac{\partial C}{\partial x} \tag{B10.5.1}$$

where C is the ion concentration in the soil solution, x is the distance from the root surface, and D_e, the effective diffusion coefficient, is given by

$$D_e = D\theta f \frac{\partial C}{\partial C'} . \tag{B10.5.2}$$

D is the diffusion coefficient in bulk water, θ is the volumetric water content, f is an impedance factor (range 0–1) to account for the tortuous pathway for diffusion, and $\partial C'/\partial C$ is the soil's *buffer power* for the nutrient. This is related to the slope of the *Q/I relation* (Section 7.2), except that C' is the amount of labile nutrient per unit volume of soil. Note that D_e is reduced when θ and f are small, and $\partial C'/\partial C$ is large. A depletion zone around a root extends a radial distance of $\sqrt{D_e t}$ from the surface, where t is the time from when the root begins to absorb the nutrient.

harsh habitats such as on sand dunes and mine spoil. The incidence of infection declines, however, with the addition of P fertilizer. Therefore little benefit is gained by cultivated crops in general because fertilizer has a more rapid effect on growth than does the mycorrhizal infection. This suppression of VAMs by P fertilizer may explain the appearance of Cu and Zn deficiencies in some P-fertilized crops, especially on high pH soils

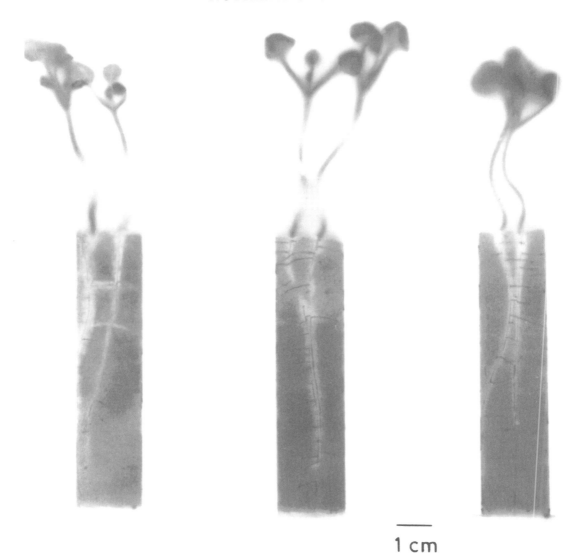

1 cm

Fig. 10.14 Autoradiograph of roots of rape in a soil labelled with
^{33}P showing a zone of depletion around the main roots, and
accumulation of P within the root axes and laterals (after Bhat &
Nye, 1973).

(Section 10.5). Members of the Cruciferae (brassicas)
and Chenopodiaceae (buckwheat) are non-mycorrhizal
and have evolved with other mechanisms for enchanc-
ing P uptake from P-deficient soils.

Leaching losses

With the exception of very sandy soils, and other situations where P has accumulated in high amounts
through the repeated application of fertilizer and/or
animal manures (Section 12.3), phosphate losses by
leaching are very small, usually < 1 kg ha^{-1} y^{-1}.

Sulphate forms sparingly soluble gypsum in arid region soils, but is readily leached from soils of humid

Table 10.6 P and S losses (kg ha^{-1} y^{-1}) in drainage from neutral and calcareous soils in southern England. After Williams, 1975 and others.

Site	PO$_4$-P	SO$_4$-S
Woburn – neutral, sandy soil over Oxford Clay	0.05	100
Saxmundham – sandy, calcareous boulder clay	0.12	86
Wytham – brown calcareous soil over Oxford Clay	0.13	78

Table 10.7 (a) Annual return (= uptake) of K, Ca and Mg (kg ha^{-1} y^{-1}) by a 40-year-old forest in West Africa. After Nye and Greenland, 1960.

Process	K	Ca	Mg
Litter fall	68	206	45
Timber fall	5	82	8
Leaf wash	220	29	18
Total	293	317	71

Table 10.7 (b) Annual return (= uptake) of K and Ca (kg ha^{-1} y^{-1}) by 36-year-old Douglas fir. After Cole *et al.*, 1967.

Process	K	Ca
Litter fall	3	11
Stem flow	1	1
Leaf wash	11	4
Total	15	16

regions. For example, where fertilizers containing SO$_4$-S are used (single superphosphate, ammonium sulphate), leaching losses of SO$_4^{2-}$ are equivalent to 30–80% of the fertilizer S in the year following application. An exception is soils high in sesquioxides and low pH, where SO$_4^{2-}$ is retained at depth (Fig. 10.2b). Data obtained from three sites on neutral and calcareous soils in southern England illustrate the contrast between leaching losses of sulphate and phosphate (Table 10.6).

10.4 Potassium, calcium and magnesium

Plant requirements

Annually, agricultural crops remove between 5 and 25 kg Mg ha^{-1}, and exceptionally up to 50–60 kg ha^{-1} for high yields of maize or oil palm. Calcium requirements range from 10 to 100 kg ha^{-1} and the K requirement is even higher at 100–300 kg ha^{-1}.

There are few estimates of the rate of cation uptake by perennial vegetation, especially forests, because it is difficult to measure the amount and composition of the annual growth increment. By assuming that in a mature forest the rate of nutrient return balances the rate of uptake from the soil, Nye and Greenland (1960) obtained the estimates of cation uptake shown in Table 10.7a. Much lower figures for K and Ca uptake were obtained by Cole *et al.* (1967) for a 36-year-old Douglas fir plantation in Washington, USA (Table 10.7b). Estimates of Ca and Mg uptake by beech and spruce forests in Sweden are slightly larger than for the Douglas fir plantation.

Soil reserves

The pool of exchangeable cations Ca^{2+}, Mg^{2+} and K$^+$ provides the immediate source of these elements for plants. Insoluble reserves may occur as calcium and magnesium carbonates, potash feldspars and interlayer K$^+$ in micaceous clays. Total reserves are therefore very variable, reflecting the conditions of soil formation, but exchangeable Ca^{2+} values usually lie between 1000 and 5000 kg ha^{-1}; Mg^{2+} between 500 and 2000 kg ha^{-1}; and K$^+$ $\sim$ 1000 kg ha^{-1}.

Atmospheric inputs

Amounts of K and Ca in all forms of precipitation and dry deposition are comparable and lie between 1 and 20 kg ha^{-1} y^{-1}. Magnesium can be much higher, especially on sea coasts (because ocean water is rich in Mg), amounting to as much as 150 kg ha^{-1} y^{-1}. Thus, Mg inputs roughly balance combined losses through plant removal and leaching, but not so for K and Ca. A striking feature of K input to the soil is the magnitude of its leaching from leaves, amply demonstrated by the figures for the tropical tree species and Douglas fir in Table 10.7. This process is one of the main reasons for the rapid cycling of K in natural ecosystems in which plants and their residues can accumulate 40 to 60% of the total labile K within the root zone.

Mineral weathering

Apart from some recent glacial materials, the reserves of weatherable primary minerals are usually small in the top 30 cm of soil where 80–90% of the plant roots are located. Nevertheless, the slow release of nutrients from the 'geologic reserve' at greater depth is often crucial for the stability of natural plant communities for which elemental losses by leaching and erosion occasionally exceed atmospheric inputs. For example, data from West Africa suggested that 58, 64 and 14 kg ha^{-1} of K, Ca and Mg, respectively, were pumped up annually from the subsoil by deep tree roots.

Leaching losses

Cation leaching is accelerated in cultivated soils, where nitrification occurs, for two reasons.
1 There is a net production of one H$^+$ per mol of NO$_3^-$ produced during ammonification and nitrification (Reactions 10.1 and 10.2). These H$^+$ ions can displace exchangeable Ca^{2+}, Mg^{2+} and K$^+$ from exchange surfaces.
2 The conversion of NH$_4^+$ to NO$_3^-$ means that the moles of anion charge in solution increases relative to the moles of cation charge. However, the extra anions are balanced by the cations, especially Ca^{2+}, displaced from the exchange surfaces. Thus, Ca^{2+} and NO$_3^-$ predominate in the water percolating through the soil.

For agricultural soils in Britain, Ca leaching losses range from 90 to 400 kg ha^{-1} y^{-1}. There is a direct correlation between Ca losses and pH, with higher losses (> 200 kg ha^{-1}) usually occuring from soils that have been limed or which contain free CaCO$_3$. Although Ca^{2+} is normally the most abundant cation in soil drainage, for acid soils close to the sea Na$^+$ will often be the dominant cation, as shown by results for a silt loam under pasture in New Zealand (Fig. 10.15).

Suppression of nitrification in soils under natural

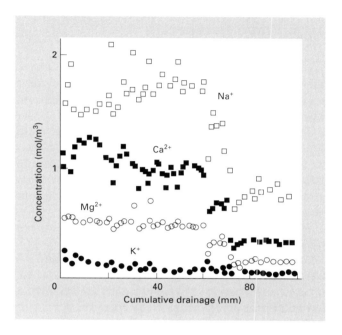

Fig. 10.15 Concentrations of Ca^{2+}, Mg^{2+}, Na$^+$ and K$^+$ in drainage from a pasture near the coast in New Zealand (after Heng *et al.* 1991).

grassland of forest will reduce cation loss by leaching. For example, under an undisturbed deciduous forest catchment in New Hampshire USA, the *net* leaching loss of K, Ca and Mg from the acid podzolic soil was 0.6, 9.2 and 2.2 kg ha^{-1} y^{-1}, respectively (Bormann and Likens, 1970). These net losses were balanced by weathering of the parent material. Nitrification appeared to be of very minor significance in the soil under the undisturbed forest, and nitrate losses in the drainage balanced inputs in precipitation. However, after the forest was clear-felled (when the timber was neither removed nor burnt), nitrification increased sharply and net N leaching losses rose to 120 kg N ha^{-1} y^{-1}, while net K, Ca, Mg losses rose to 13, 92 and 15 kg ha^{-1} y^{-1}, respectively.

10.5 Trace elements

Definitions and sources

Those elements whose total concentration in the soil is normally $< 1000\ mg\ kg^{-1}$ are called *trace elements*. They fall into three categories.

1 The essential *micronutrients* Cu, Zn, Mn, B and Mo which are beneficial to plants at normal concentrations in the plant (ranging from $0.1\ mg\ kg^{-1}$ for Mo to $100\ mg\ kg^{-1}$ for Mn), but become toxic at higher concentrations. Iron (Fe) is micronutrient that is not strictly a trace element.

2 Elements such as Cr, Se, I and Co which are not essential for plants but are essential for animals.

3 Elements such as Li, Be, As, Hg, Cd, Pb and Ni which are not required by plants or animals and are toxic to either group at concentrations greater than a few milligrams per kilogram in the organism.

With industrialization and the generation of wastes from industry and large cities, together with the input of chemicals in intensive agriculture, reports of undesirably high concentrations of many of these elements in soils, plants, animals or water are becoming more frequent. We speak of *contamination* and *pollution* of the environment (Box 10.6).

Trace elements in soil are normally derived from the parent material and natural deposition from the atmosphere (e.g. from volcanoes, dust and sea spray). In the 20th century, however, anthropogenic sources have increased markedly, especially from mining, industrial processing, motor vehicles, agricultural chemicals and urban waste. These sources have increased trace element *loadings* in soil–plant systems via the atmosphere and direct solid and/or liquid applications (e.g. spreading animal waste, sewage sludge or in landfills).

Influence of parent material

Trace elements are largely bound in mineral lattices, to be released by weathering, so that the type of parent material determines the natural abundance of the individual elements in soil. Iron, Cu, Mn, Zn, Co, Mo, Ni and Cr occur in ferromagnesian minerals, common in ultrabasic and basic igneous rocks. Fe, Mn and Mo also occur as insoluble oxides (often with Co coprecipitated in the MnO_2), and Zn, Fe, Pb and Cu as equally insoluble sulphides in sedimentary rocks. Boron occurs as the resistant mineral tourmaline in acid igneous rocks and to some extent may substitute for Al in the tetrahedral sheet of $2:1$ clay minerals. Typical trace element contents of soils formed on parent rocks of different basicity are presented in Table 10.8.

The term '*heavy metals*' is frequently used for metals and metalloids involved in contaminating or polluting the environment. By implication, these elements are considered toxic to either plants or animals if concentrations exceed a certain threshold. This terminology is vague and the terms 'micronutrient' and 'trace element' used here are to be preferred. Threshold concentrations and any adverse effects should be defined for each element.

The terms 'concentration' and 'pollution' tend to be used interchangeably. Sometimes 'contamination' is restricted to the presence of a substance not normally found in soil, which may or may not have a harmful effect. Commonly, 'pollution' is used as a generic term to cover the introduction by humans of any substance having an adverse effect on the environment. It has a pejorative connotation.

Box 10.6 Soil contamination and pollution.

Table 10.8 Trace element contents (kg ha^{-1}) of soils on parent materials of different rock type. After Mitchell, 1970 and others.

| Trace element | ← Increasing basicity | | | |
	Serpentine	Andesite	Granite	Sandstone
Co	160	16	< 4	< 4
Ni	1600	20	20	30
Cr	6000	120	10	60
Mo	2	< 2	< 2	< 2
Cu	40	20	< 20	< 20
Mn	6000	1600	1400	400
Pb	2	12	36	24

Atmospheric inputs

Trace elements enter the air as gases, aerosols and particulates and return to the soil–plant system mainly by dry deposition. Metals that are used extensively in industry – Cd, Zn, Cu, Ni and Pb – show the greatest enrichment in the air of industrialized regions, relative to metals like Fe or Ti that are naturally abundant. The amounts deposited are therefore likely to be greater in industrial regions of the northern hemisphere than in remote rural areas of the continents. This is shown in Table 10.9 where estimated annual rates of deposition of several trace elements are given. The minimum values are representative of rural areas, whereas the median values are biased towards urban areas because of the greater density of observations there.

Table 10.9 Estimated rates of trace element deposition (g ha^{-1} y^{-1}) from the atmosphere over Europe and North American. After Sposito and Page, 1984.

| Element | Europe | | North America | |
	Minimum	Median	Minimum	Median
Mo	< 0.3	—	0.2	—
Co	0.3	—	0.2	4.7
Cd	0.8	3	< 0.2	6.5
Ni	6.3	39	< 16	142
Mn	14	68	0.1	236
Cu	13	536	7.9	441
Zn	20	19	—	788
Pb	87	189	71	4257

Soil reactions and availability to plants

Free ions and complexes in solution

The reactions involved in the turnover of a trace element in the soil–plant system are summarized in Fig. 10.16. The chemistry of an individual element determines the extent to which it participates in any one of these reactions. The rate of any reaction depends not only on the concentration of reactants and products, but also on steriochemical factors (for complex formation) and environmental conditions.

Most trace elements – Fe, Cu, Mn, Zn, Co, Cd, Ni and Pb – occur as cations. Others such as Cr can occur as an anion (the chromates $Cr_2O_7^{2-}$ and CrO_4^{2-}) or cation Cr^{3+}, depending on the oxidation state. Of the remainder, Mo, As and Se occur as oxyanions – MoO_4^{2-} (molybdate), $H_2AsO_4^-$ and $HAsO_4^{2-}$ (arsenate), SeO_4^{2-} (selenate) and SeO_3^{2-} (selenite). Boron occurs as H_3BO_3 (boric acid) which dissociates to H^+ and $B(OH)_4^-$ (borate) at pH > 8, and I occurs as I^-. The cations form complexes of varying stability with inorganic and organic ligands. The most abundant ligand is H_2O so the cations are normally hydrated. The principles of complex formation are as follows

$$a M^{n+} + b L^{l-} \leftrightarrow \{M_a L_b\}^q \qquad (10.12)$$

where charge balance requires that

$$an - bl = q \qquad (10.13)$$

and q, the valency of the complex, may be $+$, 0 or $-$. The stability of the complex is given by the *conditional stability constant* $^c K_s$, defined by the expression

$$^c K_s = \frac{[M_a L_b^q]}{[M^{n+}]^a [L^{1-}]^b}. \qquad (10.14)$$

The constant is called conditional because its value usually depends on the solution composition and temperature. Metal–inorganic ligand complexes may form, depending on the pH and concentration of suitable ligands. Some examples are $CdCl^+$, $MnSO_4^0$, $NiSO_4^0$ and $ZnHPO_4^0$, all of which modify the metal's adsorption characteristics and its mobility in the soil. Com-

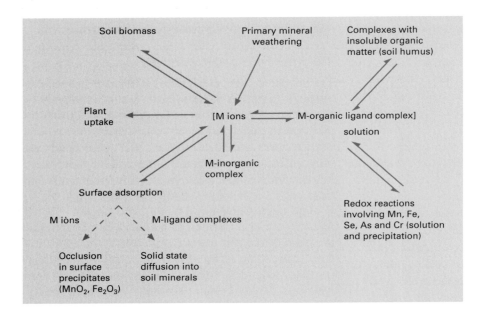

Fig 10.16 Trace element (M) equilibria in the soil.

puter programs have been developed to enable 'speciation' of elements in soil solutions (e.g. Sposito and Coves, 1995). However, many metals form much more stable complexes with organic ligands. These can be bidentate, tridentate or even quadridentate 'chelates' (Section 3.4) which contributes to their stability. The complexes exist in solution as anions or cations of low charge : mass ratio, but they can also be adsorbed (see below).

The cations Fe, Cu, Mn, Ni and Zn also hydrolyse at pH values between 3 and 10 and eventually precipitate as insoluble hydroxides. This is a special case of Equation 10.16, for example

$$Fe^{3+}(\text{soln}) + 3OH^-(\text{soln}) \leftrightarrow Fe(OH)_3^0(\text{solid}). \quad (10.15)$$

Equation 10.15 is usually considered as a dissolution–precipitation reaction for which the *thermodynamic solubility constant* K_{sp} is defined by

$$K_{sp} = \frac{(Fe^{3+})(OH^-)^3}{(Fe(OH)_3)}. \quad (10.16)$$

Taking the activity of pure solid $Fe(OH)_3$ as 1, Equation 10.16 becomes

$$K_{sp} = (Fe^{3+})(OH^-)^3. \quad (10.17)$$

Whether the insoluble hydroxide or a soluble organic complex (FeL_b^q) is thermodynamically more stable at a given pH depends on which chemical form maintains the lower activity of free Fe^{3+} ions in solution. This fact underlies the choice of organic compounds (chelating agents) to form soluble complexes with micronutrients, such as Fe, that become unavailable to plants at high pH due to the precipitation of the hydroxide (Box 10.7).

Trace elements in the solid phase

The concentration of free trace elements in the soil solution is normally very low ($< 10^{-7}$M). The surface adsorbed component shown in Fig. 10.16 accounts for about 10% of the total of any trace element in soil. The remainder is held in insoluble precipitates, primary and secondary minerals, or is complexed with soil organic matter. Metal cations are adsorbed:

• non-specifically, by electrostatic attraction to negatively charged surfaces, in competition with the macronutrient cations Ca^{2+}, Mg^{2+} and K^+, or

Box 10.7 Organic complexes and micronutrient availability.

Deficiency of Fe is manifest in plants as '*lime-induced* chlorosis'; that is, interveinal yellowing of leaves due to the lack of chlorophyll. The deficiency of micronutrients such as Fe and Mn can be corrected by supplying the elements in a soluble complex. For example, ethylenediamine tetra-acetic acid (EDTA) is a synthetic chelating agent which forms stable complexes with most trace elements. Figure 10.17 shows the effect of pH on the balance between the stability of $Fe(OH)_3$ and the complex $FeEDTA^-$. The change from the chelate to the hydroxide as the more stable phase occurs *c.* pH 9 normally, but in calcareous soils it occurs *c.* pH 8 because of the increased stability of the $CaEDTA^{2-}$ complex at high pH. Thus, chelating agents, such as EDDHA (ethylenediamine di(*o*-hydroxyphenylacetic acid)) which have a higher affinity for Fe than for Ca, should be used on calcareous soils. However, there is a evidence that plants take up more Fe from complexes of lower stability, indicating that the complex may have to be broken down before the Fe can be absorbed into the cell. Complexing agents such as EDTA and DTPA (diethylene triamine penta-acetic acid) are used to extract soils to measure the '*bio-availability*' of trace elements (Box 10.8).

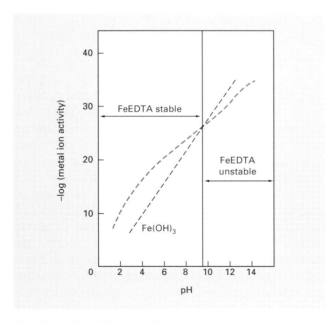

Fig. 10.17 pH stability range for FeEDTA (after Martell, 1957).

• through the formation of outer-sphere complexes (Section 7.1), and
• specifically through the formation of inner-sphere complexes on oxyhydroxide surfaces (Section 7.1).

Inner-sphere complexes are more stable than outer-sphere complexes because no water molecules separate the ion from the surface and a degree of covalent bonding is involved. Hydrated Cu^{2+} ions, for example, form outer-sphere complexes with the cavities in the siloxane sheet of 2 : 1 clay surfaces from which they can be displaced by Ca^{2+}. On the other hand, Cu^{2+} ions forming inner-sphere complexes with COOH groups on organic matter are not exchangeable to Ca^{2+}. Although stable complexes are formed between trace metal ions and organic ligands both in solution and on surfaces, because the organic C in solution (at 50–150 ppm) is very much less than the C in the solid phase, organic complex formation tends to immobilize metals rather than to solubilize them. Sewage sludge, particularly that derived from industrial wastes, often contains relatively high concentrations of metals such as Cu, Zn, and Ni in fairly labile organic combination which may pose problems when large applications are made to agriculture soils (Section 11.1).

The adsorption of metal cations increases with pH rise, with maximum adsorption occuring 3 to 4 pH units below the pK_1 for hydrolysis of the cation. The effect of pH is primarily due to the increase in negative charge of variable charge surfaces, but the presence of even low concentrations of the hydrolysed cations may also contribute. This explains why the availability to plants of most trace elements decreases with pH rise.

Anions such as MoO_4^{2-} are most strongly adsorbed at low pH on sesquioxide surfaces by a ligand-exchange mechanism analogous to that described for phosphate (Section 7.3). B is adsorbed by the same mechanism on oxide sufaces as $B(OH)_4^-$ but as this species only appears at pH > 8, B solubility decreases with increase in pH. Molybdenum is therefore the only *micronutrient* to increase in availability as the soil pH rises. The two arsenate anions are also adsorbed by ligand exchange, but in the case of Se, SeO_4^{2-} (the more stable form, selenate) is only weakly adsorbed whereas SeO_3^{2-} is adsorbed by ligand exchange. Thus, selenate can be quite mobile in soil. Both As and Se anions can be methylated in biological reactions in soil and released as gases (e.g. $(CH_3)_2HAsO_2$ and $(CH_3)_2Se$).

The adsorbed phase of trace elements comprises the 'labile' component of the soil reserve, the size of which can determine the bioavailability of the element. Measuring trace element bioavailability is important for assessing not only the deficiency or otherwise of micronutrients for crop growth, but also the potential hazard of these elements when the soil loading is high (Box 10.8).

The effect of poor drainage

The manganese and iron of insoluble conpounds are mobilized in waterlogged soils as Mn^{2+} and Fe^{2+} ions, as discussed in Section 8.4. In acid, waterlogged soils, chromate can be reduced to Cr^{3+}, but this forms very insoluble oxides and hydroxides. In addition, the solubility of Co, Ni, Mo and Cu (elements not reduced at

Table 10.10 The effect of drainage condition on trace element solubility in soil.

Soil	Extractable* element content (mg kg^{-1} soil)			
	Co	Ni	Mo	Cu
Freely drained	1.3	1.3	0.06	2.6
Poorly drained	1.9	3.4	0.19	6.6

* Extracted in CH_3COOH or EDTA.

the redox potentials attained in soils) may be increased under waterlogged conditions due to the accelerated weathering of ferromagnesians minerals (Table 10.10). The dissolution of manganese oxides can also release coprecipitated Co into the soil solution.

Crop removal and leaching losses

Estimates of the removal of micronutrients in harvested products vary widely but the range (in g ha^{-1} y^{-1}) is approximately as follows:

Mn	34–500
Mo	< 0.2–500
B	27–160
Zn	31–80
Cu	6–80.

A comparison with the minimum figures for atmospheric inputs in Table 10.9 suggests that soils in remote rural areas may suffer a net depletion of these elements by crops. On the other hand, these elements occur as trace contaminants in fertilizers and liming materials (e.g. basic slag) and unknown amounts are released through mineral weathering each year.

Because of precipitation in soluble compounds and strong retention by mineral and organic surfaces, micronutrient losses by leaching are very small, except for B, and Fe and Mn in some gleyed soils. Nevertheless, without significant atmospheric inputs, cumulative losses from very old soils eventually lead to widespread micronutrient deficiencies, as occur in the coastal areas of eastern and southern Australia. Leaching of Se from naturally rich sediments in the San Joaquin Valley of

Box 10.8 Bioavailability of trace elements.

In Section 7.2 the concept is introduced of a *quantity intensity* (Q/I) relationship which controls the availability of a nutrient ion (K^+) to plants. This concept of nutrient availability, or *bioavailability*, is developed further in Section 11.2. With respect to micronutrients and trace elements, the concentration of the free ionic species in solution at an absorbing surface (e.g. a root or algal cell) controls the rate of uptake by the organism. But because the free ion concentration is normally very low, the extent to which that concentration can be maintained by dissociation of a complex, or desorption from the solid phase, becomes the controlling factor in determining bioavailability. Thus, Tiller (1996) and others have argued that the Q factor is paramount in determining trace element bioavailability and that it should be measured as the *total potentially labile element* (per unit mass of soil). Tiller and his coworkers found that acids were not suited to measuring Q because of their partial neutralization in calcareous soils and their dissolution of soil minerals. They preferred to extract soil in 0.05 or 0.1M EDTA for periods up to 7 days, to completely remove elements with an affinity for soil surfaces and present at heavy loadings.

Unfortunately, much of the work on the trace element chemistry of soils has been empirical and led to the use of a bewildering range of complexing, extracting and dissolving reagents (Beckett, 1989). The rational Q/I approach has not often been applied so that many conflicting results have been obtained on bioavailability: so much so that legislation on soil contamination by heavy metals is framed in terms of total soil content, not bioavailable content. In many cases, therefore, critical levels for unacceptable contamination have been set quite low and the risk associated with higher total soil contents overestimated (Chaney and Oliver, 1996). Where some differentiation of the ease of release of an element from the solid phase is needed to refine the estimate of bioavailability, the following fractionation sequence is suggested:
1 weakly bound forms desorbed in dilute $CaCl_2$ or NH_4NO_3,
2 strongly bound forms extracted in 0.05 or 0.1M EDTA (prolonged if necessary), and
3 residue dissolved in strong acids or measured by X-ray fluorescence.

California, USA has caused serious pollution of surface and underground water.

10.6 Summary

The *macronutrient* elements C, H, O, N, P, S, Ca, Mg, K and Cl are essential for plant growth and reproduction and occur in concentrations > 1000 mg kg^{-1} DM. The *micronutrient* elements Fe, Mn, Zn, Cu, B and Mo occur at concentrations < 100 mg kg^{-1} DM. The micronutrients are included in the trace elements – elements normally occurring at a concentration < 1000 mg kg^{-1}

soil. Also included are elements such as Cr, Se, I and Co which are essential only for animals, and Li, Be, As, Hg, Cd, Pb and Ni which are not required by plants or animals, and are toxic at concentrations greater than a few milligrams per kilogram in the organism.

Gains and losses of nutrients in natural ecosystems, comprising the soil–plants–animals–atmosphere, are roughly in balance so that continued growth depends on the cycling of nutrients between the *biomass*, *organic* and *inorganic* stores. Removals from agricultural systems, especially of crop products, generally exceed natural inputs unless these are augmented by fertilizers and organic manures, and in the case of some trace elements, by accessions from the atmosphere.

Nutrients in the biomass store pass to the organic store by excretion from, or on the death of, living organisms. The greater part of the organic and inorganic stores is in the soil. Additions to the inorganic store occur through *mineralization* of organic residues, *weathering*, and *atmospheric inputs* (rainfall and dry deposition), and from fertilizers in agriculture systems. That part of the inorganic store on which plants and other organisms feed is called the *available* or *labile pool*, consisting of ions in solution and ions adsorbed by clays, sesquioxides and organic matter. Nutrients and potentially toxic elements held in insoluble precipitates or strongly adsorbed complexes (organic or inorganic) are considered *non-labile*. The *bioavailability* of a nutrient or potentially toxic element is best characterized by its *quantity/intensity* (Q/I) relationship, but many empirical extraction methods have been developed which obscure the definition of Q and I factors.

The ability of a minority of plants in symbiotic association and free-living micro-organisms, to reduce N_2 gas to NH_3 within the cell (*nitrogen fixation*) provides a unique input of N to the biomass and hence organic stores. But substantial losses of N can occur following *ammonification* and *nitrification*, according to

$$\text{organic-N} \rightarrow NH_4^+ + OH^- \text{ and}$$

$$NH_4^+ + 2O_2 \rightarrow NO_3^- + H_2O + 2H^+,$$

through NO_3^- leaching, and NO_3^- reduction (*denitrification*) to N_2O and N_2 under anoxic conditions. Note that ammonification/nitrification is a potentially acidifying process. Volatilization of NH_3 (from NH_4^+ at high pH) also occur, following the hydrolysis and decomposition of urea in animal excreta. Substantial amounts of N are returned to soil–plant systems from the atmosphere, dry deposition being particularly significant in urban and industrialized areas.

Atmospheric inputs of S are much greater than P, especially in industrialized regions, but are now declining. Biological transformations of P and S in soil are similar, but P cycling is more conservative than S because the ions $H_2PO_4^-$ and HPO_4^{2-} are strongly adsorbed, or form insoluble precipitates with Fe, Al and Ca. SO_2^{2-} is only weakly adsorbed and does not form insoluble precipitates (except $CaSO_4$ in aride soils). SO_4^{2-} is also reduced to sulphides under intense reducing conditions. Many plants have evolved fungus–root associations or *mycorrhizas* which enhance P absorption (and Cu and Zn) from deficient soils. In contrast to sulphate, which is readily leached (except from acid sesquioxidic soils), P losses by leaching are small because P concentrations in the soil solution are very low.

Calcium, Mg and K are held mainly as exchangeable cations, the supply of which buffers the soil solution against depletion. Cation losses are accelerated when anions such as NO_3^-, SO_4^{2-} and HCO_3^- are plentiful in the percolating water, and also under acid conditions when H^+ and Al^{3+} ions occupy exchange sites. The micronutrients cations, Fe, Mn, Cu, Zn and Co, and the trace element cations Cr, Ni, Hg, Cd and Pb, form complexes with negatively charged surfaces, those formed with organic compounds being more stable than the inorganic complexes. The ability to form stable complexes with soluble organic ligands (*chelates*) is used to prevent the micronutrients precipitating as insoluble hydroxides at alkaline soil pHs. Molybdenum, B, Se, As and I occur as anions in solution. Molybdate is most strongly adsorbed by sesquioxides, followed by As, Se, B and I.

References

Anon. (1983) *The Nitrogen Cycle of the United Kingdom.* Royal Society Study Group Report.

Beckett P. H. T. (1989) The use of extractants in studies on trace metals in soils, sewage sludges and sludge-treated soils. *Advances in Soil Science* **9**, 143–176.

Bhat K. K. S. & Nye P. H. (1973) Diffusion of phosphate to plant roots in soil. *Plant and Soil* **38**, 161–175.

Birch H. F. (1958) The effect of soil drying on humus decomposition and nitrogen availability. *Plant and soil* **10**, 9–31.

Bormann F. H. & Likens G. E. (1970) Nutrient cycles of an ecosystem. *Scientific American* **223** (4), 92–101.

Chaney R. L. & Oliver D. P. (1996) Sources, potential adverse effects and remediation of agricultural soil contaminants, in *Contaminants and the Soil Environment in the Australasia-Pacific Region* (Eds R Naidu, R. S. Kookuna, D. P. Oliver, S. Rogers, M. J. & M. J. McLaughlin). Kluwer, Dordrecht, pp. 323–359.

Cole D. W., Gessel S. P. & Dice S. F. (1967) Distribution and cycling of nitrogen, phosphorus, potassium and calcium in a second-growth Douglas fir ecosytem, in *Primary Productivity and Mineral Cycling in Natural Ecosystems* (Ed. H. E. Young). Ecological Society of America, New York.

Garrett M. K., Watson C. J., Jordan C., Steen R. W. J. and Smith R. V. (1992) The nitrogen economy of grazed grassland. *Proceedings of the Fertilizer Society*, No. 326, 3–32.

Halm B. J., Stewart J. W. B. & Halstead R. L. (1972) The phosphorus cycle in a native grassland ecosystem, in *Isotopes and Radiation in Soil–Plant Relationships including Forestry.* International Atomic Energy Agency, Vienna.

Heng L. K., White R. E., Bolan N. S. & Scotter D. R. (1991). Leaching losses of major nutrients from a mole-drained soil under pasture. *New Zealand Journal of Agricultural Research* **34**, 325–334.

Lindsay W. L. (1979) *Chemical Equilibria in Soils.* Wiley, New York.

McGrath S. P., Zhao F. J. & Withers P. J. A. (1996) Development of sulphur deficiency in crops and its treatment. *Proceedings of the Fertiliser Society*, No. 379, pp. 1–47.

Martell A. E. (1957) The chemistry of metal chelates in plant nutrition. *Soil Science* **84**, 13–41.

Mitchell R. L. (1970) Trace elements in soils and factors that affect their availability. *Geological Society of America, Special Paper* **140**, 9–16.

Nicholson T. H. (1967) Vesicular–arbuscular mycorrhiza – a universal plant symbiosis. *Science Progress, Oxford* **55**, 561–581.

Nutman P. S. (1965) Symbiotic nitrogen fixation, in *Soil Nitrogen* (Eds W. V. Bartholomew & F. E. Clark). Agronomy No. 10. American Society of Agronomy, Madison.

Nye P. H. & Greenland D. J. (1960) *The Soil under Shifting Cultivation.* Commonwealth Bureau of Soils, Technical Communication No. 51. Harpenden, England.

Nye P. H. & Tinker P. B. (1977) *Solute Movement in the Soil–Root System.* Blackwell Scientific Publications, Oxford.

Sposito G. & Coves J. (1995). SOILCHEM on the Macintosh, in *Chemical Equilibrium and Reaction Models* (Eds R. H. Loeppert, A. P. Schwab & S. Goldberg). Soil Science Society of America, Madison, pp. 271–287.

Sposito G. & Page A. L. (1984) Cycling of metal ions in the soil environment, in *Metal Ions in Biological Systems*, Vol. 18 (Ed. H. Siegel). Marcel Dekker, New York.

Tiller K. G. (1996) Soil contamination issues: past, present and future, a personal perspective, in *Contaminants and the Soil Environment in the Australasia–Pacific Region* (Eds R Naidu, R. S. Kookuna, D. P. Oliver, S. Rogers, M. J. & M. J. McLaughlin). Kluwer, Dordrecht, pp. 1–27.

White R. E. (1980) Retention and release of phosphate by soil and soil constituents, in *Soils and Agriculture* (Ed. P. B. Tinker). Critical Reports on Applied Chemistry, Vol. 2. Blackwell Scientific Publications, Oxford, pp. 71–114.

White R. E. & Sharpley A. N. (1996). The fate of non-metal contaminants in the soil environment, in *Contaminants and the Soil Environment in the Australasia–Pacific Region* (Eds R Naidu, R. S. Kookuna, D. P. Oliver, S. Rogers, M. J. & M. J. McLaughlin). Kluwer, Dordrecht, pp. 29–67.

Williams R. J. B. (1975) *The chemical composition of water from land drainage at Saxmundham and Woburn (1970–1975).* Rothamsted Experimental Station Report for 1975, Part 2, pp. 37–62.

Further reading

Alloway B. J. (Ed.) (1990) *Heavy Metals in Soils.* Blackie, Glasgow.

Harter R. D. & Naidu R. (1995) Role of metal–organic complexation in metal sorption by soils. *Advances in Agronomy* **55**, 219–263.

Haynes R. J. & Williams P. H. (1993) Nutrient cycling and soil fertility in the grazed pasture system. *Advances in Agronomy* **49**, 119–199.

Reid C. P. P. (1990) Mycorrhizas, in *The Rhizosphere* (Ed. J. M. Lynch). Wiley, Chichester, pp. 281–315.

Woomer P. L. & Swift M. J. (1994) (Eds) *The Biological Management of Tropical Soil Fertility.* Wiley, Chichester.

Part 3
Soil Management

'And earth is so surely the food of all plants that with
the proper share of the elements which each species
requires, I do not find but that any common earth will
nourish any plant.'
Jethro Tull (1733) in *Horse-hoeing Husbandry*,
republished by William Cobbett in 1829

Chapter 11
Maintenance of Soil Productivity

11.1 Traditional methods

Shifting cultivation

Shifting cultivation is widespread in the tropics, being practised on some 500 million hectares of rainforest and open forest (savanna), which is about one-quarter of the potential arable land of these regions. The cycle, which is also called 'slash-and-burn' agriculture, begins with the clearing and burning of the natural vegetation, which in the case of the rainforest releases a great store of available nutrients for the growth of crops such as maize, cowpeas, cassava and groundnuts. However, loss of nutrients from the ash by erosion and leaching can be high (Table 11.1), and the encroachment of weeds which compete with the crops is rapid. After two or three crops the land is abandoned to the regenerating forest and the farmer moves to a new site, where the cycle is repeated.

Traditionally, very little chemical fertilizer, if any, is used in shifting cultivation. Regrowth of the native vegetation, especially the forest, is vital to the stability of such an agricultural system. During this period, which has been called the *bush fallow* (Nye and Green-

land, 1960), the soil's fertility is restored by the deep-rooting trees or perennial grasses. These draw up nutrients from deep in the subsoil and weathering parent material and return large quantities of litter (up to $10 \, t \, C \, ha^{-1} \, y^{-1}$) to the soil surface. There is also an input of symbiotically fixed N in soils of the humid forest regions. Experience in the savanna regions suggests that a cycle of 2–4 years' cropping and 6–12 years' fallow can maintain fertility in the long term, but in the wet forest zones a 1–2-year cropping period and 10–20 years' fallow are preferred.

Rotational cropping

In the more closely settled temperate regions, competition for farming land has forced the adoption of intensive cropping systems much sooner than in the tropics. One of the earliest was the simple 3-course rotation of autumn cereal–spring cereal–fallow, practised in England for 2000 years. This wholly arable system was largely displaced in the 18th century by rotations, such as the Norfolk four-course, in which grass–clover pastures and root crops were alternated with cereals, thus permitting close integration of livestock and arable farming. Added advantages were the nitrogen fixed by the legume and the abundant residues left in the soil by the pasture.

However, depressed cereal prices in the 1880s and a growing belief in the value of grass as a 'soil conditioner' led to the temporary pasture or *ley* being extended for three or more years. Gradually the arable–ley rotation evolved into the practice of *mixed farming*. The essence of this system is the rearing of livestock on farm-produced grass and cereals, in the course of which

Table 11.1 Rates of soil erosion under shifting cultivation and undisturbed rainforest. After Brady, 1996.

Land use	Erosion (t ha⁻¹ y⁻¹)		
	Minimum	Median	Maximum
Undisturbed rainforest	0.03	0.30	6.2
Shifting cultivation			
Bush fallow	0.05	0.15	7.4
Cropping period	0.4	2.8	70

much of the nutrient gathered from the soil is returned in organic manures.

Arable–ley rotations became the mainstay of agriculture in the winter rainfall zone of southern Australia, following the widespread use of superphosphate and introduction of the annual pasture legume *Trifolium subterraneum* (subclover) earlier this century. Phosphorus was applied, and N was fixed by the legume during the pasture phase, and these nutrients cycled through the grazing animals. Soil organic matter slowly increased and soil structure improved. After two or more years, the pasture was ploughed in and cereals were sown, nourished mainly by the reserves built up during the pasture phase. However, on some soils in the zone of > 500 mm annual rainfall, after 40–50 years of such agriculture, problems of accelerated soil acidification (Section 11.3) and dryland salinization (Section 13.2) have emerged.

Organic manures

Farmyard manure (FYM) and deep litter (from stall-fed animals), slurry, poultry (broiler) manure, sewage sludge and compost are examples of *organic manures* (Table 11.2). Their beneficial effects lie as much in the improvement of structure from large additions of organic matter, as in the contribution made to the soil's nutrient supply. Organic manures also encourage flourishing populations of small animals, especially earthworms, as well as micro-organisms that may produce growth stimulants (indoleacetic acid, cytokinins) in the rhizosphere. The following categories of organic manure are important.

• *FYM* consists of cattle dung and urine mixed with straw. It is very variable in composition and although it is low in macronutrients, at normal application rates of 40–50 t ha^{-1}, the quantities of micronutrients supplied are roughly equivalent to those removed in a succession of four to five crops (Table 11.3).

• *Slurry* is a mixture of dung and urine and washing water from animal houses and milking parlours. Compositions of some representative slurries, liquid manures and poultry manure are given in Table 11.2.

Table 11.2 Range and median (in parentheses) element contents of organic manures (% fresh weight). After MAFF, 1976 and ADAS, 1982.

	Dry matter	N	P	K
FYM*	11–92 (23)	0.2–3.5 (0.6)	0.04–1.3 (0.13)	0.08–3.6 (0.6)
Poultry manure	5–96 (29)	0.1–6.8 (1.7)	0.04–3.4 (0.6)	0.03–3.6 (0.6)
Liquid cattle manure and slurry	1–60 (4)	0.005–4.8 (0.3)	0.0002–1.1 (0.04)	0.0008–3.5 (0.25)
Pig slurry	1–69 (4)	0.01–4.8 (0.4)	0.004–2.1 (0.09)	0.017–2.7 (0.17)
Compost	24–95 (28)	0.6–1.1 (0.9)	0.18–0.35 (0.22)	0.17–0.75 (0.33)
Sewage sludge				
Liquid digested	4	0.2	0.1	0.01
Dewatered digested	50	1.5	0.5	—

*FYM, farmyard manure.

Table 11.3 Micronutrients (kg ha^{-1}) in farmyard manure (FYM) compared with those removed in crops. After Cooke, 1982.

	Mn	Zn	Cu	Mo
FYM (45 t ha^{-1})	3.36	1.12	0.56	0.01
Total in four arable crops in succession	2.50	1.80	0.30	0.01

• *Compost* is made by accelerating the rate of humification of plant and animal residues in well aerated, moist heaps. Ideally, a compost of low C:N ratio (*c*. 10–12) and acceptable nutrient content can be made in 6–8 weeks (Table 11.2).

• *Sewage sludge.* A number of types of sludge may be produced from raw sewage. Liquid *raw sludge* from the primary sedimentation tanks is sometimes offered to farmers, but, more commonly, sludge produced after secondary treatment of sewage is used. Anaerobic fermentation reduces pathogen numbers and smell, and produces a black *liquid digested sludge* containing 2–4% DM (Table 11.2). *Dewatered sludges* (raw or digested) contain 40–50% DM and are often called 'sludge cake'. When a flocculant, such as lime or $FeSO_4$ has been used before dewatering, the sludge has better 'condition' and is of value in improving the structure of reconstituted soil on mine spoil or derelict land. However, much of the soluble N and P is lost during the dewatering treatment such that the increase in total N and P content through dewatering is not in proportion to the increase in dry matter percentage (Table 11.2). Constraints on the use of sewage sludge are discussed in Box 11.1.

Green manure

The practice of ploughing in a quick-growing leafy crop before maturity is called *green manuring*. A major benefit is the avoidance of NO_3^- leaching in susceptible soils, because the nitrate is taken up by the green manure crop and slowly released to a subsequent cash crop as the low C : N ratio residues decompose. If the green manure crop is a legume, there can be an additional boost to the soil N supply from symbiotically fixed nitrogen.

Box 11.1 The use of sewage sludge or biosolids.

To emphasize the potential benefits of the nutrients, sewage sludge is often now referred to as '*biosolids*'. Sludge can be applied at rates of 50–80 t DM ha^{-1}, provided that the acceptable levels of phytotoxic elements (Cu, Zn, Ni and B) are not exceeded. In the UK, Cu and Ni are considered to be twice and eight times as toxic as Zn, and the total *Zn equivalent* of these metals added to agricultural land should not exceed 560 kg ha^{-1} over an extended period (30 years). If one-fifth of the total permitted load is applied in a single dressing, no more should be added for another 5 years. Soil approaching the limit should be kept at a minimum pH of 6.5 for arable crops, or 6.0 for grassland. Possibly twice as much could be added to calcareous soils because of the insolubility of the metal hydroxides at high pH (Section 10.5).

There are other elements in sludge which might affect human or animal health, namely Cd, Pb, Hg, As, Se, Cr, F and Mo, the chemistry of which in soil is discussed in Section 10.5. In the UK, the maximum soil concentration of these elements (to 7.5 and 15 cm depth in grassland and arable soils, respectively), together with the upper limit for application, are given in Table 11.4. Of these elements, Cr and Pb usually occur in the highest concentrations in sludge (80–1600 mg kg^{-1} DM). Even though Cd is low in sludges (generally < 50 mg kg^{-1}), there is particular concern in Australia because the maximum permissible concentration in food (other than meat and fish products) is set at 0.05 mg kg^{-1} fresh weight. One of the reasons for this caution is that superphosphate used in Australia has traditionally been relatively high in Cd, having been made from Pacific Island phosphate rocks which are high in Cd.

As noted in Box 10.8, the maximum permissible concentrations are in terms of the total element in the soil, which is conservative because the bioavailable content will be much lower.

Table 11.4 Soil concentrations and application limits to safeguard human and animal health. After ADAS, 1982.

Element	Maximum soil concentration (mg dm^{-3})	Application limit (kg ha^{-1})
Cadmium	3.5	5
Lead	550	1000
Mercury	1	2
Molybdenum	4	4
Chromium	600	1000
Selenium	3	5
Arsenic	10	10
Fluorine	500	600

The long-term effect on soil organic matter is minimal, since the succulent resides are rapidly decomposed and contribute little to the soil humus. Similarly, effects on structure through the stimulation of soil microbial activity are ephemeral.

Recent developments

Since the 1950s, the demand for food created by a rapidly growing world population and a need to grow cash crops for export earnings have induced a swing away from traditional methods of agriculture in many parts of the world, especially in the tropics. For example, although shifting cultivation is conservative of natural resources when properly managed, it supports relatively few people (less than one-tenth of the world's population). Also, as population pressure on natural resources has increased, migrants have moved into the shifting-cultivation areas who do not have the knowledge or incentive to follow the traditional practices. Thus, the rate of forest clearance has increased, the bush–fallow period has shortened and the total stock of nutrients in the ecosystem has declined (Fig. 11.1). The consequences of these changes are serious not only for food production, but also for contributing to the enhanced 'greenhouse effect' (Box 3.1). More permanent systems of agriculture are being developed for these tropical areas which will require substantial inputs of nutrients from external sources to sustain high levels of production.

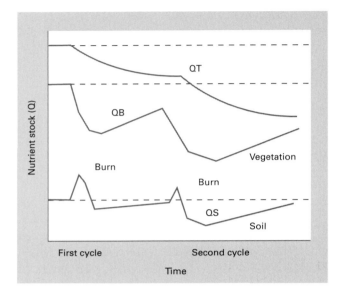

Fig. 11.1 Changes in the nutrient stock in slash-and-burn agriculture as a function of time. QT = total nutrient stock in the ecosystem; QB = nutrient stock in plant biomass, and QS = nutrient stock in the soil; QT = QB and QS (after Juo and Mann, 1996).

With the intensification of agriculture in temperate regions, especially after the Second World War, livestock and arable farming have become increasingly divorced, so that animal manures are not plentiful in areas where they are most needed. Inevitably, there has been a change-over to cereal monoculture – 'the one-course rotation' – and increasing reliance on chemical fertilizers in place of traditional manuring and fallowing. The need to make high economic returns per hectare has meant that green manure crops could not be grown on a large scale; nor is this approach feasible for profitable agriculture in countries, such as Australia, where water resources are scarce.

Nevertheless, agriculture has in general been very successful over the last three decades in producing enough food to satisfy a growing world population, now > 6 billion (Fig. 11.2). Production has increased at about 1% per annum in developed countries and 3% per annum in developing countries, although there are problem areas such as Africa south of the Sahara and

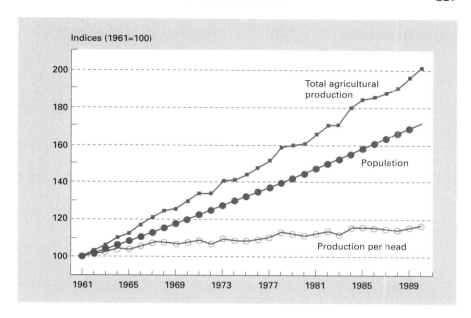

Fig. 11.2 World agricultural production, population and production per head (after FAO, 1993).

South Asia. In many of the developing countries both higher yields per hectare and more land under cultivation have contributed to the rise in production; but in the developed world and much of South and East Asia, there have been dramatic increases in yield per hectare through the introduction of better crop varieties, which require high inputs of fertilizer and water (Box 11.2). However, this achievement has not been without cost in terms of *accelerated soil acidification* (Section 11.3), *deterioration of soil structure* (Section 11.4), *accelerated erosion* (Section 11.5), *eutrophication* of water bodies (Section 12.3) and *soil salinization* (Section 13.2). Before discussing such problems, it is appropriate to review the means of assessing soil fertility.

11.2 Productivity and soil fertility

Soil, climate, pests, disease, genetic potential of the crop and man's management are the main factors governing *productivity*, as measured by the yield of crop or animal product per hectare. This book is primarily concerned with soil properties – how they interact with soil management and their effect on productivity. The *fertil-*

ity of the soil, and its manipulation by the proper use of fertilizers (Chapter 12), is probably the most important single soil factor affecting productivity. Physical properties, notably structure, are normally secondary although these become of prime importance when soils are susceptible to degradation, or once the fertility problems have been solved.

Soil nutrient supplying power

The assessment of soil fertility has two aspects:
• *qualitative* – the aim is to identify which nutrients are deficient (or in excess), and
• *quantitative* – the aim is to estimate how much of a particular limiting nutrient is required to achieve optimum growth under the prevailing environmental conditions.

Qualitative aspects – diagnosis of nutrient deficiency

Various signs of disorder (chlorosis, stunted growth or necrosis of leaf margins) indicate the deficiency of one or more essential elements in a plant. However, visual

Beginning in the 1950s, the '*Green Revolution*' was the international community's response to the need to increase per capita food production, particularly in developing countries, where in 1960 some 80% of the population of 1.7 billion people had a daily food intake < 2100 calories (FAO, 1993). The program focused on three related activities:
• the breeding of new high-yielding, early maturing varieties of the staple cereals wheat and rice (HYVs),
• promotion and distribution of packages of high inputs including fertilizers, pesticides and the regulation of water supplies, and
• implementation of these technical innovations in the most favourable agro-climatic regions and suitable rural communities.

The extraordinarily successful outcome of the program has been that food production per head is 20% higher in 1990 than 1960 (Fig. 11.2), and in developing countries only 8.5% of the population of *c.* 4 billion now has a daily food intake < 2100 calories. Notwithstanding this success, there is evidence that yields from the HYVs have in some instances reached a plateau, the causes of which have been variously ascribed to increased incidences of pests and disease, and soil degradation, associated with intensive monoculture. With the high inputs of fertilizers and pesticides, there have inevitably been losses to the environment so that the quality and quantity of water in many densely settled and farmed regions have deteriorated, especially in the Ganges Plain of India and the eastern seaboard of China. Addressing these problems to develop and implement more *sustainable* agricultural systems is the challenge for the current generation of soil scientists.

symptoms appear only after the plant has suffered a check in growth due to 'hidden hunger' for the deficient element. Growth and the concentration of the element in the plant's tissues are closely linked, as shown in Fig. 11.3, so that *plant analysis* or *tissue testing* is useful in diagnosing deficiency (or toxicity).

Note in Fig. 11.3 that the *critical value* is the element concentration in the plant below which an improvement in the supply of the element leads to increased growth. To be of use in deficiency diagnosis, such values must be determined for individual crops, preferably when no other nutrient limits growth. Under these conditions the critical value should be independent of soil type; but it should be noted that moisture supply to the plant has a significant influence on critical values.

Quantitative aspects – soil testing

The quantitative approach involves analysing a soil to measure the amount of nutrient that is 'available' for plant growth. This is called *soil testing*. Fundamentally, the concept of nutrient 'availability' is more complex than first appears because:
• plants take up nutrient ions from the soil solution in the root zone for which the total amount of nutrient present is usually insufficient to meet the plant demand,
• the concentration of a nutrient in the soil is *buffered* by the whole labile pool which includes an easily desorbable surface phase (e.g. for Ca, Mg, K, P and S) and also may include a readily mineralizable organic fraction (e.g. for N, S and P), and

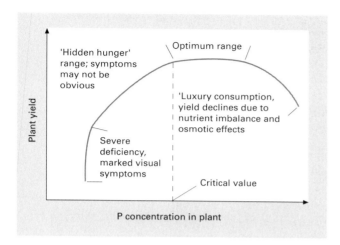

Fig. 11.3 The relationship between yield and plant tissue concentration of phosphorus.

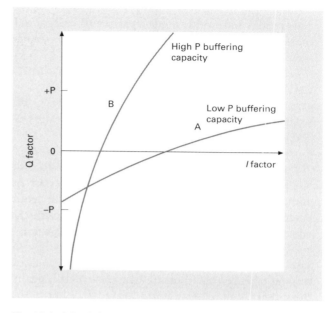

Fig. 11.4 Q/I relations for two soils showing contrasting phosphate buffering capacities.

• the *rate of replenishment* of a nutrient in solution at a root surface depends not only on the soil's buffering capacity for that nutrient, but also on the rate of movement to root surface by mass flow or diffusion (Box 10.5).

This complexity may be resolved by measuring the *quantity/intensity* (*Q/I*) relationship for a nutrient (Box 10.8). The concentration of an ion is independent of the size of the system studied and is therefore a measure of the *intensity factor* (*I*). The amount of an ion extracted by an anion or cation exchange resin, or a neutral salt solution, corresponds to the easily desorbable labile component and hence measures the *quantity factor* (*Q*). The change in *I* with change in *Q* depends on the slope of the *Q/I* relation, which defines the buffering capacity of the soil for a particular nutrient. Further, the effective diffusion coefficient for an ion such as $H_2PO_4^-$ depends largely on the reciprocal of the buffering capacity (Box 10.5). Figure 11.4 illustrates the contrasting *phosphate buffering capacities* (PBC) of two soils. Clearly, the P concentration in solution (*I*) in soil A, which has a low PBC, will decrease rapidly as P is absorbed by plants; but much less P fertilizer will need to be applied to this soil to raise *I* to a satisfactory value than for soil B, which has a high PBC.

For N, not only the total amount of NH_4^+ and NO_3^-, but also the amount of readily mineralizable organic N may need to be measured to give Q. In some cases, especially for N and P but also for other nutrients, a stable isotope (^{15}N) or radioactive isotope (^{32}P), has been used to measure Q. For example, the amount of isotopically exchangeable P (called the *E* value or sometimes the *L* value, Fig. 12.8) is calculated from the quantity of tracer added (^{32}P soil) and the specific activity of the soil solution after isotopic exchange between the solution and soil surfaces has taken place, using the equation

$$E \text{ value} = {}^{32}P\text{soil} \times \frac{{}^{31}P \text{ soln}}{{}^{32}P \text{ soln}}. \tag{11.1}$$

The implications of these concepts for practical soil testing are discussed in Box 11.3.

Quantitative assessment of fertilizer needs

The amount of fertilizer needed to correct a specificnu-

Box 11.3 Soil testing in practice.

Throughout the world, a great variety of empirical soil tests are used to measure 'available' nutrients, even for a single element such as P. Broadly, these tests employ:

- neutral salt solutions, e.g. 0.01M $CaCl_2$ (for P and K), 0.01M $Ca(H_2PO_4)_2$ (for S), 2M KCl (for mineral N), or
- anion or cation exchange resins (e.g. for P, S, K, Mg and Ca), or
- acids or alkalis of varying strength (e.g. 0.005M H_2SO_4 for P or 0.5M $NaHCO_3$ for P), or
- complexing agents (e.g. 0.005M DTPA for Cu, Zn, Mn and Fe).

The milder extractants (neutral salts) tend to reflect the I factor of a soil's nutrient supply, whereas the more severe extractants (acid and complexing agents) reflect the Q factor. Specific details of the methods are given in books such as Rayment and Higginson (1992). Irrespective of whether the Q or I factor (or both) is measured, a soil test must be *calibrated* against crop yield, usually on a variety of soils with different rates of the element supplied in fertilizer. The resultant plot of yield response (percentage increase over the control) against the test values for the suite of soils allows a *critical value* to be established (Fig. B11.3.1).

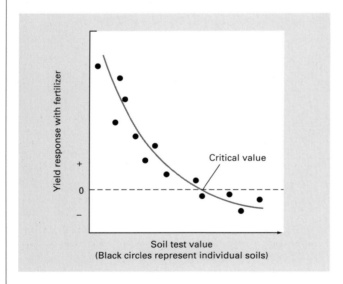

Fig. B11.3.1 Yield response to fertilizer in relation to soil test value (points represent individual soils tested).

Continued

Box 11.3 *Continued*

In many countries a solution of 0.5M $NaHCO_3$ buffered at pH 8.5 (the 'Olsen test') is the standard extractant for soil P. In the UK, no soil test is considered satisfactory for N, but in other European countries and many states of the USA, measurements of 'residual' soil mineral N (exchangeable NH_4^+ and solution NO_3^-) made at the end of cropping or in late winter have been found useful for predicting fertilizer N requirements for subsequent crops. The depth of sampling can range from 0.6 to 1.8 m.

Besides the choice of soil test for an element, the *representivity* of the sample analysed needs to be considered. Soil is naturally very variable (Chapter 5). The spatial distribution of 'available' nutrients is variable, not only because of the natural variability in soil properties but also because of localized plant removals, biological transformations, fertilizer inputs, and the heterogeneous return of dung and urine from animals. Variations in temperature and moisture with time also affect nutrient availability. Hence, to improve the accuracy of a soil test, a sufficient number of samples must be taken to cover the range of variability in the area to be tested. Analysing a large number of such samples (15 or more) for each test is usually not cost-effective, so the samples must be bulked, mixed and then subsampled to provide a representative 'composite' sample for analysis.

trient deficiency can be estimated in the course of calibrating a soil test by means of a *fertilizer rate trial*, in which the yields at different rates of fertilizer (amounts per hectare) are measured.

The response to fertilizer frequently follows a *law of diminishing returns*, which means that the yield increase for successive equal increments of fertilizer becomes steadily less. If y is the yield and x the amount of fertilizer element (e.g. N) in kilograms per hectare, this law can be expressed as

$$y = A - B \exp(-Cx). \qquad (11.2)$$

As shown in Fig. 11.5a, A is the *maximum* (asymptotic) yield attainable, $(A-B)$ is the yield without fertilizer and C is the rate of change of y with x.

Particularly with N, a decline in yield at high fertilizer rates is sometimes observed which has led to the fitting of quadratic or linear intersecting models to such fertilizer–response curves. However, whichever model is chosen, the precision in predicting the amount of

fertilizer needed to attain maximum yield is poor because of factors other than N, such as climate, affecting the yield. More important is the prediction of fertilizer required to produce an *optimum yield*, which may be defined as the yield at which profit per hectare is greatest (Fig. 11.5b). For example, when the value of the extra yield due to fertilizer (curve AB) is compared with the fertilizer cost line (AC), the vertical XY indicates the fertilizer rate giving the biggest value–cost differential per hectare. If the crop's value, or the fertilizer's cost changes, XY slides to the right or left to a new point of maximum profit. Other approaches to estimating fertilizer requirements are outlined in Box 11.4.

11.3 Soil acidity and liming

Sources of acidity

A soil becomes acid if Ca^{2+}, Mg^{2+}, K^+ and Na^+ ions are leached from the profile faster than they are released

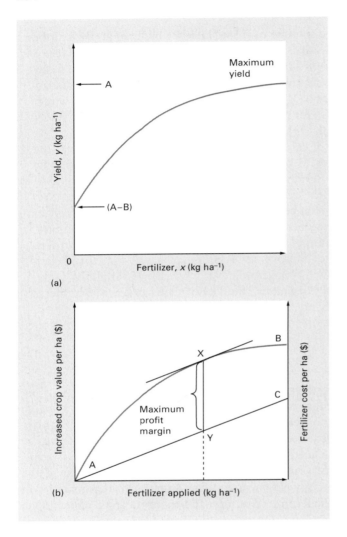

(a)

(b)

Fig. 11.5 (a) A typical fertilizer response curve obeying Equation 11.2 and (b) maximum profit per hectare from fertilizer investment.

by mineral weathering, and H^+ and Al^{3+} ions become predominant on the exchange surfaces (Section 7.2 and Box 9.2). The intensity of the acidity, measured by the soil pH (Box 7.4), is the result of an interaction between soil mineral type, climate and vegetation. The main sources of net acid inputs are as follows:
• The dissolution of CO_2 in the soil water to form

carbonic acid which dissociates according to

$$CO_2 + H_2O \leftrightarrow H_2CO_3 \leftrightarrow H^+ + HCO_3^- \leftrightarrow H^+ + CO_3^{2-}. \tag{11.3}$$

Pure rain water in equilibrium with CO_2 in the atmosphere has a pH of 5.65. However, soil respiration increases the partial pressure of CO_2 in the soil air, so driving reaction 11.3 to the right and producing more H^+.
• The accumulation and humification of soil organic matter, producing humic residues with a high density of carboxyl and phenolic groups that dissociate H^+ ions. The balancing alkalinity is not released until the organic matter decomposes. Removal of C in plant and animal products has a similar acidifying effect.
• Nitrification of NH_4^+ ions, producing H^+ ions, and NO_3^- which is susceptible to leaching (Equation 10.2). Removal of N in plant and animal products also has an acidifying effect because there is no opportunity for this organic N to be ammonified and release OH^- ions to the soil (Equation 10.1).
• Inputs of H_2SO_4, HNO_3 and $(NH_4)_2SO_4$ from the atmosphere or 'acid rain' (Sections 10.2 and 10.3). The acidifying effect of the H_2SO_4 is not necessarily eliminated by reaction with NH_3 to form $(NH_4)_2SO_4$, because H^+ ions are released again when NH_4^+ is oxidized in the soil. Even when the NH_4^+ ions are taken up by plant roots there can be an acidifying effect on the *rhizosphere* because H^+ ions are released to balance any net surplus of cations over anions absorbed, when N is supplied mainly as NH_4^+ and not NO_3^-. And with atmospheric accessions of NH_4^+, volatilization of NH_3 becomes an acidifying process because there are no neutralizing OH^- ions present (*cf.* Equation 10.8).
• In soils formed on marine muds, or coal-bearing sedimentary rocks, the oxidation of iron pyrites FeS_2 (formed under intense reducing conditions in the original deposit) gives rise to *acid sulphate soils* (Box 9.4).

Plant response

Under natural conditions, plants have evolved and

Box 11.4 Alternative approaches to estimating fertilizer requirements.

The requirement for a particular nutrient in a pasture or cropping system can be estimated using a *nutrient balance* approach. This requires a knowledge or an estimation of:

1 external inputs of the nutrient to the soil (in residues and atmospheric inputs),
2 transformations in the soil leading to net gains or losses to the labile pool,
3 losses by leaching, erosion, in gas form and by product removal, and
4 the desired crop yield or number of grazing animals to be supported.

Point 4 quantifies the demand for the nutrient, which must be balanced against the sum of points 1–3. The shortfall determines how much of the nutrient must be supplied in fertilizer. This is the principle behind recommendations for P, K and S fertilizer for pasture in New Zealand (Cornforth and Sinclair, 1984), and also behind the SUNDIAL model (Simulation of Nitrogen Dynamics in Arable Land) for N fertilizer recommendations for crops in the UK (Smith *et al.*, 1996). In the New Zealand system, when the pasture is in the establishment phase and fertility needs to be built up, a soil test value is used to modify the fertilizer recommendation obtained from the model.

These approaches produce a 'blanket' fertilizer recommendation for an area (a field or perhaps a farm). Another approach is to map the variation in crop yield accurately at a scale of a few metres, as the crop is being harvested, and to use this information to adjust subsequent fertilizer application. If yield maps are constructed over a number of years (to take account of year-on-year variability in climate, pests and disease), those areas of a field which consistently yield higher or lower than the field average can be identified. The amount of fertilizer applied to the 'high' areas can be decreased and that applied to the 'low' areas increased to achieve an overall increase in yield, and minimize any off-site environmental impacts through excess fertilizer use. This is the basis of *site-specific* fertilizer management or *precision farming*.

adapted to the prevailing soil pH. The *calcifuges* are tolerant of acidity when compared with the *calcicoles*, which are intolerant of acidity or 'lime-loving'. Note that sensitivity to soil acidity is primarily a sensitivity to exchangeable Al^{3+} (Section 7.2). Cultivated plants also exhibit considerable diversity in their response to acidity, as the selection of crop and pasture species in Table 11.5 shows. Within the legume family, the tropical species are more tolerant of acidity than most medics and clovers of temperate regions.

Accelerated soil acidification

Soil acidification is a natural process, but the rate can be accelerated by human activities, as indicated in 'Sources of acidity' above. For example, in large areas of southern and eastern Australia the productivity of agriculture has been raised by ley–arable farming and improved pastures based on subclover (Section 11.1). At the same time, soil organic matter contents have risen and the removal of C and N in crop and animal

Table 11.5 Sensitivity of cultivated plants to soil acidity (exchangeable Al^{3+}).

Pasture species	Crop species	Tolerance to aluminium
Cocksfoot	Rye Potato Oats Triticale	High tolerant
Ryegrass Subclover White clover	Wheat Maize	Tolerant
Phalaris	Canola (rape)	Sensitive
Lucerne Trefoil	Barley Sugar beet	Highly sensitive

products, and the leaching of NO_3^-, have increased. Thus, net acid inputs to the soils have increased by approximately 1–2 kmol H^+ ha^{-1} y^{-1}, compared to the unimproved systems.

The effect that increased acid inputs have on the soil pH is determined by the soil's *pH buffering capacity* (Section 7.2). While the additional input of 1–2 kmol H^+ ha^{-1} y^{-1} is relatively small compared to atmospheric acid inputs in the northern hemisphere, the podzolic soils of southern Australia are naturally acid, and have low pH buffering capacities, in the A horizon. Thus, the pH of soil A horizons has declined at the rate of 1 pH unit per 30–50 years and some 21 million hectares of agricultural soils in southern Australia are now classed as strongly acidic (pH (H_2O) < 5.6 or pH ($CaCl_2$) < 4.8) (Anon., 1995). Neutralization of this extra acidity would require an annual application of 50–100 kg $CaCO_3$ ha^{-1}. However, lime is expensive, particularly for pastoral enterprises, so alternative strategies such as the use of acid-tolerant species and the sowing of deep-rooted perennial grasses to minimize nitrate leaching are being employed. Such strategies can only slow down the rate of acidification and in the end, lime will be needed to neutralize the extra acidity (see below).

In southern England, the estimated atmospheric acid input is *c.* 5 kmol H^+ ha^{-1} y^{-1} which is well above the *critical load* (the level at which adverse environmental effects might occur) of *c.* 1 kmol H^+ ha^{-1} y^{-1} (Blake *et al.*, 1994).

Liming

The problems associated with soil acidity – slow turnover of organic matter, poor nodulation of some legumes, Ca and Mo deficiencies, Al and Mn toxicities – can be remedied by liming. Ground limestone, chalk, marl and basic slag are used as liming materials, the active constituent being primarily $CaCO_3$, with some burnt lime (CaO) and hydrated lime ($Ca(OH)_2$).

When lime is added to moist soil, it slowly dissolves by hydrolysis to produce an alkaline pH, according to the reaction

$$CaCO_3 + 2H_2O \leftrightarrow Ca(OH)_2 + H_2CO_3 . \qquad (11.4)$$

However, the strong base $Ca(OH)_2$ reacts with dissolved CO_2 from the soil air to form $Ca(HCO_3)_2$ so that the net reaction is

$$CaCO_3 + CO_2 + H_2O \leftrightarrow Ca(HCO_3)_2 . \qquad (11.5)$$

The final pH attained can be predicted from the expression

$$pH = K - \tfrac{1}{2}\log(P_{CO_2}) - \tfrac{1}{2}\log(Ca), \qquad (11.6)$$

where P_{CO_2} is the partial pressure of CO_2, (Ca) is the activity of Ca^{2+} ions in solution, and $K = 4.8$ if the carbonate has the solubility of pure calcite. For soil carbonates, which are more soluble than pure calcite, $K \simeq 5.2$. The predicted pH for a soil containing free $CaCO_3$, which is in equilibrium with various partial pressure of CO_2, is shown in Table 11.6.

Assessing the amount of $CaCO_3$ required to neutralize soil acidity or raise the soil pH to a desired value of crop growth involves measuring the soil's *lime requirement* (Box 11.5).

Table 11.6 Predicted soil pHs for calcareous soils.

Ca activity (M)	CO_2 partial pressure (bar)		
	0.00036	0.0036	0.01
0.001	8.42	7.92	7.70

Box 11.5 Lime requirement.

The *lime requirement* depends on the pH buffering capacity of the soil (Section 7.2), and is expressed as the quantity of ground limestone or chalk (t ha^{-1} to a depth of 15 cm) required to raise the soil pH to a desired value. Methods of measuring the lime requirement range from
- equilibrating a soil sample with a single buffer solution, to
- titration with $Ca(OH)_2$, to
- calculation from a model based on estimated rates of Ca loss in the field.
- Details of laboratory methods are given in Rayment and Higginson (1992). Laboratory analysis is also necessary to measure the *neutralizing value (NV)* of different liming materials. The standard in the UK is pure CaO. The *NV* of liming materials ranges from 90 for burnt lime to *c.* 50 for ground limestone to *c.* 40 for basic slag.

A pH (H_2O) of 6.5 is recommended for temperate crops, but liming of pastures to pH (H_2O) 6 is preferred, because grasses are more tolerant of acidity and the possibility of inducing deficiencies of Mn, Cu and Zn is minimized. Tropical species are more acid tolerant and, furthermore, liming to pH 6–6.5 may reduce P availability in soils high in exchangeable Al^{3+} (Section 7.3) and destabilize the structure of kaolinitic clay soils (Section 7.4). In such soils, liming to hydrolyse the exchangeable Al^{3+}, normally achieved at pH 5.5, is preferable (Section 7.2).

Finely ground $CaCO_3$ made to adhere to legume seeds – a process called *lime pelleting* – greatly improves nodulation of sensitive species in acid soils, especially when an effective strain of *Rhizobium* bacteria is included in the coating.

11.4 The importance of soil structure

Structure dependent properties

Good soil management should aim to create optimum physical conditions for plant growth, as shown by:
1 adequate aeration for roots and micro-organisms,
2 adequate available water,
3 ease of root penetration, permitting thorough exploitation of the soil for water and nutrients,
4 rapid and uniform seed germination, and
5 resistance of the soil to slaking, surface sealing and accelerated erosion by wind and water.

Two of the most useful indices of structure are *bulk density*, which is inversely related to total porosity (Section 4.5), and *aggregate stability*, which is assessed by the coherence of soil aggregates in water (Box 4.4). Bulk density measurements are more relevant to points 1, 2 and 3 above, whereas aggregate stability is more relevant to points 4 and 5. Organic matter, texture and land use modify these properties. For example, it has been found that bulk density increases as organic matter content decreases, and is highest in the sandy textural classes. Aggregate stability, on the other hand, decreases with declining organic matter and is least for soils in the silty and fine sandy loam classes.

Aeration and water supply depend on the soil's pore size distribution, the key indices being the *air capacity* C_a (the air-filled porosity at the soil's *field capacity* (*FC*) – Section 4.5) and *available water capacity* (*AWC*) (Section 6.4). Using these two properties, classes of soil droughtiness at one extreme and susceptibility to water-

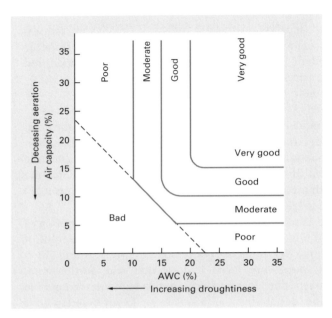

Fig. 11.6 Classification of the structural quality of topsoils (after Hall *et al.*, 1977).

logging at the other may be separated, as illustrated for surface soils in England and Wales which regularly experience a soil moisture deficit (SMD) > 100 mm (Fig. 11.6). Soils with < 10% *AWC* are droughty, and those with C_a < 5% are likely to be waterlogged and anaerobic. The limiting *AWC* value is lowered slightly and the limiting C_a value raised to 10% for soils of wetter regions. Irrespective of the individual values of C_a and *AWC*, a storage pore space value (the sum of *AWC* and C_a) < 23% is undesirable; hence the truncation of the curve separating the 'poor' and 'moderate' classes in Fig. 11.6.

The effect of land use

Grassland vs arable

Soil physical conditions are usually best under permanent grassland (or forest) and deteriorate at a rate dependent on climate, soil texture and management as the soil is cultivated.

When land is ploughed, disruption of peds exposes previously inaccessible organic matter to attack by micro-organisms, and populations of structure-stabilizing fungi and earthworms decrease markedly. Removal of the protection of a permanent canopy of vegetation exposes the soil to the direct impact of rain, and the loss of surface litter reduces the detention time of water before surface runoff and consequent erosion begins. Shearing forces generated by tractor wheels and ploughs can rupture macro- and microaggregates (Section 4.4). Structure is most vulnerable to damage when the soil is wet (> *FC*) because the distance between clay particles in the domains and quasi-crystals is greatest then. The external force may cause deflocculation and reorientation of the deflocculated clay particles to form dense layers of low permeability to gases and water called *plough pans*. The detection of a plough pan or other compacted layers in a soil profile is discussed in Box 11.6.

Measurements of water-stable aggregates show that the deterioration in structure is most rapid in the first year after grassland is ploughed (Fig. 11.8). Total porosity and *AWC* in the plough layer also decline as cultivation is prolonged (Fig. 4.3b). These trends are reversed, however, by the introduction of a pasture phase (a ley) into the cropping cycle. The restorative effect of the ley depends on its duration – this is illustrated by the comparison of measurements made on a continuously cultivated (fallow–wheat) soil put down to pasture for various lengths of time (Fig. 11.9). Note that the improvement in aggregate stability is largely confined to the surface 10 cm or so where the grass roots are concentrated. Improved aggregation following short leys (< 4 years) is primarily due to the effect of polysaccharides produced by grass roots and associated rhizosphere micro-organisms (Fig. 4.12). The resulting aggregation is sometimes called 'soft' because it does not survive wet sieving. Longer leys, with an increase in the contribution from humified organic matter, are needed to produce the more stable 'hard' aggregation. For British soils, 4% organic matter ($\simeq$ 1.6% C) has been suggested as the critical lower limit for aggregate stability under cultivation.

Box 11.6 Detecting soil compaction.

Compaction results in reduced porosity and so can be detected from an increase in soil *bulk density* (Section 4.4). The bulk density might increase from the acceptable range of 1.0–1.3 Mg m^{-3} to > 1.5 Mg m^{-3}. Such measurements can be easily made on surface soil by driving in a steel cylinder of known volume, extracting the intact soil core, oven-drying and weighing. The diameter of the cylinder should be at least 10 times the thickness of the cylinder wall to avoid additional compaction of the soil core.

It is less convenient to measure bulk density at depth in soil because a hole must be excavated. An alternative approach is measure the soil's penetration resistance with a *penetrometer*. This consists of a steel rod about 1 m long, with a conical tip and a proving ring and strain gauge at the top (Fig. 11.7). As the conical tip is pushed into the soil, the ring is distorted and the force is recorded by the gauge. The rod is graduated so that the depth of the tip in the soil can be measured. The readings are most reliable when the rate of entry of the rod is constant, which can be achieved using a tractor-mounted hydraulic system. Penetration resistance can be very variable spatially, so a large number of readings are required. Penetrometer measurements also depend on soil moisture content (Box 4.2) and are best used in a comparative rather than absolute sense. Used in this way, such measurements can identify areas of compacted soil near gateways and headlands in arable fields, which are correlated with areas where crop yield is depressed (Box 11.3).

Tillage methods

The traditional method of cultivation or tillage in Western agriculture has involved thorough disturbance of the top 20–25 cm of soil, usually with inversion of the furrow slice by mouldboard ploughing (Fig. 11.10). Residues of the previous crop and volunteer weeds are buried, and after repeated passes with discs and harrows the soil is reduced to a fine tilth suitable for seeding. This method was exported to North America, Africa and Australasia, often with disastrous results in terms of soil structural deterioration and erosion.

In contrast, traditional methods of cultivation in the tropics were very different. Following the clearing of the natural vegetation (with or without a burn), the first crop was sown with the minimum of soil disturbance (Section 11.1). Plant debris left on the surface protected the soil from erosion until the crop was well established.

Intercropping with tall and short crops was common so that little bare ground was exposed to the erosive tropical rain.

In developed countries over the past few decades, however, various factors have led to a reappraisal of traditional ploughing methods, notably:

• experiments at Rothamsted in the 1930s had shown that a fine seedbed tilth did not produce higher yields than less elaborately prepared seedbeds, provided that weeds were controlled,

• manpower and fuel became very costly, and

• the adverse effects of soil erosion, not only on agricultural productivity but also in polluting the environment, gave increasing cause for concern.

New tillage methods were devised which were a blend of modern and ancient practices. They combine *minimum* disruption of the natural structure of the soil with the use of new herbicides to control weeds. These prac-

Fig. 11.7 An example of a hand-held penetrometer (after Davies *et al.*, 1993).

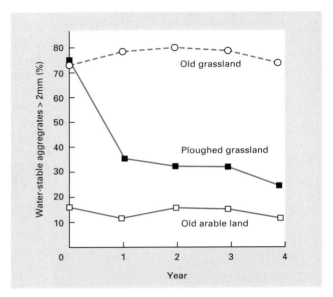

Fig. 11.8 Aggregation of the fine earth fraction of a soil under different land use (after Low, 1972).

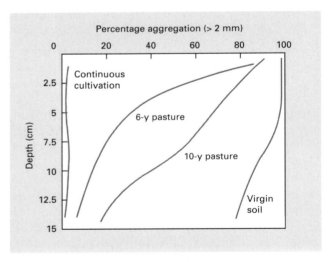

Fig. 11.9 The effect of leys of different duration on the water-stable aggregation of a soil under cultivation in southern Australia (after Graecen, 1958).

tices range from *zero tillage, no-till* or *direct drilling*, where the only mechanical operation is seed sowing, to *reduced* or *minimum tillage*, which involves fewer and shallower passes with tines or discs than does conventional ploughing. A more inclusive concept is that of *conservation tillage* which does not necessarily involve less tillage, but is broadly conceived of as minimizing soil and water loss relative to conventional cultivation. This is achieved primarily through the retention of crop residues (> 30% soil cover) to protect the soil surface and reduce evaporative losses (Sections 6.5 and 11.5).

Direct drilling or no-till

This practice has gained widespread popularity in the USA, particularly in the southern and mid-western corn

(maize) belt, and also in tropical Africa, South America and Australia. For row crops, such as maize, the seed is almost always sown directly into a grass sod that has

Fig. 11.10 Inverted furrow slices following mouldboard ploughing (courtesy of D. E. Patterson).

Fig. 11.11 Direct-drilling into wheat stubble.

• surface runoff and erosion are greatly reduced. This effect is now very well documented in many countries, as illustrated by the data in Table 11.7.

There is also an energy saving of 7–18% for no-till when compared with conventional ploughing, the higher figure occurring with direct-drilled legumes: slightly higher amounts of N fertilizer are required for many direct-drilled non-legumes, which offsets some of the energy saving on fuel.

It has been predicted that about 65% of the seven major annual crops in the USA will be under no-till by the year 2000, whereas in Britain only about 30% of the cereal land is considered suitable for this system. The main reasons for this difference lie in the *climate* and the *soil*, and it is appropriate to examine some of the

been killed with a herbicide, or into the residues of a previous crop. Soybeans, for example, are frequently sown into the stubble of a winter cereal (Fig. 11.11). In the countries cited, the preservation of plant residues on the soil surface is a major benefit of no-till agriculture because:

• the mulch formed shades the soil and reduces water loss by evaporation,

• the temperature of the surface is kept at tolerable levels, and

Table 11.7 Average soil losses under different soil management in Illinois, USA (annual precipitation *c.* 1300 mm). After Gard and McKibben, 1973.

Management system	Average annual soil loss (t ha^{-1})	
	5% slope	9% slope
Conventional tillage, maize and wheat	7.6	21.5
No-till, maize and wheat	0.9	1.3
No-till, continuous maize	0.6	0.9

changes in soil properties that occur when no-till re-places ploughing.

Changes in soil properties

Organic matter gradually increases in the top 5 cm and the water stability of the surface peds is enhanced under direct drilling. On many soils, bulk density is increased down to the old plough depth so that total porosity decreases, mainly at the expense of macropores. Never-theless, water infiltration and percolation do not neces-sarily suffer because earthworms multiply two- or threefold, especially the deep burrowing *Lumbricus ter-restris*, and the predominantly vertical orientation and continuity of their holes provides a pathway for rapid water flow. Furthermore, the higher bulk density im-proves the soil's 'trafficability'; that is, its load-bearing capacity when wet.

Changes also occur in the chemical properties of the soil. Phosphorus and K become concentrated in the 0–5 cm layer, which is no disadvantage provided the soil stays moist (e.g. under a mulch); but these elements may become less available to the plant if the soil dries out. Total N follows the trend in organic C, but soil nitrate levels are often lower. Nitrate leaching is not increased because the NO_3^- is mainly generated within soil aggregates where it is better protected from leach-ing. But there tends to be a 'trade-off' between leaching and denitrification losses in soils of humid regions (Fig. 10.7), so denitrification generally increases within the larger aggregates of direct-drilled soils. Lower ex-changeable Ca and pH values in the 0–5 cm depth of some direct–drilled soils may be a function of the in-crease in surface organic matter, so that additional lime is required. Other possible disadvantages are:
• slow warming of the wet soil in spring, which retards crop growth (particularly in Britain and northern USA and Canada),
• the production of volatile fatty acids around ferment-ing crop residues in the drill silt, which may inhibit seed germination (Section 8.3), and
• an increase in perennial weeds, especially grass weeds in cereals, which are difficult to control with herbicides alone (in some cases due to the development of herbi-cide resistance in the weeds).

An illustration of the beneficial surface mulch on a silt loam after several years of no-till maize in Ken-tucky, USA is shown in Fig. 11.12. Overall, the advan-tages of no-till agriculture can be summarized as:
• improved surface soil structure, with less surface crusting and better infiltration of water,
• control of erosion by wind and water,
• conservation of soil moisture,
• lower surface soil temperatures (warm and hot cli-mates only), and
• lower energy requirements.

Fig. 11.12 Wheat stubble residues under no-till maize in Kentucky.

Soil conditioners

Because plant and microbial polysaccharides are known to improve aggregation (Section 4.4), synthetic macro-molecules, called *soil conditioners*, have been used for the same purpose. The main types of soil conditioners are shown in Table 11.8. Application is at the rate of 0.2 g (soluble polymer) or 10 g (emulsifiable polymer) per kilogram of soil. Moist conditions are necessary to enable the soluble polymers to diffuse into the larger pores (> 30 μm diameter) where they are most effective. With the bitumen emulsions, the soil should be allowed to dry to achieve maximum aggregate stability. A problem with all soil conditioners is that only a shallow depth of soil can be easily treated. For this reason and reasons of cost, they tend to be used in special situations, for example:

• stabilizing sand dunes against wind erosion,
• stabilizing the bare surface of rehabilitated mine spoil, road cuttings and over landfills to protect the surface from erosion until vegetation is established, and
• protecting the fine surface tilth of seedbeds, especially that of high-value vegetable crops and sugar beet.

The beneficial effect of a soil conditioner on surface is demonstrated in Fig. 11.13.

11.5 Soil erosion

Influence of climate, human activities and scale

The term *erosion* describes the transport of soil constituents by natural forces, primarily water and wind. As indicated in Section 5.2, past phases of geologic erosion led to the formation of sedimentary deposits which are

Fig. 11.13 Stabilization of surface soil structure – the darker plots have been treated with PVA solution (courtesy of J. M. Oades).

the parent materials of many present-day soils; but in the short term, accelerated erosion due to human activities, especially agriculture, is of greater concern. Note that *erosion* and *deposition* are complementary processes because soil removed from the steeper slopes is deposited to a variable extent on footslopes and flood plains. Similarly, soil eroded from one area by wind is usually deposited in another part of the landscape (unless blown out to sea). Thus, soil erosion rates expressed per unit area are highest for individual fields, and decrease progressively when calculated on a whole farm to

Table 11.8 Summary of the main types of soil conditioners.

Soluble polymers (hydrophilic)	Emulsifiable polymers (hydrophobic)
Polyvinyl alcohol (PVA)	Bitumen
Polyacrylamide (PAM)	Polyvinylacetate (PVAc)
Polyethyleneglycol (PEG)	Polyurethane

small catchment to whole landscape scale. Erosion rates for agricultural land are mainly field-scale and tend to be high; for example, for cropland in the USA:

- 20% of the area loses > 20 t ha^{-1} y^{-1},
- 50% loses between 7.5 and 20 t ha^{-1} y^{-1}, and
- 30% loses < 7.5 t ha^{-1} y^{-1},

whereas the rate under natural vegetation (catchment scale) is 0.03–3 t ha^{-1} y^{-1} (Morgan, 1986). Similarly in Australia, long-term natural rates are < 1 t ha^{-1} y^{-1} (Anon., 1995) compared to agricultural land where losses range from:

- good pasture < 0.5 t ha^{-1} y^{-1},
- summer crops *c.* 10 t ha^{-1} y^{-1}, and
- recently cleared crop land in northern Australia from 10 to 50 t ha^{-1} y^{-1}.

These losses may be compared with stream sediment yields < 1–4 t ha^{-1} y^{-1}, based on a global study of catchments (Fig. 11.14). The solid line in this figure shows the general relationship between erosion by water and *effective precipitation* – defined as the precipitation re-

quired to produce a known amount of runoff under specific temperature conditions. In arid and semi-arid regions, erosion increases as effective precipitation increases to a peak between 400 and 500 mm – the vegetation is sparse and the rainfall tends to show great variability, with flooding rains often following prolonged dry periods. As the climate becomes more humid and vegetation more abundant and stable, erosion losses decreases (although there is some evidence that at effective precipitation > 1500 mm, erosion begins to increase again even under natural vegetation). Certainly, the erosive power of high rainfall is attested by the broken curve in Fig. 11.14, which applies to subhumid and humid regions when the natural vegetation is removed. This also demonstrates the vulnerability of land in tropical regions to erosion by water when the natural rainforest is clear-felled, for timber harvesting, or felled and burnt for agriculture (Section 11.1).

Erosion is very damaging to soil fertility because it is mainly the nutrient-rich surface soil that is removed, and of that, predominantly the fine and light fractions – clay and organic matter – leaving behind the more inert sand and gravel. The fine and light fractions remain in suspension longer (in air or water) and so are carried further. Suspended stream sediment is eventually deposited in lakes, reservoirs or marine estuaries where it is lost to the land and may cause water-quality problems (Section 12.3).

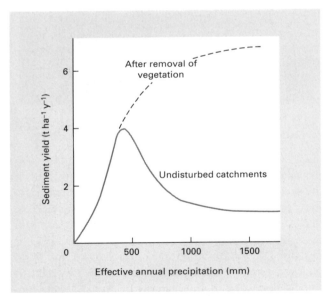

Fig. 11.14 Sediment loads in rivers in relation to annual precipitation (adjusted for differences in temperature) (after Langbein and Schumm, 1958).

Erosion by water

Soil loss by erosion depends on:

- the potential of rain to erode – the rainfall *erosivity*, and
- the susceptibility of soil to erosion – the soil *erodibility*.

Rainfall erosivity

To erode soil, work must be done to disrupt aggregates and move soil particles. The energy is provided by the kinetic energy (*KE*) of falling raindrops and flowing water. The principal effect of raindrops is to *detach* soil particles, whereas that of surface flow is to *transport* the detached particles. Calculations show that the *KE* of rain is ∼ 200 times greater than the *KE* of runoff, and field experiments confirm that erosion from a protected surface can be < 1/100th of that from a bare surface under the same rainfall (Hudson, 1981).

Raindrop impact initiates *splash* erosion (Fig. 11.15). On a sloping surface the downhill momentum transmitted to the soil particles is greater than the uphill component so there is a net downslope splash erosion. Slope also accelerates runoff velocity and the two effects combine to produce *wash* or *interrill* erosion. Where the water is concentrated, for example in cultivation furrows or tractor wheel marks, *rill erosion* is initiated (Fig. 11.16). Rills can be eliminated by normal cultivation methods, but they may deepen to form gullies so large that they cannot be crossed by farm machinery.

Raindrops vary in size up to *c.* 5 mm diameter. The *median drop diameter*, that is the diameter such that half the rain in a storm falls in drops of a smaller size, increases with rate of rainfall or *intensity* up to 25–50 mm h^{-1} above which it does not change much. Raindrop *KE* is given by the equation

$$KE = 1/2 \text{ mass} \times (\text{terminal velocity})^2. \qquad (11.7)$$

Terminal velocity increases with drop size (mass) so the

(a)

(b)

Fig. 11.15 (a) Soil surface just before raindrop impact. (b) Soil surface immediately after raindrop impact showing splash erosion (courtesy of D. Payer).

Fig. 11.16 Rill erosion caused by surface runoff on bare soil (courtesy of A.J. Low).

KE of the rain increases sharply up to moderate intensities (25–50 mm) and then more gradually, as illustrated in Fig. 11.17.

Another important characteristic of rainfall is the frequency of high-intensity storms, for which there is a striking difference between tropical and temperate regions. Approximately 95% of temperate rain, but only 60% of tropical rain, falls at intensities < 25 mm h^{-1}, which are considered non-erosive. Furthermore, the maximum intensity of tropical rainfall may exceed 150 mm h^{-1}, which is twice that of temperate rainfall. The high average *KE* and generally greater amounts of tropical rainfall make erosion by water potentially a very serious problem in these regions, whereas it is of less importance in humid temperate lands.

Soil erodibility

Intrinsically, a soil's *erodibility* must depend on the complex interaction of factors such as clay content and type of exchangeable cations, organic matter content, structure, infiltration capacity and shear strength (Box 13.7). However, no simple erodibility index based on soil properties measured in the laboratory or field

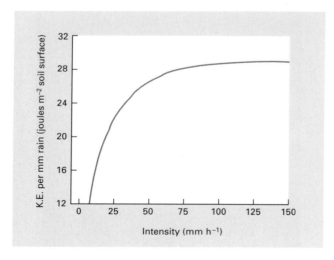

Fig. 11.17 Kinetic energy per mm of rain in relation to rainfall intensity (after Hudson, 1981).

has yet been devised. Instead, the susceptibility of a soil to erosion is defined in practical terms by the equation

$$\frac{A}{R} = K \tag{11.8}$$

where *A* is the mean annual soil loss in t ha^{-1}, *R* is a *rainfall erosivity* index (Box 11.7), and *K* is the *soil erodibility* factor.

Given that the value of *A* and *R* for a particular soil and site are known, Equation 11.8 can be solved to give the value of *K* under standard conditions. The standard conditions are a slope length of 22.6 m and gradient of 9% for bare soil that is ploughed up and down the slope. If the actual conditions of slope, surface coverage and land management differ from the standard, the consequent effect on *A* is obtained by rearranging Equation 11.8 and adding extra terms to give

$$A = R \times K \times L \times S \times P \times C \tag{11.9}$$

where the symbols *L*, *S*, *P* and *C* represent, respectively, a length factor, slope factor, conservation practice factor and crop management factor. Equation 11.9 is called the *universal soil loss equation* (ULSE). The values of *L*, *S*, *P* and *C* are unity when the standard conditions apply; if not, they change in a way that reflects the effect of the changed conditions in either *increasing* or *decreasing* erosion. The effects of these factors on the value of *A* are briefly examined as follows.

Soil factor K. Of the intrinsic soil properties affecting *K* (see above), structure is the most integrative and is also amenable to manipulation, through methods of tillage and the effect of leys or permanent pasture (Section 11.4). For example, the dramatic effect of ploughing old grassland on the percentage of water-stable aggregates in the soil was shown in Fig. 11.8. Experiments in the USA on different soil types, but under the standard conditions of *L*, *S*, *P* and *C*, produced a range of *K* values between 0.03 and 0.69. Unfortunately, such detailed information is not commonly available in other countries.

Box 11.7 Rainfall erosivity indices.

One index of rainfall erosivity which as been widely used is the EI_{30} value, calculated as the product of the KE of a storm and its maximum 30-min intensity. The latter is the maximum rainfall in any 30-min period, expressed in millimetres per hour. An alternative is the $KE > 25$ index which sums the KE values of a storm for those 15-min intervals when the rainfall intensity is > 25 mm h^{-1}. Both indices are empirical and must be tested by correlation against actual soil losses for individual soils under a range of rainfall intensities. Such testing suggests that the EI_{30} value should be calculated only for storms yielding $\geqslant 12.5$ mm of rain with a maximum 5-min intensity > 25 mm h^{-1}. Further, the $KE > 25$ index seems best suited to tropical regions and a lower threshold intensity of 10 mm h^{-1} is more appropriate in temperate regions (Morgan, 1986). The unmodified EI_{30} index is used in the *universal soil loss equation*.

Slope factors LS. Both steepness (*S*) and length (*L*) of slope affect erodibility: the former because of the greater potential energy of soil and water high on a slope, which can be converted to kinetic energy; the latter because the longer the slope, the greater the surface runoff and its downhill velocity. When soil conservation measures, such as contour terracing or strip cropping, are introduced, the distance *L* of vulnerable soil between structures is adjusted according to the slope *S*, so that a composite factor *LS* is used in the USLE.

Conservation practice P. Ploughing straight up and down a slope is the worst soil management practice (*P* = 1). Ploughing along contours can reduce *P* by 10–50%, the greatest effect being achieved on slopes between 2 and 7%. More elaborate procedures range from *contour listing* – small temporary ridges thrown up by cultivation, to *contour ridges* or *contour banks* – permanent banks with graded channels above for water flow, to *bench terraces*, which convert a steep slope into a series of steps. Generally these structures are designed to:

• reduce the length of slope from which runoff is generated,

• hold water on the slope longer so that more of it infiltrates the soil, and

• divert surplus water into safe spillways down which it may escape with minimum destructive effect, such as the grassed waterways shown in Fig. 11.18.

Much earth movement and grading may be required, so that the final design is a compromise between the cost of the work, including the loss of cropping flexibility incurred, and the value of the land and its produce.

Crop management C. Manipulation of this factor offers the greatest scope for reducing erosion loss because *C*

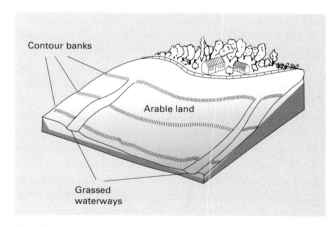

Fig. 11.18 Perspective of crop land with contour banks and waterways.

Fig. 11.19 Trash blanket on soil after 'green cane harvesting' in north Queensland.

can vary from 0.001 for well-kept woodland, to 0.05 for continuous forage cropping, to 1.0 for continuous bare fallow. Practices aimed at reducing C must focus on not leaving the soil bare when the rainfall erosivity is high, and growing high-density, healthy crops so that the soil surface is protected and the binding of soil by the roots is strong.

Permanent pasture, leys, minimum tillage, direct drilling and conservative tillage all have merit for erosion control. A specific example of conservation tillage is *stubble mulching*, whereby the residues of a previous crop (ofter a cereal crop) are chopped up and scattered over the soil surface. A striking example is the *trash blanket* applied to ratoon crops of sugar cane in north Queensland whereby the cane is harvested green and most of the leaf material chopped up and spread several cm deep over the field (Fig. 11.19). Other conservative crop management practices are:
- *strip cropping* – strips of arable land and pasture alter-

nate down a slope, thereby reducing the length of slope most vulnerable to erosion, and
- *alley cropping* – advocated for steep slopes (> 10%) under high rainfall in the tropics, whereby shrubs or trees are grown in thin strips (*c*. 1 m wide) across the slope and cash crops are grown in the 'alleys' between (*c*. 5 m wide). If possible, the tree/shrub is leguminous so that N is fixed, and the prunings from the strips and stubble from the crop (e.g. maize, pulses) are retained as a mulch on the soil surface.

The use of USLE and other more physically based soil erosion models is discussed further in Box 11.8.

Erosion by wind

Soil movement by wind is prominent in arid areas and on coastlines, where the wind is strong and consistent and the vegetation is adversely affected by salinity. By analogy with water erosion, the severity of wind erosion depends on the *potential* of the wind to erode and the *susceptibility* of the soil to erosion. Although not as widespread as water erosion, the effects of wind erosion

Box 11.8 Models for predicting soil erosion.

Because the USLE is an empirical model, it works best in the regions for which it was originally developed (the USA east of the Rocky Mountains). If the maximum average soil loss that can be tolerated in a production system is known (set equal to A in Equation 11.9), various combinations of L, S, P and C can be tried such that the product of the terms on the RHS of the equation does not exceed A. When used outside the USA, the appropriate values of L, S, P and C for soil conservation and cropping practices on the local soils, under local rainfall conditions, should be established.

USLE applies to agricultural land only and can not be used to predict sediment yields and other off-site effects of erosion. Elements of the USLE have been combined with physically based concepts of runoff, soil detachment/deposition and transport to produce the CREAMS model – Chemicals, Runoff and Erosion from Agricultural Management Systems. This is a dynamic model with a daily time step – average annual erosion rates are obtained by summation – which has given reasonably good predictions in several countries (Morgan, 1986). However, CREAMS is being displaced in the USA by WEPP (Water Erosion Prediction Project), a new model designed for all land surfaces, which is entirely process based and designed to predict, for single rainfall events, runoff and erosion and sediment concentrations in receiving channels. WEPP is currently the subject of on-going developing and testing (Blackburn *et al.*, 1994)

are often more dramatic in that the quantities of soil moved per hour across the edge of 1 ha can be of the same order as that moved by water erosion in 1 year.

Wind action

For given soil conditions (moisture and particle density ρ_p), erosion by wind depends on the wind speed and the roughness of the surface. Roughness alters the frictional drag on the wind and the likelihood of turbulence, which can impart an upward velocity component to loose particles or small aggregates (Fig. 11.20a). The air-borne particle gains a forward velocity, and because of the increase in wind velocity away from the soil surface (Fig. 11.20b), the higher the jump, the greater the particle's momentum. However, surface roughness, through its drag effect, also decreases the gradient in

forward wind velocity over the soil and this affects its ability to erode.

Particle or aggregate size influences surface roughness and the density obviously determines how much energy must be expended to raise a particle of a particular size off the ground. The interaction of these factors determines in a complex way the *threshold wind velocity* (i.e. the minimum velocity required to move a particle) as illustrated in Fig. 11.21. This diagram shows that particles or aggregates around 0.1–0.15 mm diameter are the most vulnerable, irrespective of their density. Once a particle has been moved, there are three possible consequences.

• Particles < 0.05 mm in diameter that rise more than approximately 30 cm into the air may stay suspended for some time. The smaller the particle, or the lower its density, the more likely are wind currents to be able to

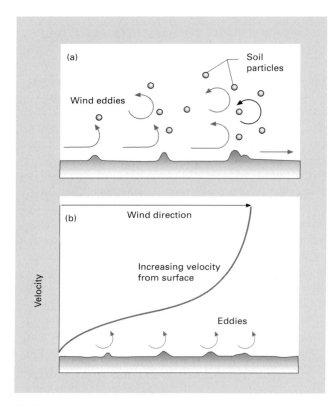

Fig. 11.20 (a) Turbulent wind eddies over a roughened soil surface. (b) Vertical profile of wind velocity over a soil surface (after Hudson 1981).

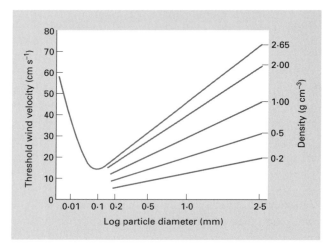

Fig. 11.21 The effect of particle size and density ρ_p on the threshold wind velocity for particle movement (after Marshall, 1972).

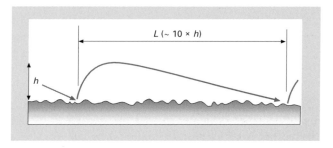

Fig. 11.22 Jump and glide path of saltating soil particles (after Hudson 1981).

counteract the force of gravity and keep the particle suspended. The resulting *aerial dispersion*, if sufficiently concentrated ($> 0.2\,\mathrm{g\,m^{-3}}$), becomes a dust storm. In the past, especially during the Pleistocene glaciations, the strong winds of periglacial regions dispersed vast amounts of these fine particles (0.02–0.05 mm) to form the extensive *loess* deposits described in Section 5.2.

• Particles and aggregates between 0.05 and 0.5 mm in diameter do not usually rise more than 30 cm and their jump is followed by a long flat trajectory to earth. On striking the surface, the particle either bounces off or comes to rest, transmitting its *KE* to other particles, which are knocked upwards (Fig. 11.22). This self-sustaining process is called *saltation*. Saltating particles

also impart energy to the smaller or lighter particles which go into suspension.

• On landing, the saltating particle may strike larger, heavier particles which do not rise up but merely roll or *creep* along the surface. Particles and aggregates of a diameter between 0.5 and 2 mm move in this way.

Movement by saltation accounts for more than 50% of wind erosion, with 3–40% of movement occurring in suspension and 5–25% by surface creep. Little can be done to lessen the erosive potential of wind other than by reducing its velocity by means of *wind breaks*, and by shortening its *fetch*, that is the distance of its unhindered travel over erodible land.

Soil and crop factors

Moist soil supporting healthy vegetation is the most effective deterrent to wind erosion. Cultivated soil is vulnerable when the surface is dry and the structure weak, especially if most of the aggregates are < 1 mm in diameter. Often coarse- and fine-textured soils are more erodible than loams, which have the right proportions of clay, slit and sand to form stable aggregates of a non-erodible size. The fen peat soils of eastern England are particularly vulnerable to wind erosion once drained and cultivated because of the lightness of the organic-rich surface horizon.

With respect to surface roughness, although slight irregularities initiate saltation as the wind speed rises, very rough or cloddy surfaces protect against erosion by reducing the average wind velocity at ground level and trapping saltating particles as they land.

Vegetation, even if dead, has a similar effect, which is directly proportional to its coverage of the soil surface and height above the soil. Key crop management practices that are valuable in controlling wind erosion include:
- *stubble mulching*, pioneered on the wind-swept High Plains of the USA, and recommended for the light-textured cropping soils of the Mallee district in Australia,
- *trash farming*, whereby large one-way disc ploughs cut through the stubble but do not bury it, and
- *direct drilling* of a new crop through the residues of a previous crop.

11.6 Summary

Clearing a soil of natural vegetation interrupts the cycling of nutrients and may lead to a decline in fertility. The traditional method of farming in the wet tropics – *shifting cultivation* or *slash-and-burn agriculture* – can be sustainable if the fallow period, when the forest is allowed to regrow, is long enough for nutrients to be replenished from weathering and atmospheric sources. In temperate regions, traditional *rotational cropping* and *mixed farming* – involving arable–ley pasture rotations – will generally maintain fertility provided that *fertilizers* and/or *organic manures* are used to replace nutrients removed in farm produce. However, because of population pressure, the cost of labour and the demand for land for non-agricultural use, these traditional methods have gradually been displaced by intensive monocultural systems, such as continuous cereal cropping, and livestock farming has become separated from cropping. This has resulted in varying degrees of soil nutrient depletion, soil structure decline, salinization, accelerated acidification and erosion in many countries.

Modern intensive farming requires substantial inputs of fertilizers and/or organic manures (including sewage sludge) to maintain soil fertility. *Green manuring* may also be advocated to prevent leaching of surplus soil NO_3^- during winter or the non-cash crop period. Qualitatively, soil nutrient deficiencies can be assessed by *soil testing* or *plant analysis*, with the identification of *critical values* for a nutrient element below which plants respond to an increase in that nutrient's supply. A rational approach to soil testing is to use the *quantity/intensity* (*Q/I*) concept to measure a soil's nutrient supplying power; but in practice a variety of empirical extractants are used which provide a composite estimate of *Q* and *I* factors. Whichever approach is adopted, soil or plant tests must be calibrated against crop response to the nutrient in question using a *fertilizer rate trial*, carried out on a range of soils. Care must be taken to ensure the soil or plant sample analysed is *representative* of the area tested. Fertilizer requirements can also be assessed using a *nutrient balance* method. Crop yield maps can be used so that fertilizer application is adjusted to the specific need of parts of individual fields. This is called *precision farming*.

Soil acidity and the consequent problems of low pH, Al and Mn toxicity, Ca and Mo deficiency, are remedied by adding lime. The amount needed (the *lime requirement*) depends on the pH buffering capacity of the soil. Accelerated soil acidification resulting from high atmospheric H^+ inputs, removal of C and N in farm products, and increased NO_3^- leaching, affects many modern farming systems. It requires management to reduce both the inputs and losses, as well as liming to correct the acidity problem.

Deterioration of *soil structure* is a more insidious problem that is only revealed through inadequate soil aeration, decreased available water, impedance to root penetration and surface sealing. A grass ley following arable cropping is very effective in restoring the water stability of aggregates, especially in the surface 5 cm or so. The improvement is time-dependent, at least 4 years being required in temperate regions to achieve an improvement in 'hard' aggregation. However, the effect of the ley is relatively short-lived once it is ploughed. A similar improvement in aggregation can be obtained with synthetic organic polymers called *soil conditioners*, but these are expensive.

New tillage methods, which are a blend of *modern* (use of rapid-kill, non-persistent herbicides) and *ancient* (sowing seed with minimum soil disturbance) agricultural practices, have gained widespread popularity, particularly on soils where high surface temperatures, high evaporation, surface sealing, and water and wind erosion pose problems. These methods range from *reduced* or *minimum tillage* (fewer passes with cultivating implements) to *direct drilling or no-till*, where the only disturbance is sowing the seed. *Conservation tillage* is a generic term denoting conservation of water and soil, not necessarily by less tillage, but by keeping at least 30% of the surface covered by crop residues.

Erosion is very damaging to soil fertility because it is mainly the nutrient-rich surface soil that is removed. *Water erosion* occurs in almost any environment when the soil is bare and the structure unstable, the magnitude of the loss depending on rainfall *erosivity* and soil *erodibility*. This relationship is quantified by assigning values to variables in the *universal soil loss equation* (*USLE*)

$$A = R \times (K, LS, P, C).$$
(Soil loss) (Erosivity) (Erodibility)

Conservation measures aim at decreasing slope *steepness S* and *length L*; also modifying the *soil manage-ment factor P* (through contour banks and terraces) and the *crop management factor C* (by direct drilling, stubble mulching, strip cropping and alley farming). Although the USLE has been very successful in identifying management practices which will decrease soil loss per hectare to an acceptable value, it is not applicable to non-agricultural land nor can it be used to predict the off-site effects of erosion. For this reason, USLE is being superseded by process-based erosion models, such as the Water Erosion Prediction Project (WEPP) model in the USA, which apply to any form of land use.

Wind erosion is confined to arid regions or coastal areas where wind speeds are consistently high. The susceptibility of soil to wind erosion depends on the moisture content of the surface, the diameter and density of the surface particles (those of diameter 0.05–0.15 mm are the most vulnerable), and the degree of protection afforded the surface by vegetative cover.

References

Agricultural Development & Advisory Services (1982) *The Use of Sewage Sludge on Agricultural Land*. Booklet 2409. MAFF Publications, Northumberland.

Anon (1995) *Data Sheets on Natural Resources Issues*. Land & Water Resources Research & Development Corporation Occasional Paper No. 6/95.

Blackburn W. H., Pierson F. B., Schumann G. E. & Zartman R. (1994) *Variability in Rangelend Water Erosion Processes*. Soil Science Society of America Special Publication No. 38. Soil Society of America, Madison, Wisconsin.

Blake L., Johnstone A. E. & Goulding K. W. T. (1994) Mobilization of aluminium in soil by acid deposition and its uptake by grass cut for hay – a chemical time bomb. *Soil Use and Management* **10**, 51–55.

Brady N. C. (1996) Alternatives to slash-and-burn: a global imperative. *Agriculture, Ecosystems and Environment* **58**, 3–11.

Cooke G. W. (1982) *Fertilizing for Maximum Yield*, 3rd edn. Granada, London.

Cornforth I. S. & Sinclair A. G. (1984) *Fertiliser and Lime Recommendations for Pastures and Crops in New Zealand*, 2nd edn. Ministry of Agriculture and Fisheries, Wellington.

Davies D. B., Eagle D. J. & Finney J. B. (1993) *Soil Management*, 5th edn. Farming Press, Ipswich.

FAO (1993) *Agriculture: Towards 2010*. Food and Agriculture Organization of the United Nations, Rome.

Gard L. E. & McKibben G. E. (1973) No-till crop production

proving a most promising conservation measure. *Outlook on Agriculture* **7**, 149–154.

Graecen E. L. (1958) The soil structure profile under pastures. *Australian Journal Agricultural Research* **9**, 129–137.

Hall D. G. M., Reeve M. J., Thomasson A. J. & Wrights V. F. (1977) *Water Retention, Porosity and Density of Field Soils*. Soil Survey of England and Wales, Technical Monograph No. 9, Harpenden.

Hudson N. W. (1981) *Soil Conservation*, 2nd edn. Batsford, London.

Juo A. S. R. & Mann A. (1996) Chemical dynamics of slash-and-burn agriculture. *Agriculture, Ecosystems and Environment* **58**, 49–60.

Langbein W. B. & Schumm S. A. (1958) Yield of sediment in relation to mean annual precipitation. *Transactions of the American Geophysical Union* **39**, 1076–1084.

Low A. J. (1972) The effect of cultivation on the structure and other physical characteristics of grassland and arable soils (1945–1970). *Journal of Soil Science* **23**, 363–380.

Marshall J. K. (1972) Principles of soil erosion and its prevention, in *The Use of Trees and Shrubs in the Dry Country of Australia*. Forestry and Timber Bureau, Canberra, pp. 90–107.

Ministry of Agriculture, Fisheries and Food (1976) *Organic Manures*. Bulletin 210. HMSO, London.

Morgan R. P. C. (1986) *Soil Erosion and Conservation*. Longman Scientific and Technical, Harlow.

Nye P. H. & Greenland D. J. (1960) *The Soil under Shifting Cultivation*. Commonwealth Bureau of Soils, Technical Communication No. 51. Harpenden. England.

Rayment G. E. & Higginson F. R. (1992) *Australian Laboratory Handbook of Soil and Water Chemical Methods*. Australian Soil and Land Survey Handbook. Inkata Press, Melbourne.

Smith J. U., Bradbury N. J. & Addiscott T. M. (1996) SUNDIAL: a PC-based system for simulating nitrogen dynamics in arable land. *Agronomy Journal* **88**, 38–43.

Further reading

Carter M. R. (Ed.) (1994) *Conservation Tillage in Temperate Agroecosystems*. Lewis, Baton Rouge, Florida.

De Boodt M. (1979) Soil conditioning for better management. *Outlook on Agriculture* **10**, 63–70.

Havlin J. L. & Jacobsen J. S. (Eds) (1994) *Soil Testing: Prospects for Improving Nutrient Recommendations*. Soil Science Society of America Special Publication No. 40. Soil Science Society of America. Madison, Wisconsin.

Troeh F. R. & Thompson L. M. (1993) *Soils and Soil Fertility*, 5th edn. Oxford University Press, New York.

Ulrich B. & Sumner M. E. (Eds) (1991) *Soil Acidity*. Springer, Berlin.

Chapter 12
Fertilizers and Pesticides

12.1 Some definitions

The prime purpose for the use of *agricultural chemicals* (sometimes referred to as *agrochemicals*) is to increase the quantity and improve the quality of agricultural products. The broad categories of agricultural chemicals are:

• *fertilizers*, introduced in Chapter 11, which are used to correct nutrient deficiencies and imbalances in plants and animals and so improve growth and product quality, and

• *pesticides* (sometimes called *plant-protection* chemicals) which are used to protect crops from pests and disease, or to eliminate competition from aggressive weeds. This protection extends to the crop products after harvest and during storage.

Much of the increase in world agricultural production, and production per capita (Fig. 11.2), is due to expansion in the use of fertilizers, up from 14 Mt (N, P and K) in 1950 to 146 Mt in 1989, and pesticides (> 2.6 Mt of active ingredient used in 1989). To achieve their purpose, fertilizers and other chemicals must be used in accordance with best scientific principles so that the response of the soil, plant or animal is optimal, and their impact on the wider environment is minimal. These principles are discussed in previous chapters of this book and are further developed here.

Fertilizers and pesticides may be applied as solutions, suspensions, emulsions or solids to soils and plants. When applied as a foliar spray, the properties of the chemical (such as its volatility), the nature of the plant surfaces and the atmospheric conditions determine the efficacy of the application and the subsequent fate of the chemical in the environment. In other cases, the type of chemical, its method of application and the reactions it undergoes in the soil are of prime importance. In this chapter, emphasis is placed on chemical forms and soil reactions, starting with the N, P, K and S fertilizers.

12.2 Nitrogen fertilizers

Forms of N fertilizer

Soluble forms

The composition of the common water-soluble N fertilizers is given in Table 12.1. Some of these, such as NH_4NO_3 and urea, supply a single macronutrient ele-

Table 12.1 Forms of soluble N fertilizer.

Compound	Formula	N content (%)
Solids		
Sodium nitrate	$NaNO_3$	16
Calcium nitrate	$Ca(NO_3)_2$	16
Ammonium sulphate	$(NH_4)_2SO_4$	21
Calcium cyanamide	$CaCN_2$	21
Ammonium nitrate	NH_4NO_3	34
Urea	$(NH_2)_2CO$	46
Monoammonium phosphate (MAP)	$NH_4H_2PO_4$	11–12
Diammonium phosphate (DAP)	$(NH_4)_2HPO_4$	18–21
Liquids		
Aqua ammonia	NH_3 in water	25
Nitrogen solutions – urea ammonium nitrate	NH_4NO_3, $(NH_2)_2CO$ in water	30
Anhydrous ammonia	liquefied NH_3 gas	82

ment (N); others, such as $(NH_4)_1SO_4$, monoammonium phosphate (MAP) and diammonium phosphate (DAP), provide more than one macronutrient and are classed as *multinutrient* or *compound* fertilizers. *Mixed fertilizers*, which may be solid or liquid (fluid), are made by mixing single or multinutrient fertilizers and are usually identified by their N:P:K ratio. For example, a mixed fertilizer of NPK 5–5–8* can be made by dissolving NH_4NO_3, urea, ammonium polyphosphate and KCl in water.

Highly soluble solid fertilizers can pose handling and storage problems because of their hyroscopicity (absorption of water vapour). This difficulty is largely overcome by *granulation* which involves heating the fertilizer slurry in a rotating drum at the time of manufacture. As excess water is driven off, the small fertilizer particles coagulate to form granules which harden further on air-drying. Ideally, a granulated fertilizer has uniform granules, roughly spherical in shape and *c.* 3–4 mm in diameter. The ultimate granules are the very regular, spherical *prills* of NH_4NO_3 or urea. The process of granulation is intended to:

• reduce the contact area between individual particles and also the area available for water absorption, and
• prevent the fertilizer from breaking down into very fine dust particles which make the fertilizer difficult to spread, and which may create a fire hazard (e.g. during aerial application) or health hazard.

When solid mixed fertilizers are made by mixing the ingredients in a slurry before granulation, they are homogeneous. Other fertilizers made by mixing individual granulated fertilizers – a process of *bulk blending* – may not be so homogeneous because granules of different size and density can segregate in the mixture. The chemical incompatibility of some fertilizers, such as KNO_3 and urea, also precludes their bulk blending.

*In the trade, fertilizer analyses are generally given as the ratio of the percentages of $N:P_2O_5:K_2O$. For simplicity in this book, analyses are given as percentages of the elements N, P, K and S unless otherwise stated.

The choice of the most suitable N fertilizer depends on a balance of factors: the cost per kilogram of N (including transport and application costs), the effects on plant growth (both beneficial and detrimental) and the magnitude of N loss through leaching, volatilization and denitrification. A high N content and ease of handling as prills have made *urea* the most popular form of solid N fertilizer in the world. However, in the USA anhydrous NH_3 is very popular because of its high N analysis and the precision with which its application can be controlled. The advantage of a very high N content per unit weight afforded by anhydrous NH_3 is partly offset by higher costs of application, since it must be kept under pressure and injected at least 10 cm below the soil surface. Other effects are discussed below.

Slow-release fertilizers

To obviate problems arising from the high solubility of many N fertilizers and their potential vulnerability to leaching, especially in the of NO_3-N form, various sparingly soluble N compounds have been developed (Table 12.2). These *slow-release* fertilizers include both synthetic and natural *organic fertilizers*. Plant response to these fertilizers is usually inferior, per kilogram of N, when compared with soluble forms, but they are especially useful for the establishment of vegetation on difficult sites, such as reclaimed mine workings, where a slow but assured release of N over an extended period is required. An alternative solution to the leaching problem is discussed in Box 12.1.

Table 12.2 Forms of slow-release N fertilizer.

Fertilizer	Composition	N content (%)
Shoddy	Wool waste	2–15
Dried blood	By-product of meat processing	~ 13
Hoof and horn meal	By-product of meat processing	7–16
Ureaform	Ureaformaldehyde polymers	21–38
IBDU	Isobutylidene diurea	32
SCU	Sulphur-coated urea	37–40

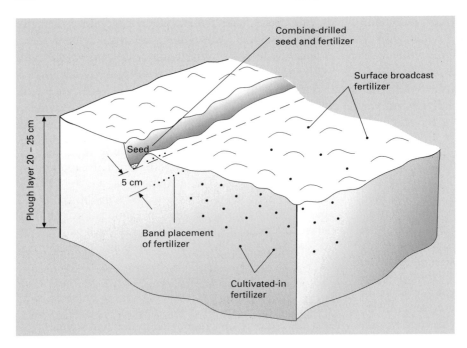

Combine-drilled
seed and fertilizer

Surface broadcast
fertilizer

Plough layer 20 – 25 cm

Seed

5 cm

Band placement
of fertilizer

Cultivated-in
fertilizer

Fig. 12.1 Ways of applying fertilizer to soil.

Reactions in the soil

Scorch and retarded germination

There is a possibility of retarded germination, injury to young roots and leaf scorch* because of the concentrated salt solution around dissolving fertilizer granules, particularly if they are drilled into the soil with the seed (Fig. 12.1). Generally, fertilizers containing NO_3-N are more harmful than NH_4-N fertilizers; for example $(Ca(NO_3)_2$ is satisfactory at < 75 kg N ha^{-1} while $(NH_4)_2SO_4$ is safe up to 125 kg N ha^{-1}. Urea is exceptional in being injurious at rates as low as 35 kg N ha^{-1}. The best method of applying urea is to place it in a band, slightly below and to one side of the seed (Fig. 12.1), where damage due to high salt concentration and losses by volatilization of NH_3 are minimized.

*Chlorosis followed by leaf death, usually extending from the tip backwards.

Mixed fertilizers containing N are less injurious to plants than straight N fertilizers at comparable rates.

pH effects

Ammonia solutions and anhydrous NH_3 produce an alkaline reaction in soil. Urea, whether from animal excreta or fertilizer, is rapidly hydrolysed to $(NH_4)_2CO_3$ (Equation 10.6), which in turn hydrolyses to NH_4OH, CO_2 and H_2O (Equation 10.7). The soil pH around granules of dissolving urea may therefore rise above 9 at which level NH_4OH is unstable and decomposes to NH_3 and H_2O (Equation 10.8, for which the $pK_a \simeq 9.5$). As the NH_3 volatilizes, the H^+ ions released neutralize the OH^- ions from the hydroxide so that the pH stabilizes and then slowly falls.

A pH rise to 8 and above in the vicinity of ammonium or urea fertilizer granules will inhibit *Nitrobacter* organisms (Section 8.3). *Nitrosomonas* is not so sensitive and functions well up to pH 9 so that NO_2^- accumulates in the soil. Although oxidation of NH_4^+ to NO_2^- causes the pH to fall gradually, the high concentration of NO_2^- may continue to inhibit *Nitrobacter*.

Much of the N from ammonium fertilizers, including urea, that remains in the soil is eventually oxidized to nitrate. Nitrification is a potentially acidifying reaction (Equation 10.2). When NH_4^+ is produced by ammonification of soil organic N (Equation 10.1), 1 mol net H^+ is produced per mole of NH_4^+ that is oxidized. But if the NH_4-N is supplied by a fertilizer, 2 mol H^+ are produced per mole of NH_4^+ oxidized. The acceleration of acidification in agricultural soils when N inputs are increased, and significant amounts of NO_3-N are leached from the root zone, is discussed in Section 11.3. A similar effect occurs in soils heavily fertilized with NH_4-N fertilizers, when the recovery of N by the crop is only *c.* 50% (see below), so that regular applications of lime are required to prevent an undesirable fall in soil pH.

According to the reaction

$$CaCO_3 + 2H^+ + SO_4^{2-} \leftrightarrow CaSO_4 + H_2CO_3. \qquad (12.1)$$

the neutralization of 2 mol H^+ ion would require 1 mol $CaCO_3$. Thus, the *theoretical* lime equivalent of a fertilizer such as $(NH_4)_2SO_4$, where all the N is in ammonium form, is 100 kg $CaCO_3$ per 14 kg N, or 7 kg $CaCO_3$ per kg N. However, because some of the NO_3-N is taken up by the crop, in practice it is found that the lime equivalent for $(NH_4)_2SO_4$ is *c.* 5.4:1 and that of NH_4NO_3 is *c.* 1.8:1. One way of counteracting the acidifying effect of ammonium fertilizers has been to produce mixed fertilizers of NH_4NO_3 and $CaCO_3$ such as 'Nitrochalk' (26% N) and limestone ammonium nitrate (20% N), but these are not as popular as the higher analysis, straight N fertilizers or compound N fertilizers. An alternative approach is to use a *nitrification inhibitor* to slow down the rate of oxidation of NH_4^+ in the soil (Box 12.1).

Nitrification and consequent N losses

Nitrate produced by nitrification of NH_4-N fertilizers is susceptible to loss by leaching out of the root zone (Section 10.2) and by denitrification (Section 8.4) In addition to biological denitrification (Section 10.2), there is the possibility of *chemo-denitrification* in N-fertilized soils. This appears to involve the reaction of nitrous acid (HNO_2) with soil organic matter to release gaseous N compounds. The most likely reactive organic components are phenolic compounds, which are known to form nitrosophenols by reacting with HNO_2. Under acid conditions (excess HNO_2), these are decomposed to N_2O and N_2. Direct leaching of soluble N fertilizer is only likely to occur when heavy rain falls soon after fertilizer application, which has been observed to occur with arable crops and N-fertilized grassland in spring in the UK. Research in southern England with winter wheat on several soil types suggests a direct relationship between the percentage of the fertilizer N lost and rainfall in the 3 weeks after application (Fig. 12.2). However, it is likely that only about one-third of this loss was by leaching and the remainder by denitrification. Overall, it can be concluded from experiments on cereal crops that the *direct* loss of fertilizer N by leaching is unlikely to exceed *c.* 6% (Addiscott *et al.*, 1991). Losses from row crops such as potatoes can be higher.

But even after the fertilizer N has entered the soil biomass–organic matter pool, it becomes vulnerable to leaching as the organic N is mineralized and especially when a 'flush' of mineralization follows the rewetting of a dry soil by rain or irrigation (Section 10.2). The leaching loss of soil-generated NO_3^- (and also NO_3^- from surface-applied fertilizer) can be exacerbated in such situations when heavy rain falls on well-structured soils, because of preferential flow of water and NO_3^- down the macropores into the drainage water (Fig. 10.8b). Thus, whenever the productivity of agricultural systems has been increased over several decades through the use of N fertilizers, there has been an *indirect* effect of rising nitrate concentrations in surface waters and underground aquifers (Fig. 12.3). In surface waters, high NO_3-N concentrations can accelerate *eutrophication* (Section 12.3) and in both surface and groundwaters high concentrations may affect human health (Box 12.2).

Box 12.1 Nitrification inhibitors.

Interest in nitrification inhibitors followed the development of the compound called 'N-serve' or nitrapyrin (2-chloro-6 trichloromethyl pyridine) which has been shown to inhibit the nitrification of NH_4-N fertilizers, including urea. The degree of inhibition is highly variable, but under field conditions it is rarely greater than 25%. Nitrapyrin has to be applied at least annually (at a rate of $c.$ 1% of the N applied) because it is decomposed by soil micro-organisms. In regions where the winters are very cold, nitrapyrin has some effect in delaying the nitrification of NH_4-N fertilizer, applied in the autumn, until the crop begins to grow rapidly in spring.

Other nitrification inhibitors that have been used include ATC (4-amino-1,2,4-triazole), DCD (dicyandiamide), C_2H_2 (acetylene) and CS_2 (carbon disulphide). With gases such as CS_2 and C_2H_2 the problem is to keep a sufficiently high concentration ($c.$ 1% by volume) in the soil air at the site of the fertilizer granule. A novel way of solving this problem has been to make *encapsulated* calcium carbide (CaC_2) pellets which can be mixed with the fertilizer. CaC_2 particles (1–2 mm diameter) are successively coated with waxes containing CaC_2 to produce a pellet of 1:1 ratio of carbide and coating. On contact with water, the pellet slowly disintegrates and C_2H_2 is released according to the reaction

$$CaC_2 + 2H_2O \rightarrow C_2H_2 + Ca(OH)_2.$$ B12.1.1

This technique has proved effective in partially inhibiting the nitrification of NH_4-N fertilizers in paddy rice and irrigated wheat. When nitrification is inhibited, there is a greater chance of NH_4^+ being immobilized by micro-organisms, or lost by volatilization of NH_3 in calcareous soils. For this reason it is recommended that the NH_4-N fertilizer and inhibitor are placed at least 5 cm below the soil surface.

Efficiency of utilization

Rate and method of application

The recovery of fertilizer N in the crop or animal product depends on the climate, the kind of crop, and on the rate, method and timing of fertilizer application. Potentially, the best recovery is achieved when temperature and other factors allow the crop to be grown during the wetter part of the year, as can be done in much of the tropics. For a given climate, N recoveries are usually lowest for plantation crops and short-season arable crops, and highest for permanent grass. Bananas in the Ivory Coast, for example, recovered only 14% of the 430 kg N ha^{-1} applied, but grass recovered 53% at rates up to 700 kg N ha^{-1}. Maize grown at different locations in the USA, Central America and Brazil recovered about 56% of fertilizer N at application rates up to 150 kg N ha^{-1}. In Britain, where leaching losses in the wet winter months are appreciable, crop recovery of fertilizer N ranges between 50 and 80%; recovery by grass fertilized at rates up to 400 kg N ha^{-1} y^{-1} ranges between 50 and 65%. However, when grass is grazed by domestic animals, recovery of N in animal products is only 5–15% of the input – the other 85–95% is returned to the soil in dung and urine, and, under intensive

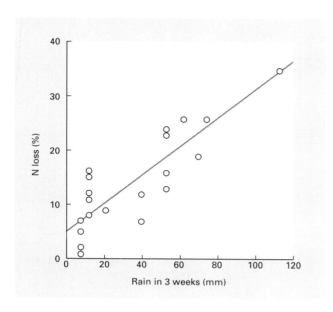

Fig. 12.2 The relationship between the loss of nitrogen fertilizer and the amount of rain in the 3 weeks after application (after Powlson *et al.* 1992).

grazing management, much is lost by a combination of leaching, denitrification and volatilization.

Timing of N application is crucial for best utilization by a crop. Spring applications in Britain give better results than those in autumn, especially in areas with annual precipitation > 675 mm (and excess winter rainfall* > 125 mm). Little nitrate remains in the soil in spring when the excess winter rainfall exceeds 150–250 mm, depending on the soil type. During the grass-growing season, N is best applied as a split dressing, say four or five times during the season, or after each cut for hay or silage. In this way, the supply of N is better matched to the growth cycle of the grass and losses by leaching are minimized. Similarly, in the southern maize belt of the USA it has been found that delaying N application for 3–4 weeks after germination avoids

*Defined as the excess of precipitation over evaporation when the soil is at field capacity during winter. Excess winter rainfall is a substitute for measured soil drainage.

leaching losses from late spring rains, and provides N at the stage of growth when the plant's demand is highest.

Residual effects

A residual effect occurs when sufficient of a fertilizer element applied to one crop remains available in the soil for the growth of a succeeding crop. For N fertilizer, residual effects depend on the rate of N applied, the crop, the rain falling between cropping periods, and the soil type. Provided that the fertilizer N applied is reasonably well matched to plant demand, the proportion of actual fertilizer N remaining in *mineral* form in the soil by harvest time is very low. However, plant growth (except under irrigation) is affected by the vagaries of the weather and farmers have a tendency to minimize the risk of diminished yield by applying extra fertilizer (as 'insurance'). Also, a significant proportion of the N fertilizer is converted into labile organic N from which it is readily mineralized (Box 10.1). Thus, field and lysimeter studies in many countries have reported nitrate concentrations in drainage from cropped soils and grazed pasture in the range 5 and 50 mg N L^{-1} (up to 100 mg N L^{-1} for intensively managed pasture). It is therefore possible to construct a schedule of probable N leaching losses as shown in Table 12.3. As excess winter rainfall (or drainage) increases, one moves across the table from left to right and slightly upwards, because nitrate concentration usually falls as the cumulative drainage volume exceeds 100 mm.

Table 12.3 Estimated N losses by leaching in relation to excess winter rainfall (or drainage).

Mean NO_3^- concentration in drainage water (mg N L^{-1})	Quantity of N (kg ha^{-1}) leached by an excess winter rainfall (mm) of:				
	100	200	300	400	500
5	5	10	15	20	25
10	10	20	30	40	50
15	15	30	45	60	75
20	20	40	60	80	100
30	30	60	90	120	150
50	50	100	150	200	250

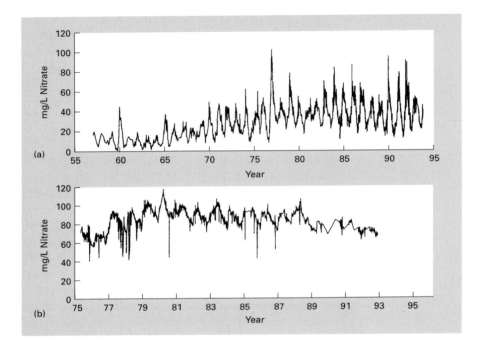

Fig. 12.3 (a) Variations with time in the nitrate concentration of the Great Ouse at Bedford, UK (after Croll, 1994). (b) Variations with time in the nitrate concentration in groundwater in limestone at Waneham Bridge, UK (after Croll, 1994).

Thus, for an excess winter rainfall of 250–300 mm and allowing for crop uptake, there will be little residual N from a fertilizer application of 150 kg N ha^{-1} to a preceding crop.

Residual effects have been observed when deep-rooting cereals, such as winter wheat, follow crops such as potatoes to which heavy N dressings have been applied (Table 12.4a). In other cases, such as cereal following cereal, part of any residual effect is undoubtedly due to the larger residue of N-rich plant material in the soil, which decomposes to release mineral N. When the preceding crop is a legume, the stimulatory effect of the residues is generally greater than that of even heavily fertilized grass (Table 12.4b).

12.3 Phosphate fertilizers

Forms of phosphate fertilizer

Phosphate fertilizers may be subdivided into:
• the water-soluble *orthophosphates*,
• the condensed orthophosphates or *polyphosphates*, and

Table 12.4 Residual effects of N-fertilized crops and legumes. After Cooke, 1969.

(a)	Wheat yield (t grain ha^{-1}) following potatoes given		
	0	and	185 kg N ha^{-1}
	2.84		3.91
(b)	Wheat yield (t grain ha^{-1}) following N-fertilized ryegrass		Clover ley
	4.83		6.12

• the *insoluble mineral* and *organic phosphates* (Table 12.5)
The natural rock phosphates, consisting of minerals of the *apatite* type with calcite, silica and other impurities, are the raw material from which *acidulated* and *partially acidulated* P fertilizers are made.

Ortho-P fertilizers

The active constituents are *monocalcium phosphate*

Box 12.2 Nitrate and human health.

Children under 3 months are most at risk from high nitrate in water (> 100 mg L^{-1} NO$_3^-$ or 22.6 mg L^{-1} NO$_3$-N) because they may develop *methaemoglobinaemia* or 'blue-baby disease'. This condition is induced by the reduction of NO$_3^-$ to NO$_2^-$ in the infant's stomach, and the absorption of excess NO$_2^-$ in the blood where it interferes with the O$_2$ carrying capacity of haemoglobin. It has also been argued that NO$_3^-$ (reduced to NO$_2^-$) could combine with amines in food to produce carcinogenic nitrosoamines. However, NO$_2^-$ in the stomach of anyone other than a very young baby is rapidly reduced to NO which kills pathogenic bacteria. Nor is there any sound epidemiological evidence of a positive correlation between nitrate intakes and stomach cancer in humans. Nevertheless, health authorities and government agencies have reacted conservatively to rising NO$_3^-$ concentrations in potable water and set low values for the maximum permissible concentration. These are 10 mg L^{-1} NO$_3$-N in the USA and 11.3 mg L^{-1} NO$_3$-N (50 mg L^{-1} NO$_3$) in the European Union. Similarly, the World Health Organization has adopted the concentration of 11.3 mg L^{-1} NO$_3$-N as its guide value for drinking water based on long-term exposure. This is not as strict as the EU requirement that the maximum concentration may never be exceeded.

The response to these limits in the UK has been broadly twofold.
- Changes in land management. Currently, under the Nitrate Sensitive Areas Scheme, farmers in sensitive areas (mainly over the chalk, limestone and sandstone aquifers which provide 40% of the potable water) receive financial incentives to make changes to their management to minimize the NO$_3^-$ available for leaching in autumn and winter.
- Treatment of the water. Here the preferred option is to blend high concentration water with low concentration water.

The rising trend in NO$_3^-$ concentrations in rivers and groundwater in the UK has now flattened out through changes in land management over the past decade or so (Fig. 12.3), but the concentrations in many sources are still above the EU limit and may remain so for some time.

(MCP), formula Ca(H$_2$PO$_4$)$_2$.H$_2$O, and *phosphoric acid*. The former compound is the basis of the *superphosphates* formed by the dissolution of rock phosphate in H$_2$SO$_4$, or H$_3$PO$_4$, for example

$$Ca_{10}(PO_4)_6F_2 + 7H_2SO_4 + 3H_2O \rightarrow 3Ca(H_2PO_4)_2.H_2O$$
$$+ 7CaSO_4 + 2HF. \qquad (12.2)$$

(single superphosphate, SSP)

If rock phosphate is reacted with excess H$_2$SO$_4$, wet-

process H$_3$PO$_4$ is produced, which is then used to dissolve fresh rock phosphate and give higher grades of superphosphate, such as *triple superphosphate*

$$Ca_{10}(PO_4)_6F_2 + 14H_3PO_4$$
$$+ 10H_2O \rightarrow 10Ca(H_2PO_4)_2.H_2O + 2HF. \qquad (12.3)$$

(triple superphosphate, TSP)

If insufficient H$_3$PO$_4$ is used to convert all the rock

Table 12.5 Forms of phosphate fertilizer.

Fertilizer	Composition	P content (%)
Ortho-P		
Phosphoric acid	H_3PO_4	23
Normal or single superphosphate	$Ca(H_2PO_4)_2$; $CaSO_4$	8–10
Concentrated or triple superphosphate	$Ca(H_2PO_4)_2$	19–21
Monoammonium phosphate	$NH_4H_2PO_4$	21–26
Diammonium phosphate	$(NH_4)_2HPO_4$	20–23
Monopotassium phosphate	KH_2PO_4	23
Poly-P		
Superphosphoric acid	$H_4P_2O_7$ and higher MW polymers	> 33
Ammonium polyphosphate	$(NH_4)_4P_2O_7$ and higher MW polymers	23
Insoluble phosphates		
Phosphate rocks	$Ca_{10}(PO_4)_6(F, OH)_2$ with variable SiO_2,$CaCO_3$ and sesquioxide impurities	6–18
Basic slag	Basic Ca, Mg phosphates, Fe_2O_3 and $CaSiO_3$	3–10
Organic phosphates	Bone meal, guano	5–13

phosphate to MCP, a partially acidulated phosphate rock (PAPR) fertilizer is produced. Single superphosphate has the advantage of containing 11% SO_4-S which is of additional value on sulphur-deficient soils.

Roasting rock phosphate ore in an electric furnace with quartz and coke reduces the phosphate minerals to elemental P, which when burnt in oxygen forms the oxide P_2O_5. This oxide combines with water to form phosphoric acid according to the reaction

$$P_2O_5 + 3H_2O \rightarrow 2H_3PO_4. \tag{12.4}$$

Phosphoric acid normally contains 52–54% P_2O_5 (23–24% P). Partial neutralization of the acid with NH_3 gives the multinutrient fertilizers MAP and DAP, discussed earlier (Section 12.1). Alternatively, ordinary superphosphate can be mixed with aqua or anhydrous NH_3 to form *ammoniated superphosphate* (*c.* 4% N and 7% P); and TSP can be ammoniated to give a product containing 8% N and 14% P. Unlike MAP and DAP, the ammoniated superphosphates contain some S. About half the P present is water soluble, but all of it is soluble in neutral ammonium citrate (Box 12.3).

The relationship between these *ortho*-P fertilizers is summarized in Fig. 12.4.

Poly-P fertilizers

Ordinary H_3PO_4 can be upgraded to *superphosphoric acid* (72% P_2O_5) by decreasing the ratio of water to P_2O_5. This acid consists of a mixture of pyrophosphoric acid ($H_4P_2O_7$) and higher molecular weight polymers, for example

$$3P_2O_5 + 5H_2O \rightarrow 2H_5P_3O_{10}. \tag{12.5}$$
$$\text{(tripolyphosphoric acid)}$$

Ammoniation of superphosphoric acid produces a very soluble fertilizer – *ammonium polyphosphate* (12% N and 23% P). This fertilizer is mainly used in liquid form, a form increasingly popular for all fertilizers in the USA and Europe. Compared to solids, *liquid fertilizers* offer the advantages of:
• precise application which is important for P and the micronutrients,
• more uniform application, especially for N,
• excellent carriers of pesticides and micronutrients, and
• well adapted to irrigation systems.
The use of soluble fertilizers in irrigation water is referred to as *fertigation*.

Insoluble phosphates

Phosphate rock (PR) ores are found in certain igneous and sedimentary rocks. The minerals present include apatites, crandallites, millisites, silica and calcite. The most important mineral *apatite* ranges in composition from *fluorapatite* $Ca_{10}(PO_4)F_2$ to *francolite*

$$Ca_{10-\alpha}(Na,Mg)_\alpha(PO_4)_{6-\beta}(CO_3)F_\gamma$$

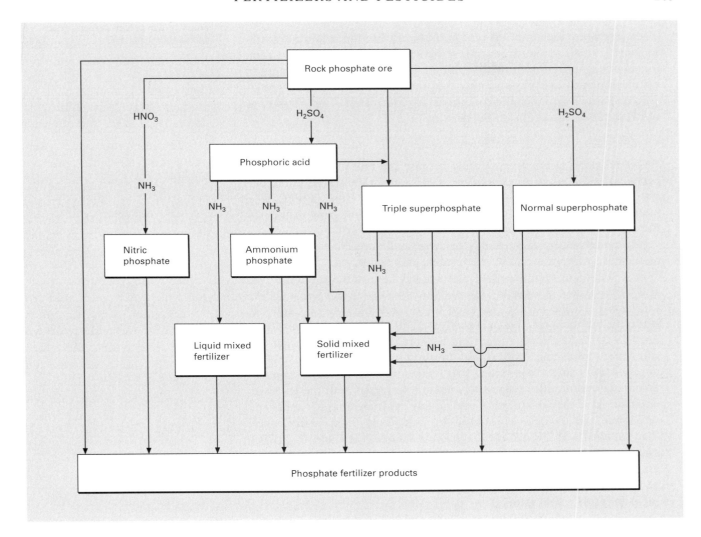

Fig. 12.4 Flow diagram of P fertilizer manufacture (after Slack, 1967).

where α and γ are functions of β, and $\beta = 1.5$ in the most substituted form. In general, the greater the degree of carbonate substitution, the smaller the crystal size and the more water soluble is the apatite. *Beneficiation* – the removal of much of the impurities by flotation and washing in water – is the first step in raising the P content of the crushed ore (to 13–18%). Finely ground

($< 150\ \mu m$) beneficiated PR can be used as a fertilizer under certain conditions (Box 12.3).

Other PR fertilizers are made by heating finely ground ore to 1200°C and above, a process called *calcination*, or by heating the ore with soda ash and silica (the product is called Rhenania phosphate in Germany). The roasting ignites organic impurities and converts fluorapatite to the more soluble β-*tricalcium phosphate* $Ca_3(PO_4)_2$.

Sometimes ground PR is mixed with *basic slag*, a

There are many sources of PR ores in the world, but the main reserves lie in North America, North and West Africa and the Middle East. The *agronomic effectiveness* of these materials when directly applied depends on the chemical and physical nature of the PR, soil properties, type of crop or pasture, and climatic conditions. A critical step is the dissolution of PR in the soil according to the reaction

$$Ca_{10}(PO_4)_6F_2 + 12H^+ \leftrightarrow 6H_2PO_4^- + 10Ca^{2+} + 2F^-. \qquad (B12.3.1)$$

This reaction is shown as reversible because the rate of dissolution depends not only on the concentration of H^+ ions in the soil solution, but also on the concentration of the main products Ca^{2+} and $H_2PO_4^-$. The interaction of soil, plant and climatic variables in determining the rate of PR dissolution is summarized in Fig. 12.5.

PRs are arbitrarily divided into 'reactive' and 'unreactive' (hard) rocks on the basis of their solubility in chemical extractants, e.g. 2% citric acid (New Zealand), or M ammonium citrate at pH 7 (Australia and the USA), or 2% formic acid (European Union). Hard rocks such as Florida (USA) and Nauru (Pacific Islands) are used for the manufacture of soluble P fertilizers, but more reactive phosphate rocks (RPRs) such as North Carolina (USA), Gafsa (North Africa) and Sechura (Peru) are used for partial acidulation (producing PAPRs), and for direct application. Given suitable crop, soil and climatic conditions, PAPRs and most RPRs can be comparable in agronomic effectiveness to soluble P fertilizers, and the RPRs in particular have the advantage of costing less per kilogram of P supplied. Agronomic effectiveness is assessed by comparing plant performance when supplied with PR against that when a soluble P fertilizer (e.g. TSP) is used at the same rate of P per hectare. For Australian and New Zealand agriculture, Bolan *et al.* (1990) suggested that suitable conditions for RPR use were:
- acid soils (pH < 6 in water),
- adequate moisture (> 800 mm rain, reasonably well distributed through the year), and
- permanent pastures on soils that are not severely deficient in P.

Box 12.3 Phosphate rocks as direct-application fertilizers.

by-product of steel manufacture from phosphatic iron ores. Basic slag must also be finely ground and is best used on acid soils where its Neutralizing Value (equivalent to approximately two-thirds its weight of ground limestone) and its micronutrient impurities are beneficial, especially for legumes.

Reactions in the soil

Dissolution of water soluble fertilizers

When a granule of MCP, the active constituent of the superphosphates (SSP and TSP), is placed in dry soil it

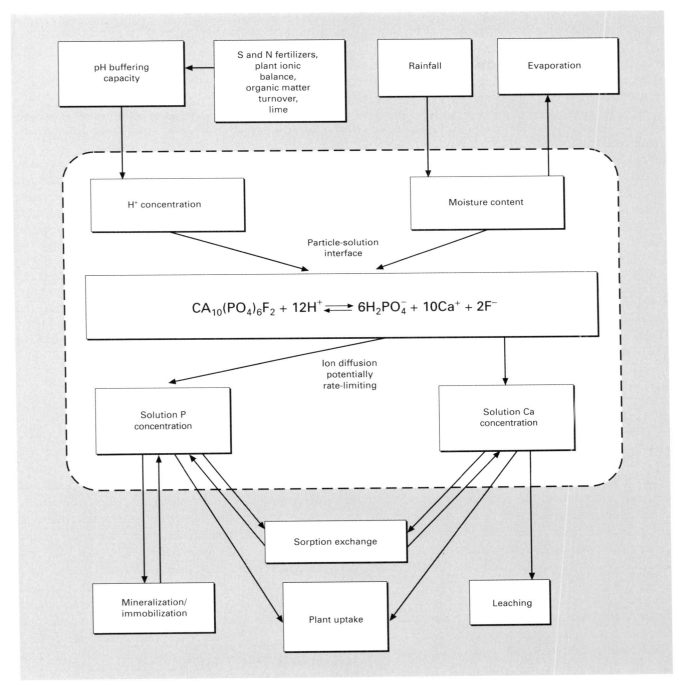

Fig. 12.5 Diagram showing the rate-determining factors (boxes in shaded areas) for PR dissolution in soil, and the variables (boxes outside shaded area) which determine the magnitude and degree of interaction of the rate-limiting factors (after Bolan *et al.* 1990).

takes up water by vapour diffusion until a saturated salt solution forms which flows outwards from the granule. In moist soil, liquid water is drawn into the dissolving granule by osmosis and the nutrient ions move out by diffusion. In dry soil, the saturated solution is very acid (pH ~ 1.5) and concentrated in P (c. 4M) and Ca (c. 1.4M): less so as the soil becomes moister. This acid solution reacts with the soil minerals, dissolving large amounts of Ca, Al, Fe and Mn, some of which are subsequently precipitated as new compounds – the *soil–fertilizer reaction products* – as the pH slowly rises and solubility products are exceeded. An MCP granule of 5 mm diameter dissolves incongruently in 24–36 h to leave a residue of dicalcium phosphate dihydrate (DCPD) and anhydrous dicalcium phosphate (DCP) containing about 20% of the original P. The sequence of events is summarized in Fig. 12.6.

The rapidity of this reaction means that a plant feeds not so much on the fertilizer itself, but on the fertilizer reaction products, which are metastable and revert slowly to thermodynamically more stable (but less soluble) products. Depending on whether the soil is calcare-ous or rich in sesquioxides, and on the presence of other salts such as NH_4Cl, $(NH_4)_2SO_4$ or KCl in the fertilizer granule, a range of intermediate products are formed including:

• potassium and ammonium *taranakites* ($H_6K_3Al_5(PO_4)_8.18H_2O$ and $H_6(NH_4)_3Al_5(PO_4)_8.18H_2O$),
• complex calcium–aluminium and calcium–iron phosphates, and
• DCPD.

The complex Al and Fe compounds hydrolyse to amorphous $AlPO_4$ and $FePO_4$, respectively, which may in very acid soils slowly revert to the stable end forms *variscite* and *strengite*, respectively. In acid soils, DCPD dissolves congruently, but at pH > 6.5 it hydrolyses to form a less soluble residue, probably *octacalcium phosphate* (OCP), which in more basic soils may revert to *tricalcium phosphate* (TCP) or *hydroxyapatite* (HA). The availability of P in the Al and Fe reaction products, relative to MCP, is illustrated in Fig. 12.7. Note also the effect of soil pH on the dissolution of the main Ca, Al and Fe fertilizer reaction products, as shown in Fig. 10.12.

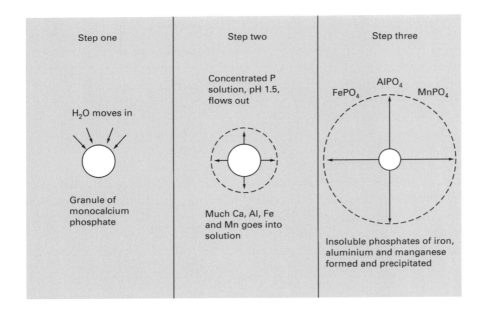

Fig. 12.6 Successive stages in the dissolution of a MCP granule in soil (after Tisdale and Nelson, 1975).

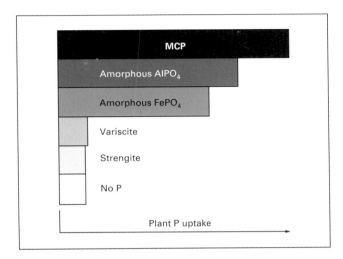

Fig. 12.7 Relative availability of different P sources to a crop (after Huffman, 1962).

Powdered vs. granular forms

Insoluble fertilizers, such as the basic calcium phosphates and PRs, are most effective when used in a finely divided form. Hygroscopic absorption of water does not cause the problem that it does with soluble fertilizers, but because fine powders are difficult to apply and create a dust problem, a form of weak aggregation or '*mini-granulation*' is sometimes employed. By contrast, granulation of soluble P fertilizers has the advantage, in addition to ease of handling, of confining the reaction between soil and fertilizer to small volumes around each granule. The rate of phosphate fixation in the soil is therefore slowed (Section 10.3).

Residual effects

Only 10–20% of fertilizer P may be absorbed by plants during the first year: the remainder is nearly all retained as fertilizer reaction products, which become less soluble with time. A rough guideline is that two-thirds of the water and citrate-soluble P remains after one crop, one-third after two crops, one-sixth after three crops and none after four crops.

Larsen (1971) has suggested that a more precise way

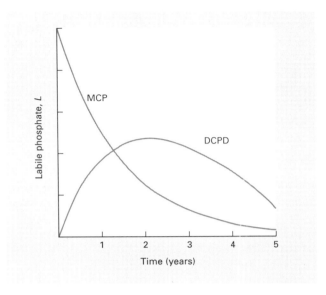

Fig. 12.8 Changes in soil labile phosphate after the addition of P fertilizer (after Larsen, 1971).

of assessing the residual effect is to measure the rate at which the *labile pool* of soil phosphate (the L value) declines after the addition of fertilizer, and to calculate the half-life for the 'decay' of the fertilizer's value to the crop. Half-lives in neutral and alkaline soils vary from 1 to 6 years. Larsen also pointed out that the less soluble P fertilizers, in contrast to soluble ones, require time to reach peak effectiveness, as well as having a half-life for reversion to an unavailable form (Fig. 12.8).

Eutrophication

Cause and effects

Eutrophication is a natural process whereby surface waters are gradually enriched with organic and inorganic nutrients carried in runoff from a surrounding catchment. One end-point of such natural eutrophication is the formation of fen peats, in both temperate and tropical climates, where base-rich waters and sediments collect in drainage depressions. Nevertheless, eutrophication can become a problem when it is accelerated by increased loads of sediment and dissolved nutrients in the streams draining urban and rural areas. The higher concentrations of P, N and organic matter promote the growth of aquatic plants, from minute *blue-green algae* (Cyanobacteria) to rooted higher plants. The most serious effects occur when there is a sudden and unpredictable surge in the growth of blue-green algae to create an '*algal bloom*' (Box 12.4).

Sources of nutrient input

Dissolved and particulate nutrient loads arise from *point* and *diffuse* sources. Examples of point sources are effluents from factories, sewage works and intensive livestock units, such as cattle feedlots, piggeries and dairies. Diffuse sources comprise the drainage from the remaining agricultural and non-agricultural land – grassland, arable and forest – which by its very nature is difficult to monitor accurately. Although the concentration of dissolved N and P may be critical for inducing an algal bloom, the total load of a nutrient entering a water body determines whether it becomes eutrophic or not.

In most countries, the bulk of the N-enrichment of surface waters comes from agricultural land. In the UK, for example, the drainage from 1 ha may carry between 5 and 250 kg N, depending on the climate, land use and N inputs (Table 12.3). In contrast to the mobile NO_3^- ion, orthophosphate ions are strongly adsorbed (Section 7.3) and the reaction products formed by P fertilizers in soil are relatively insoluble so that leaching (and surface runoff) from agricultural land usually contributes < 1 kg P ha^{-1} y^{-1} (Table 10.6). Phosphorus loads to surface waters are usually much higher from intensive animal industries and urban areas, the latter through sewage effluent. In the UK, for example, 65% of the P load to surface waters is from non-agricultural sources. However, there is accumulating evidence in several countries that, as soil P contents increase through regular inputs of fertilizer and manure P that exceed off-take in crop and animal products, P losses from agricultural land are increasing. Much of this loss is via surface runoff (Table 12.6), but losses of

Table 12.6 Phosphorus loss (kg ha^{-1} y^{-1}) from agricultural land in surface runoff. After White and Sharpley, 1996.

Land use and form of P input	Phosphorus loss	
	Dissolved P	Total P
Cereals or grass, P fertilizer	0.02–2.8	0.2– 5.5
Corn or lucerne, dairy manure	0.1–4.8	0.1– 7.4
Grass, pig manure	0.1–4.8	0.1– 4.8
Grass, poultry manure	0.0–4.3	0.1–12.4

1–3 kg P ha^{-1} y^{-1} by leaching (of soluble and colloidal P) have also been recorded.

12.4 Other fertilizers including micronutrient fertilizers

Potassium fertilizers

Potassium occurs naturally as KCl (*muriate of potash*) in salt deposits (mines and salt lakes), which also include some NaCl, K_2SO_4 and $MgSO_4$. The KCl is separated by flotation or by recrystallization, which depends on the differential response of the solubility of these salts to a change in temperature. Nevertheless, retention of a low Na and Mg content in K fertilizers is an advantage for the nutrition of some crops and also grazing animals (see below).

KCl is the most widely used K fertilizer, accounting for *c.* 95% of world use. It is especially favoured in liquid mixed fertilizers because of its high solubility. Other K fertilizers, representing a wide range of K content and water solubility, are listed in Table 12.7. Potassium sulphate is generally made by reacting SO_3

Table 12.7 Forms of potassium fertilizer.

Fertilizer	Composition	K content (%)
Potassium chloride	KCl	32–51
Potassium sulphate	K_2SO_4	41
Potassium nitrate	KNO_3	39
Kainit	$KCl.MgSO_4.NaCl.nH_2O$	12

Box 12.4 Algal blooms and water quality.

Blue-green algae such as *Anabaena, Microcystis* and *Nodularia* are natural components of all freshwater ecosystems and can also occur in saline waters. However, they only grow excessively (bloom) in nutrient-enriched waters, said to have a high trophic status or to be *eutrophic*. Waters with very low concentrations of nutrients are of low trophic status and are called *oligotrophic*. Algal blooms can have several adverse effects, for example:

• the production and release of toxins which can poison fish, waterfowl, animals and humans;

• the production of foul tastes and odours which make water less palatable to drink;

• the formation of surface scums which are visually displeasing and may emit offensive odours on decomposition. They also pose problems for water treatment works because of the fouling of filter beds;

• as the bloom subsides, dead vegetation sinks to the bottom and decomposes. Large quantities of dissolved oxygen are consumed, thus creating anoxic conditions which can result in the death of fish and other aquatic fauna.

Algal blooms are predisposed by:

• high concentrations of dissolved nutrients, especially P. Threshold P concentrations considered critical for algal blooms in several countries are given in Table B12.4.1. As water becomes more eutrophic, N becomes the limiting nutrient. Eutrophication of most marine and coastal waters is limited by N;

• high light and high temperature conditions in summer;

• low water turbidity which allows greater penetration of light;

• low flow rates which mean that nutrient concentrations can remain high. Most of the P entering surface waters is carried on suspended sediment. Much of this sediment settles to the bottom and acts as a long-term reserve for the overlying water into which P may be released by reduction of Fe–P compounds under anaerobic conditions (Section 8.4), or by desorption when the sediment is resuspended as a result of mechanical disturbance.

Table B12.4.1 Critical concentrations for P (mg L^{-1}) for the incidence of algal blooms.

Country	Dissolved reactive P	Total P (dissolved and particulate)
USA – selected States	0.05	0.1
Australia (Victoria)	–	0.05
European Union (guideline)	–	0.3

gas with water and KCl, that is

$$2KCl + SO_3 + H_2O \rightarrow K_2SO_4 + 2HCl. \quad (12.6)$$

It is less hygroscopic than KCl and therefore easier to handle. Despite its higher cost per kilogram K, it is the preferred source of K for crops that are sensitive to high concentrations of Cl, such as tobacco, potatoes and many vegetables grown in glasshouses. Potassium nitrate is also favoured for tobacco because it contains no Cl and all the N is present as NO_3^-; the preferred form of N for this crop. Kainite is an insoluble slow-release K fertilizer.

K balance in cropping

Because the K immediately available to plants is held as an exchangeable cation which is not readily leached, it is feasible to attempt to balance K removed by crops and in animal products with K applied in fertilizers and released from non-exchangeable sources in the soil.

K release

Slow release of non-exchangeable K depends on the content of potash feldspars and micaceous clay minerals. Soils with such reserves, even though low in exchangeable K, may supply between 20 and 80 kg ha^{-1} y^{-1} for many years, especially to grasses which are efficient in taking up K at low concentrations, but sandy soils lacking micaceous clays are soon exhausted of K by continuous cropping. On many soils where K is adequate initially, regular N and P fertilizing may so stimulate growth that K-deficiency symptoms appear, more often on pasture land where fodder is conserved as hay or silage.

K uptake by crops

Recovery of K fertilizer during one season varies from 25 to 80%, being highest for grassland, which may remove up to 450 kg K ha^{-1} y^{-1} when highly productive. Clovers and root crops require higher soluble K concentrations in soil than grasses, so that the ratio of N:K in fertilizer applied to such crops should be 2:3, compared to 1.5:1 for temperate grasses and 2:1 for tropical grasses. If too much K is used on pasture, the natural content of Mg in the fertilizer may be insufficient to prevent a K–Mg imbalance developing in the herbage, which may induce *grass tetany* or *hypomagnesaemia* in the grazing animals.

The Na content of K fertilizers is beneficial to animals, and Na also improves the yield and sugar content of sugar beet. The standard K fertilizer recommended for sugar beet in Germany, for example, contains 6% MgO and 8% Na$_2$O.

Sulphur fertilizers

Sulphur for fertilizers is produced by:
- roasting iron pyrites FeS$_2$ or as a by-product of the processing of non-ferrous metal sulphides,
- recovery from natural gas and gases released during oil refining which contain H$_2$S, or
- mining as elemental S.

The first two processes produce *'by-product' S*, most of which is used to make H$_2$SO$_4$ for the fertilizer industry. For use on its own, S may be made into prills from molten S and clay (up to 90% S). In the soil, S is slowly oxidized to sulphate by *Thiobacillus* bacteria according to the reaction

$$S + 3/2 O_2 + H_2O \rightarrow 2H^+ + SO_4^{2-}. \quad (12.7)$$

Another source of S for agriculture is *gypsum* (CaSO$_4$.2H$_2$O) which can be mined (from salt deposits), or produced during *flue-gas desulphurization* (Box 12.5), or as a by-product in the manufacture of triple and concentrated superphosphates (*phosphogypsum*). (NH$_4$)$_2$SO$_4$, containing 24% S, is the most soluble sulphate salt in fertilizers and is used in liquid mixed fertilizers.

Crop requirements

Crop requirements for S average 20–30 kg S ha^{-1}, and exceptionally up to 50 kg S ha^{-1} for some of the brassicas (e.g. oilseed rape). With the decrease in S emissions

The emission of S to the atmosphere from the combustion of fossil fuels and the effect of this S on the magnitude of acid inputs to the soil are discussed in Sections 10.3 and 11.3. The decrease in S emission achieved in recent years is partly due to a switch to lower S content fuels, but mainly to emission control by 'scrubbing' the flue gases from power stations through beds of crushed limestone, which produces gypsum. Gypsum produced in this way is mainly used a the raw material for plaster board manufacture, but can be used in agriculture if competitive in price with other forms of S.

Box 12.5 Sulphur emissions and flue-gas desulphurization

over western Europe in recent years (Box 10.3), S deficiency in crops is becoming more widespread and is bound to increase. This is also true of grassland that is heavily fertilized with N, because the N:S ratio in the herbage should be kept near 10:1 for the optimum nutrition of ruminant animals. Similarly, in areas of the tropics and subtropics where the soils are highly leached and atmospheric inputs of S low (e.g. northern Australia), the extent of S-deficient soils is widespread. For soils under pasture in high rainfall areas, SO_4-S is readily leached and in this case, supplying part of the fertilizer S as finely divided elemental S (a slow-release form) has merit for decreasing the rate of sulphate leaching. In New Zealand, for example, superphosphate fortified with elemental S (up to 50%) is used on soils that are naturally very deficient in S and subject to high rates of leaching.

Micronutrient fertilizers

The micronutrients Fe, Mn, Zn, Cu, B and Mo may need to be applied individually, or in mixed fertilizers, to correct a soil deficiency. Cobalt may also be required to correct a vitamin B_{12} deficiency in ruminants and for effective N_2 fixation by legumes. Selenium and I supplements to animal diets may be needed in deficient areas.

These elements can be added to the soil as sparingly soluble salts or oxides, or as chelates (Section 10.5). When used in the last mentioned form, or applied as foliar sprays of soluble salts, such as $ZnSO_4.7H_2O$, they are intended for immediate uptake by the plant. In their less soluble forms they maintain a very low concentration in the soil solution for several years, thus providing a prolonged *residual effect*. An increasingly popular slow-release form is the *frit*, made by fusing the element in a glass that can be crushed and mixed with NPK fertilizers. Table 12.8 gives the normal concentrations of these elements in mixed fertilizers and the maximum rate of application considered safe for most crops.

12.5 Pesticides in the soil

Environmental impact

The term 'pesticide' embraces the many natural and synthetic chemical compounds, which have biocidal or biostatic effects. They can be more specifically designated *insecticides*, *miticides*, *nematicides*, *fungicides*, *herbicides*, *rodenticides* or *molluscicides* according to the group of organisms they are intended to control. Of these, the herbicides, insecticides and fungicides account for the bulk of agricultural usage, in that order.

Table 12.8 Application rates for micronutrients in fertilizer. After Jones, 1982.

Element	Content (%)	Safe maximum (kg ha^{-1})
Mo	0.005	0.2
B	0.025	1.0
Mn	0.15	10
Cu	0.15	10
Zn	0.25	10

The *ideal pesticide* is one that controls only the target organism and persists long enough to complete this purpose before degrading into harmless products. It must also be of low toxicity to mammals. In practice, however, the ideal is not always achieved and chemicals have been used in agriculture (and in public health programmes, such as for the eradication of malaria-carrying mosquitoes) that are 'broad spectrum' in their activity; that is, they kill harmless and beneficial organisms as well as the target organisms. Furthermore, there are pesticides, such as the *organochlorine insecticides*, that are very stable and have the added property of high lipid solubility. They therefore accumulate in the fatty tissues of animals, particularly predators high up in natural food chains. Extensive use of these and other persistent chemicals has in the past led to their residues and breakdown products becoming widely disseminated, and there have been many instances of their detrimental effect on beneficial insects and plants, domestic animals and wildlife.

As a result of public concern over the environmental effects of these persistent and sometimes toxic chemicals, pesticide manufacturers voluntarily, or in response to government legislation, now undertake extensive research into the development and testing of new products before their release as pesticides, as shown in Fig. 12.9. These procedures are designed to evaluate not only the efficacy of a chemical for the control of specific pests but also its *environmental impact*, that is:
• the effect of the chemical on non-target organisms – beneficial insects, indigenous flora and fauna, especially mammals, and
• its persistence, defined as the *residence time* of the pesticide in a defined compartment (soil, water, organisms or air) of the environment (Fig. 12.10).

The soil plays a key role in determining the fate of a pesticide in the environment. Nematicides, many insecticides and fungicides are deliberately applied to soil to control soil-inhabiting pests. There are also soil-applied herbicides (e.g. pre-emergent herbicides) and an increasing number of systemic insecticides and fungicides intended for absorption by roots and transport to the aerial parts of plants. An even wider range of compounds reaches the soil unintentionally because of rainwash from leaves and the incorporation of pesticide-treated plant residues. Much attention has therefore

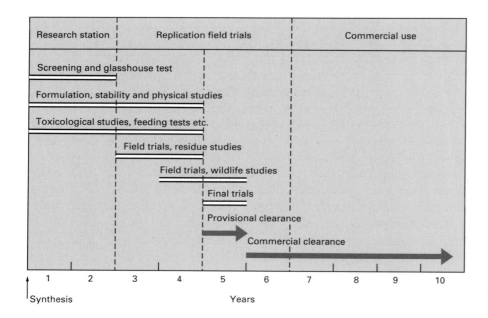

Fig. 12.9 Pesticide testing procedures in the UK (after Glasser, 1976).

been paid to improving the efficiency of pesticide application to target organisms which can be achieved through:
- improvement in spray technology, particularly to deliver sprays of optimum drop size and maximum chemical content per unit volume,
- minimizing spray drift, and
- microencapsulation to delay the release of an active constituent (controlled-release technology).

Pesticide persistence in soil

The main factors governing pesticide stability in soil and the loss of pesticide to surface waters, groundwater and air may be summarized as follows:
- volatilization,
- adsorption by soil minerals and organic matter,
- chemical and biological transformation,
- transport in the air, liquid or solid phase,
- absorption by plants and animals.

The relative importance of each process depends very much on:
- *the pesticide's properties* – its water solubility, volatility, affinity for organic or mineral surfaces,
- its *formulation* – usually the pesticide is dissolved in a suitable oil or organic solvent (with an emulsifier, and diluted with water as required), or supplied as a wettable powder (the pesticide is adsorbed on finely divided clay particles with a wetting agent), or as granules (the pesticide is mixed with inert fillers and pelletized),
- its *mode of application* – whether applied to crop surfaces or cultivated into the soil, and
- the *environmental conditions* (temperature and rainfall), soil type and cropping system.

Volatilization

Volatilization can be a major mechanism of pesticide loss from plant and soil surfaces, and for very persistent chemicals, such as the organochlorines, the means whereby they become widely dispersed in the environment. For example, for a relatively volatile compound like *lindane* (Table 12.9), more than 90% of the pesti-

Table 12.9 Description and key properties of four important organochlorine insecticides. After Guenzi, 1974 and others.

Common name	Chemical name	Solubility in water ($mg\ L^{-1}$)	Vapour pressure at 30°C (mmHg)
DDT	1,1-bis(4-chlorophenyl)-2,2,2-trichloroethane	0.001–0.04	7×10^{-7}
Dieldrin (HEOD)	1,2,3,4,10,10-hexachloro-cyclopentadiene	0.1–0.25	1×10^{-5}
Lindane (γ-BHC)	γ-1,2,3,4,5,6-hexachloro-cyclohexane	7.3–10.0	13×10^{-5}
Toxaphene	Chlorinated camphene containing *c.* 68% chlorine	0.4	No data

cide exposed on moist soil or plant surfaces can be lost to the air within 48 h of application. Incorporating the pesticide into the soil can decrease vapour losses substantially because of adsorption, but the factors controlling the volatilization of a soil-incorporated pesticide are complex (Box 12.6).

Adsorption–desorption

Pesticide molecules in the soil solution reach equilibrium with the solid phase by a variety of adsorption–desorption mechanisms, which are described for simple inorganic and organic ions in Chapter 7. Broadly, the position of this equilibrium depends on the chemical nature of the pesticide and the type of adsorbent (Box 2.5). The adsorbent in soil comprises a mixture of organic and inorganic surfaces which are predominantly negatively charged. For this purpose, pesticides may be grouped as:
- hydrophobic – water solubility $< 10^{-3}$M or approximately 300 mg L^{-1},
- sufficiently polar that the solubility $> 10^{-3}$M, but not a cation or weak acid or base,
- weak bases, i.e. $RH_2^+ \leftrightarrow RH$ (pesticide) + H^+,

Box 12.6 Volatilization of soil-incorporated pesticide.

In soil, the pesticide is distributed through the soil solids–water–air continuum. Partitioning between the soil solution and soil air is governed by Henry's law, that is

$$VD = hC, \hspace{4cm} \text{B12.6.1}$$

where VD is the vapour density (mass of vapour per unit volume), C is the concentration in solution and h is the *Henry's law coefficient*. The concentration in solution is governed by the adsorption–desorption equilibrium with the solid phase (see below). In moist soil, as the pesticide volatilizes from the surface a diffusion gradient develops, and the diffusive flux may limit the rate of surface volatilization for a compound with a high h value ($> 2.65 \times 10^{-5}$). However, if water is evaporating from the surface, pesticide is also delivered to the surface by mass flow, described as the '*wick effect*'. For relatively soluble chemicals such as lindane, mass flow will support a volatilization rate up to five times faster than when soil movement is by diffusion alone. The volatilization of pesticides with a h value considerably $< 2.65 \times 10^{-5}$ is controlled not by soil movement but by the boundary layer conditions above the soil surface.

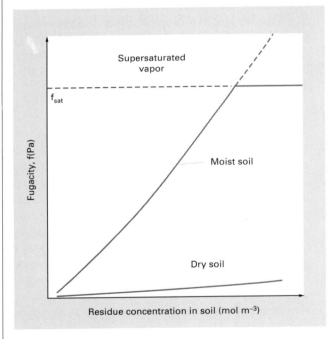

Fig. B12.6.1 The relationship between fugacity and pesticide residue concentration in moist and dry soil (after Taylor and Spencer, 1990).

Continued

Box 12.6 *Continued*

Irrespective of whether movement through the soil is by mass flow or diffusion, once the soil surface has become very dry (less than the equivalent of a monolayer of water molecules adhering to the solid surfaces), a pesticide is much more strongly adsorbed and the vapour density in the soil air (and hence volatilization) is decreased accordingly. The interpretation of pesticide volatilization in terms of *fugacity*, or the escaping tendency from the micro-environment of moist or dry soil, is illustrated in Fig. B12.6.1. For a moist soil, the fugacity increases as the pesticide concentration increases up to the maximum set by the saturated vapour pressure for the compound at that temperature. The slope of the line is inversely related to the K_d value for the pesticide in that soil. The figure also shows that the fugacity in dry soil is very low and changes little with increase in pesticide concentration.

- weak acids, i.e. RH (pesticide) $\leftrightarrow$ R$^-$ + H$^+$, and
- cations.

For the non-polar, hydrophobic compounds of low water solubility (e.g. the organochlorines), soil organic matter is the most important adsorbent. For pesticides that are weak bases or weak acids, the strength of adsorption depends on the soil pH because this influences the charge on the adsorbing surfaces and the charge on the molecule. Pesticides that are cations are the most strongly adsorbed. Some examples follow.

Weak bases. The *triazine* and *pyridione* herbicides are very weak bases with pK_a values ranging from 1.7 for atrazine to 4.3 for prometon. The molecules can protonate at pHs up to 2 units above the pK_a. The triazine *simazine*, for example,

protonates in acid media at one of its amino groups. This enhances its adsorption by clays.

Weak acids. Other herbicides, such as the *phenoxyalkanoic acids* and substituted *phenols*, are very weak acids. They exist as undissociated molecules or anions at normal soil pHs and hence are not appreciably adsorbed, except by non-specific adsorption or on positively charged sesquioxides. The more common representatives of this group are

2,4-D   2,4-dichlorophenoxyacetic acid

2,4,5-T   2,4,5-trichlorophenoxyacetic acid

MCPA   (4-chloro-o-tolyl)phenoxyacetic acid.

The herbicide *glyphosate* is an interesting example of a chemical that at normal soil pH values probably exists as the zwitterion

$$^-OOCCH_2\overset{+}{N}H_2CH_2PO_3H^-.$$

Because the charged groups are all pH-dependent, the net charge on the molecule changes with soil pH, but it is strongly adsorbed over a range of pH.

Cations. The bipyridyl herbicides *diquat* and *paraquat* are divalent cations which have the chemical formulas

Box 12.7 Sorption isotherms for pesticides.

The term *sorption* is used to describe retention of an adsorbate by an adsorbent, irrespective of the mechanism. The distribution of a pesticide between solid and liquid phases in soil can be described by *sorption isotherm*. If the isotherm is linear, or approximately so, the distribution coefficient K_d is given by the equation

$$Q = K_d C \tag{B12.7.1}$$

where Q is the amount sorbed per unit mass of soil and C is the pesticide concentration in solution at equilibrium. For non-ionized, non-polar molecules of low water solubility, organic matter provides the most important sorbing surface in the soil. The sorption characteristics of these chemicals is therefore better described by the equation

$$Q = K_{oc} C \tag{B12.7.2}$$

where the distribution coefficient K_{oc} is defined as K_d/soil organic C content. If not measured directly in soil, K_{oc} for different pesticides can be estimated with reasonable confidence from their octanol–water partition coefficients, that is

$$K_{ow} = \frac{[\text{pesticide concentration}]_{octanol}}{[\text{pesticide concentration}]_{water}}, \tag{B12.7.3}$$

where terms in brackets are the concentration of the pesticide in octanol and water, respectively. K_{oc} values are best used when the ratio of clay content to organic C is < 40. If the sorption isotherm is not linear, it may often be described by the non-linear *Freundlich equation*

$$Q = kC^n \tag{B12.7.4}$$

where k and n are coefficients.

Because of their positive charge and size, these molecules are very strongly adsorbed. Adsorption renders paraquat inert and also induces a shift in the wavelength of maximum light adsorption into the range of sunlight so that the molecule becomes unstable and is photolysed.

The interaction between adsorbing pesticides and soil is discussed in Box 12.7.

The combined effects of vapour pressure, water solubility and adsorption affinity on the distribution of several pesticides between the solid phase, water and air is exemplified in Table 12.10, assuming the soil has 50% pore space which is half-filled with water. This demonstrates that even for a volatile fumigant, such as ethylene dibromide, the major part of the chemical is retained by the solid phase. This has important consequences for the activity and biological transformations of a pesticide, as discussed below.

Chemical and biological transformations

With the increase in potency of pesticides, the quantity

which is needed for effective pest control has decreased dramatically. For example, at the beginning of this century inorganic pesticides, such as sodium chlorate, lead arsenate and flowers of sulphur, were applied at rates of $10–500 \text{ kg ha}^{-1}$. After the Second World War, with the advent of organochlorines, organophosphates and selective herbicides, rates fell to $0.5–5 \text{ kg ha}^{-1}$. At present, the synthetic pyrethroid insecticides are effective at rates of 10 g ha^{-1} or less. Thus, the concentration of pesticide in the soil is very small when compared with the other reactive constituents, such as clay, organic matter etc., so that the decay of the chemical usually follows first-order kinetics (*cf.* Equation 10.3).

Transformations of pesticides can occur *abiotically* or *biotically*. Note that these transformations can result in a relatively harmless chemical being converted into a compound more toxic to target and non-target organisms, as well as to the break down of the chemical to less harmful products. The former process is called *activation*; the second *detoxication* (Alexander, 1994). The versatility of micro-organisms is such, however, that eventually any pesticide will be metabolized or decomposed in some environmental niche; but the rates of decomposition can vary enormously. In this section, we are primarily concerned with transformations that lead to detoxication and decomposition.

Abiotic processes. These include:
• hydrolysis (e.g. the organophosphate and organophosphorothioate esters, carbamates, amides and anilides – mainly by alkaline hydrolysis),
• reduction under anoxic conditions, especially in sedi-

ments (but difficult to distinguish from biotic reduction – Section 8.4), and
• photolysis through exposure to sunlight (UV and visible radiation), most commonly in water. Some examples are paraquat (in the adsorbed state) and the pyrethroids. The natural pyrethroids and the early synthetic compounds decompose within a few hours of exposure to sunlight. On the other hand, the later synthetic compounds, such as *deltamethrin*, are not only photostable but are also potent at very low concentrations (Table 12.11). The stable synthetic pyrethroids are now being used against many pests previously controlled by DDT because of their potency and low environmental impact.

Biotic processes. The major process of pesticide decomposition in soil is through microbial activity. This is true of all groups of pesticides, with the notable exception of DDT and the cyclodienes, such as *dieldrin* and *lindane*, which are particularly recalcitrant. The processes involved are:
• biodegradation, in which the pesticide is a substrate for microbial growth,
• cometabolism, in which the pesticide is changed by the micro-organisms but does not serve as a direct energy source for growth,
• polymerization, in which pesticide molecules are linked with other molecules either of pesticide or naturally occurring organic molecules,
• accumulation, in which the pesticide is incorporated into the micro-organisms, and
• secondary effects in which the pesticide is changed

Table 12.10 Distribution of pesticides in a soil of 50% porosity, half-filled with water. After Hartley and Graham-Bryce, 1980.

Compound	K_d (L kg^{-1})	Amount in each phase (%)		
		Air	Liquid	Solid
Ethylene dibromide	0.5	0.7	28.4	70.9
Dimethoate	0.3	1.6×10^{-7}	40	60
Simazine	1.9	1.3×10^{-7}	9.5	90.5
Monuron	2.2	2.0×10^{-7}	8.3	91.7
DDT	5.0×14^4	1.2×10^{-6}	4.0×10^{-4}	100

Table 12.11 Activity of some chemical insecticides. After Graham-Bruce, 1983.

Chemical	Target organism	Median effective dose* (mg kg^{-1})
DDT	Housefly	10
Parathion	Housefly	2
Synergized deltamethrin	Housefly	0.002

*Dose that kills 50% of the population.

because of changes in pH, redox potential or reactive surfaces brought about as a result of microbial activity.

These are discussed further by Bollag and Liu (1990). Sorption of a chemical usually makes it physiologically inactive, although if it is reversibly adsorbed, its removal from solution by decomposition will cause more to be desorbed and become active. The long persistence of the bipyridyls is due to their irreversible sorption by clays when used at normal rates. But it is also possible that sorption of a chemical, especially on organic surfaces, may place it in closer proximity to colonies of bacteria or extracellular enzymes, thus hastening its decomposition. Therefore, no sound generalization on the effect of sorption on the rate of biotic decomposition of pesticides in soil can be made. Maximum persistence times of the major groups of pesticides, measured at the time for 90% or more of the chemical to disappear from the site of application, are shown in Fig. 12.10.

Enrichment and bioremediation. Biodegradation of the phenoxyalkanoic acid herbicides illustrates a phenomenon also observed with the phenylureas, organophosphates and carbamate insecticides when they are added to a soil not previously exposed to the compound in question. An initial lag phrase during which the herbicide concentration remains constant is followed by a period of rapid disappearance (curve A, Fig. 12.11). Soil micro-organisms, principally bacteria which are capable of biodegrading the foreign molecule, multiply during the lag phase. These 'adapted' organisms then enjoy a competitive advantage for substrate over non-adapted organisms, resulting in their proliferation and rapid breakdown of the herbicide. The soil is then said to be *enriched* with adapted organisms, a condition persisting for several months in the absence of fresh herbicide. Herbicide reapplied to an enriched soil is detoxified without any lag phrase (curve B, Fig. 12.11). The enhanced biodegradation of pesticides due to enrichment effects in soil is quite common and can lead to an apparent decline in the effectiveness of a pesticide in controlling its target organisms.

The phenomenon of enrichment can be used to advantage to accelerate the biodegradation of pesticides (and other undesirable organic chemicals) which are serious contaminants in soil, waste disposal sites and groundwater. This is the process of *bioremediation* where the source of contamination is treated *in situ* by stimulating the growth of organisms that can decompose the contaminants.

Absorption and transport of pesticides

Uptake of pesticides by plant roots depends on their water solubility; absorption through the cuticle of leaves depends more on their lipid solubility. Transport in the liquid phase mainly occurs by mass flow (Box 12.6), and therefore depends on the chemical's water solubility and the rate of water flow. Herbicides, like picloram, the phenylureas (monuron, fenuron) and the phenoxyalkanoics, can be transported in surface runoff or leached. Surface runoff losses are potentially

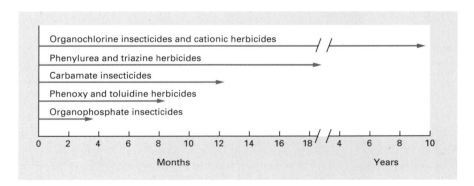

Fig. 12.10 Maximum persistence of pesticides in soil under a mild climate (after Stewart *et al.*, 1975).

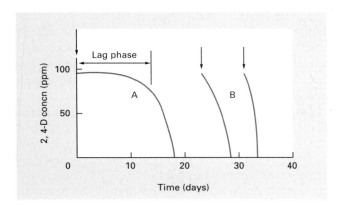

Fig. 12.11 Time course of microbial decomposition of 2,4-D in soil (curve A–initial enrichment, curve B–addition of herbicide to enriched soil; arrows indicate time of addition of the herbicide) (after Audus, 1960).

more serious than leaching loss in their off-site effects because once in the soil, these compounds are usually rapidly biodegraded. Strongly adsorbed pesticides (DDT, paraquat) can be transported when soil particles are removed during erosion. This can lead to residues of these persistent chemicals building up in stream and lake sediments. However, there is evidence that DDT

is decomposed more rapidly to the less toxic DDE under anaerobic conditions, in the presence of plant residues, so that its disappearance from sediments may be accelerated.

A diagrammatic summary of modes of pesticide input, reactions in soil and pathways of loss to the environment is shown in Fig. 12.12.

12.6 Summary

Substances recognized as *fertilizers* and *pesticides* have been used for centuries. But in the last 50 years there has been a phenomenal increase in the quantity and variety of agricultural chemicals used to meet the need for greater production and improved quality of the crops harvested.

The main nutrients supplied are N, P, K and S either singly in *straight* fertilizers or in various combinations in *compound* or *multinutrient* fertilizers. Most of the simple compounds of N – NH_4NO_3 $Ca(NO_3)_2$, $(NH_4)_2SO_4$ and $(NH_2)_2CO$ – are soluble solids, but N is also used as solutions of NH_4NO_3, urea, or NH_3 dissolved in water and as liquefied NH_3 gas. The N of these compounds is immediately available to the plant,

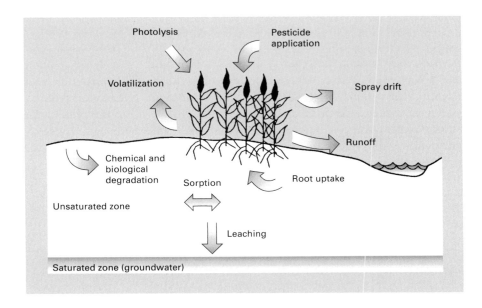

Fig. 12.12 Pesticide inputs, reaction in soil and pathways of loss (after Severn and Ballard, 1990).

but is vulnerable to leaching and denitrification in the NO_3^- form. Provided the amount and timing of the N-fertilizer input are matched to the crop's requirements, little of the fertilizer remains in *mineral* form in the soil by harvest time. But repeated high N inputs as fertilizer or in manures will build up the pool of labile soil organic N which can mineralize rapidly. Thus, leaching losses of NO_3^- from intensively farmed land can range up to $250\ kg\ N\ ha^{-1}\ y^{-1}$, depending on the rate of fertilizer (or manure) applied, the land use and the excess winter rainfall. Loss by leaching (and denitrification) can be decreased by the use of slow-release organic N fertilizers, or by nitrification inhibitors applied with a soluble N fertilizer.

Naturally occurring *phosphate rocks* (PRs) are insoluble in water and for most purposes must be treated with strong acids (H_2SO_4 or H_3PO_4) to convert the bulk of the P to a soluble form. An exception are the PRs consisting of francolite minerals (the reactive phosphate rocks RPRs) which may be used for direct application to permanent pastures on soils of pH < 6 (in water), under humid climates. Monocalcium phosphate $Ca(H_2PO_4)_2.H_2O$ is the active constituent of *single* and *triple superphosphates* (SSP and TSP, respectively), whereas higher analysis P fertilizers (the *polyphosphates*) are made from *superphosphoric acid*, formed by decreasing the amount of water in which P_2O_5 is dissolved. Soluble P fertilizers react quickly with the soil to form less soluble fertilizer reaction products on which plants feed. These fertilizer reaction products can have a long residual effect. P lost from agricultural systems can have a major effect on the *eutrophication* of surface waters and the outbreak of undesirable blue-green algal blooms.

The main K compounds KCl and K_2SO_4 are very soluble, but because K^+ is retained as an exchangeable cation in most soils, leaching losses are not large. Up to the last 20 years or so, inputs of S to the atmosphere from the combustion of fossil fuels, mainly in industrialized countries, were substantial and subsequent deposition on the land was sufficient to satisfy most crop requirements. With emission controls, however, deposition from the atmosphere has declined dramatically and S deficiency is becoming more widespread, necessitating the use of more S fertilizer – as $(NH_4)_2SO_4$, SSP, elemental S and gypsum $CaSO_4.2H_2O$. The micronutrients Fe, Mn, Zn, Cu, B, Mo and Co are applied at low concentrations in mixed fertilizers, or individually as chelates or soluble salts.

Ideally, a pesticide should be lethal for the target organism and remain active only long enough to control that pest. Pesticide chemicals are inactivated in the soil by sorption on to clays and organic matter, or undergo chemical or biological transformations. The strength of adsorption depends on the properties of the pesticide molecule and the nature of the adsorbent. Transformations can be *abiotic* or *biotic*. Biotic processes may cause chemicals to be converted to more active (and toxic) compounds – the process of *activation*; or cause the chemical to be rendered less active and toxic – the process of *detoxication*. The most important detoxication process in soil is decomposition when a pesticide becomes the substrate for microbial growth. Repeated use of a pesticide can result in soil becoming *enriched* with micro-organisms that rapidly decompose the pesticide. Although such an effect may decrease the pesticide's potency in controlling pests, it is of use for decontaminating sites *in situ* by the process of *bioremediation*. Of the more stable compounds, the *organochlorine insecticides* have become widely disseminated in the environment due to volatilization from the soil and their accumulation in animal fat tissues. More soluble compounds, such as the phenoxyacid and urea herbicides, may be leached, but fortunately most of them are decomposed rapidly by soil micro-organisms.

References

Addiscott T. M., Whitmore A. P. & Powlson D. S. (1991) *Farming, Fertilizers and the Nitrate Problem.* CAB International, Wallingford.

Alexander M. (1994) *Biodegradation and Bioremediation.* Academic Press, San Diego.

Audus L. J. (1960) Microbiological breakdown of herbicides in soils, in *Herbicides and the Soil* (Eds E. K. Woodford & G. R. Sagar). Blackwell Scientific Publications, Oxford, pp. 1–18.

Bolan N. S., White R. E. & Hedley M. J. (1990) A review of the use of phosphate rocks as fertilisers for direct application in

Australia and New Zealand. *Australian Journal of Experimental Agriculture* **30**, 297–313.

Bollag J. M. & Lius S. Y. (1990) Biological transformation processes of pesticides, in *Pesticides in the Soil Environment: Processes, Impacts and Modeling* (Ed. H. H. Cheng). No. 2 Soil Science Society of America Book Series. Soil Science Society of America, Madison, Wisconsin, pp. 169–211.

Cooke G. W. (1969) Prediction of nitrogen requirements of arable crops in mainly arable cropping systems, in *Nitrogen and Soil Organic Matter*. MAFF Technical Bulletin No. 15, pp. 40–60.

Croll B. T. (1994) Nitrate – best agricultural practice for water – the UK experience. *Proceedings of the Fertiliser Society* No. 359. The Fertiliser Society, London.

Glasser R. F. (1976) Pesticides: the legal environment, in *Pesticides and Human Welfare* (Eds D. L. Gunn & J. G. R. Stevens). Oxford University Press, Oxford, pp. 228–239.

Graham-Bryce I. J. (1983) Biting the hand that feeds. *Chemistry and Industry* No. 19, pp. 733–739.

Guenzi W. D. (Ed.) (1974) *Pesticides in Soil and Water*. Soil Science Society of America, Madison, Wisconsin.

Hartley G. S. & Graham-Bryce I. J. (1980) *Physical Principles of Pesticide Behaviour*. Academic Press, London.

Huffman E. O. (1962) Reactions of phosphate in soils: recent research by TVA. *Proceedings of the Fertiliser Society*. The Fertiliser Society, London, pp. 5–35.

Jones U. S. (1982) *Fertilizers and Soil Fertility*, 2nd edn. Prentice Hall, Virginia.

Larsen S. (1971) Residual phosphate in soils, in *Residual Value of Applied Nutrients*. MAFF Technical Bulletin No. 20, pp. 34–40.

Powlson D. S., Hart P. B. S., Poulton P. R., Johnstone A. E. & Jenkinson D. S. (1992) The influence of soil type, crop management and weather on the recovery of ^{15}N-labelled fertilizer applied to winter wheat in spring. *Journal of Agricultural Science, Cambridge* **118**, 83–100.

Slack A. V. (1967) *Chemistry and Technology of Fertilizers*. Wiley Interscience, New York.

Severn D. J. & Ballard G. (1990) Risk/benefit and regulations, in *Pesticides in the Soil Environment: Processes, Impacts and Modeling* (Ed. H. H. Cheng). No. 2 Soil Science Society of America Book Series. Soil Science Society of America, Madison, Wisconsin, pp. 467–491.

Stewart B. A., Woolhiser D. A., Wischmeier W. H., Caro J. H. & Frere M. H. (1975) *Control of Water Pollution from Cropland*, Vol. 1. Agricultural Research Service and Environmental Protection Agency, Washington DC.

Taylor A. W. & Spencer W. F. (1990) Volatilization and vapour transport processes, in *Pesticides in the Soil Environment: Processes, Impacts and Modeling* (Ed. H. H. Cheng). No. 2 Soil Science Society of America Book Series. Soil Science Society of America, Madison, Wisconsin, pp. 213–269.

Tisdale L. S. & Nelson W. L. (1975) *Soil Fertility and Fertilizers*, 3rd edn. Macmillan, London.

White R. E. & Sharpley A. N. (1996) The fate of non-metal contaminants in the soil environment. In: *Contaminants and the Soil Environment in the Australasia–Pacific Region* (eds R. Naidu, R. S. Kookune, D. P. Oliver, S. Rogers & M. J. McLaughlin). Kluwer Dordrecht, pp. 29–67.

Further reading

Bacon P. E. (Ed.) (1995) *Nitrogen Fertilization in the Environment*. Marcel Dekker, New York.

Bockman O. C., Kaarstad O, Lie O. H. & Richards I. (1990) *Agriculture and Fertilizers. Fertilizers in Perspective*. Norsk Hydro, Oslo.

Huffman E. O. (1968) The reactions of fertilizer phosphate with soils. *Outlook on Agriculture* **5**, 202–207.

Ministry of Agriculture, Fisheries and Food (1994) *Fertiliser Recommendations for Agricultural and Horticultural Crops*. Reference Book 209. HMSO, London.

Richardson M. L. (Ed.) (1991) *Chemistry, Agriculture and the Environment*. The Royal Society of Chemistry, Cambridge.

Chapter 13
Problem Soils

13.1 A broad perspective

Nutrient deficiencies, soil acidity, structural instability or a soil's susceptibility to erosion can affect the growth of natural vegetation and agricultural crops (Chapter 11). Other conditions which can severely limit the use of land for agriculture are a high concentration of soluble salts (*salinity*) and excess soil water (*waterlogging*).

As discussed in Sections 9.2 and 9.5, an excess of precipitation over evaporation for several months each year, impermeable subsurface layers and high groundwater tables, separately or combined, induce soil *waterlogging* and the attendant problems of:
- inadequate aeration for root growth and soil microbial activity,
- poor trafficability of the soil for machinery and animals, with the danger of surface structural collapse and 'poaching',*
- invasion by flood-tolerant weeds (e.g. sedges and rushes) and an increase in animal parasites and diseases favoured by wet conditions, and
- slow warming of the soil in spring.

The causal climatic factors are beyond control, but an improvement in soil drainage (Section 13.4) does much to mitigate these ill effects.

On the other hand, in areas of high evaporation, serious problems occur when the watertable rises to within 2 m of the soil surface and *salinization* occurs due to the upward movement and evaporation of saline groundwater (Sections 6.5 and 9.6). Such a situation may arise when the steady-state equilibrium of the soil's hydrology is disturbed by natural events, such as earth

*Rutting and damage to the wet soil surface by the hooves of animals; sometimes called 'pugging'.

movements (faulting) or long-term climatic change (which may extend over thousands of years), or by man's intervention through changing the land use or supplying irrigation (usually a time span of tens of years). Problems associated with hydrological disturbances, and their solutions, fall into two categories:
- soils not initially saline, with deep watertables, which become saline under irrigation or due to a change in land use, and
- soils already salinized and of limited cropping potential, but requiring careful management to avoid chemical and physical deterioration under irrigation.

13.2 Water management for salinity control

Permeation of saline groundwater

Soil salinization is a natural process. This is evidenced by the existence of salt pans and salt lakes in the inland areas of large land masses, where regional drainage collects naturally in low-lying parts of the landscape and salts accumulate following the evaporation of water. For example, in the southwest of Western Australia, many of the valley bottoms are naturally salinized and support a scrubby heath vegetation tolerant of high salinity. Groundwater levels are kept low under the valley slopes by the deep-rooted indigenous eucalyptus trees (Fig. 13.1a). But following forest clearance and the planting of short-rooted pasture and crop species, the rate of deep percolation through the soil has increased from *c.* 5 to 10 mm y^{-1}, resulting in a rise in the regional watertable. With the increase in hydraulic gradient, more saline water moves to the valley bottoms

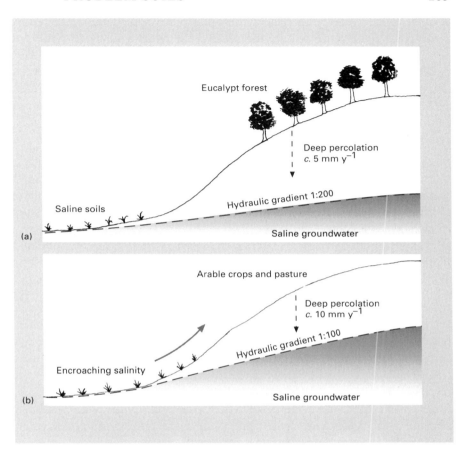

Fig. 13.1 (a) Natural landscape in south-western Australia in steady-state hydrological equilibrium. (b) Same landscape with raised groundwater table after forest clearance (after Holmes, 1971).

and gradually the area of saline soils creeps upslope (Fig. 13.1b). This sequence of events has been repeated in other parts of Australia, and is an example of *dryland salinization* (Box 13.1).

Soils under irrigation

Regulating the water supply

Some soils, notably those of the Nile Valley in north Africa, have been irrigated for centuries without adverse effects. During the annual Nile flood, water was led on to the land and impounded in basins of 400–16 000 ha for up to 40 days, after which the water receded and crops were planted. The soils were naturally well drained and

the annual flooding leached out any accumulated salts. More recently, in order that irrigated crops could be grown throughout the year, dams and barrages have been built to regulate the river's flow. Productivity has increased, but not without cost, because with the application of 1.5–2 m of water annually, most of which is lost by evapotranspiration, there has been an inevitable build-up of salts in the soils.

Abundant and cheap supplies of water also encourage farmers to apply more water than a crop can use, as insurance against the risk of any water stress which might reduce yield. The excess water drains to the groundwater, which slowly rises. Another problem is *seepage loss* from the canals and ditches that conduct water to the fields, which is especially serious in Pakis-

Box 13.1 Dryland salinity in Australia.

Although dryland salinity occurs naturally in Australia, the clearing of land for agriculture has caused existing areas to expand and salinity to appear in areas previously unaffected. *Induced* dryland salinity occurs as:
- *salt scalds* in arid and semi-arid regions, where surface vegetative cover is lost, exposing saline subsoils which are relatively impermeable, or
- *saline seepages* – typically caused by the sequence of events illustrated in Fig. 13.1.

The hill tops and upper slopes comprise the *recharge areas* for groundwater. The groundwater may be naturally saline, or becomes saline as the watertable rises and salts in the soil and parent material are dissolved and mobilized. Where the groundwater emerges at low points in the landscape (*discharge areas*), dryland salting occurs. Streams draining the land also become more saline.

Dryland salinity can be prevented (or at least minimized) by controlling recharge, through the retention of trees on recharge and discharge areas.

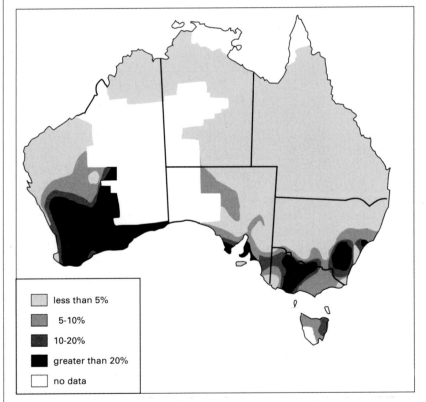

less than 5%

5-10%

10-20%

greater than 20%

no data

Fig. B13.1.1 Proportion of Australian farms with a salinity problem.

Continued

Box 13.1 *Continued*

However, where agriculture is already established, control (and amelioration) of dryland salinity can be attempted by:

- replacing short-rooted annual pastures and crops with deep-rooted perennial species such as the grass *Phalaris aquatica* and lucerne (*Medicago sativa*), combined with better grazing management to maintain the perennials;
- groundwater pumping (where aquifers are reasonably permeable), or surface drainage works to minimize accessions to groundwater. Groundwater pumping is expensive and only feasible in areas of high-value horticultural crops;
- planting of salt-tolerant species such as *Melaleuca* and tall wheat grass can be combined with surface drainage measures to reclaim land that is salinized.

In 1992, the area of salt scalds and seepage salting was estimated as 3.8 and 1.2 million hectares, respectively (Anon., 1995a). A further 1.6 million hectares are considered to be at risk. Estimates from Western Australia are more pessimistic, with 2.9 million hectares estimated to be affected in that State alone by the year 2010 (Anon., 1995b). An indication of the extent of dryland salinity in Australia is given in Fig. B13.1.1.

tan where extensive waterlogging of the permeable soils of the Indus Valley occurs alongside unlined irrigation canals (Fig. 13.2). A similar problem occurs in Australia, where water for agriculture and domestic use flows northwards into the semi-arid Wimmera–Mallee District through thousands of kilometres of unlined irrigation channels. Losses by evaporation and seepage exceed 90% of the water leaving the sources in the Grampian Mountains to the south. Seepage is prevented by lining canals with concrete – the best, but most expensive material. Asphalt or plastic sheeting are cheaper but of limited durability (Fig. 13.3).

Irrigation scheduling

Water application should be adjusted according to the *soil moisture deficit* (*SMD*) that develops under a crop (Section 6.4). The loss of soil water can be measured directly using a neutron probe (Box 6.4). Alternatively, it can be calculated from measured soil water suctions (using tensiometers, or gypsum resistance blocks for suctions > 85 kPa – Box 6.3) and a knowledge of the soil's moisture characteristic curve (Section 6.4). More appropriate for large areas is the prediction of *SMD* from evapotranspiration losses (*ET*), calculated from the Penman equation (Section 6.5). Thus

predicted *SMD* per metre depth $= ET$ (mm d^{-1}) × days from last watering.   (13.1)

Given that available water capacities (*AWC*) normally range from *c.* 80 mm m^{-1} for sandy soils to *c.* 200 mm m^{-1} for silty clays and clay loams, an *ET* rate of 6 mm d^{-1} (Fig. 6.13) would remove just over half the available water in 7–17 days, depending on the soil texture. When this point is reached, re-irrigation is recommended to avoid a check to crop growth.

Irrigation methods

The method of applying water also influences the likelihood of soil salinization. Control of infiltration and deep percolation is essential to prevent watertables rising above the critical depth of 2 m from the soil

Fig. 13.2 Soils in the background that have been waterlogged by lateral seepage from a large irrigation canal in Pakistan (courtesy of H. van Someren).

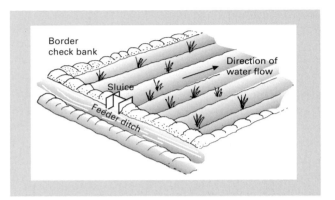

Fig. 13.4 Border check flood irrigation.

surface. Near level (< 4% slope) or uniformly graded land can be irrigated by *flood* or *furrow* methods, for example:

• *border check* or *border strip* flooding, in which water is distributed along bays separated by levees or mounds running parallel to the direction of flow (Fig. 13.4). Ideally, the bays should be constructed after the land has been levelled by laser-guided earth moving equipment. Sufficient depth of water (> 50 mm) must be applied at each irrigation to ensure that water reaches the bottom end of each bay before infiltrating the soil. To avoid overwatering at the input end, bays should be

Fig. 13.3 Lined irrigation channel in Victoria.

shorter in permeable soils (< 90 m) than in less permeable clays (up to 270 m). On the lowest gradients ($< 1\%$), construction of levees across the slope at vertical intervals of approximately 0.05 m to form rectangular basins gives more uniform water application, and therefore better control of infiltration. However, this method of *basin flooding* is more labour intensive and usually practical only on highly productive orchard and cereal crops;

• *furrow irrigation* is suited to row crops and can be used on land too steep or uneven to be flooded, provided that the furrows follow the contours. But there is the problem of salt redistribution in the soil as water moves by capillarity from the wet furrows to evaporate from the ridges where the plants are growing. The salt concentration at the apex of the ridge can be 5–10 times higher than in the body of the soil, creating an unfavourable environment for seed germination and seedling growth. This problem can be avoided by sowing seeds in double rows on broad sloping ridges (Fig. 13.5a), or in single rows on the longer slope of asymmetric ridges (Fig. 13.5b).

• *Sprinklers*, *sprays* or *trickle* emitters are suitable for

irregular terrain and more uniform (and precise) infiltration of water than is achievable by flood or furrow methods.

• *Spray* or *sprinkler* irrigation allows greater flexibility in the use of terrain and depth of water applied than either flood or furrow methods, but it requires more costly equipment. The water is delivered through upright nozzles fitted to pipes that are fixed or movable. Of the latter, self-propelled waterguns and motor-driven 'centre-pivot' sprays systems permit a uniform distribution of water and hence improved efficiency of application. This is important for keeping the leaching fraction under irrigation to the minimum necessary for proper control of soil salinity (see below). One drawback of spray irrigation is the evaporative losses, especially on windy days, which reduce the efficiency of application.

• *Trickle* or *drip* irrigation is an adaptation of spray irrigation that minimizes evaporative losses and is especially suited to high-value orchard and row crops on soils of low *AWC*. Water is delivered through flow-regulating 'emitters', which are placed close to the plants at regular intervals. Salts in the soil are leached to the periphery of the wetted zone, the shape of which depends on the soil's permeability and the rate of water application (Fig. 13.6). The irrigation rate can be adjusted to create a volume of low-salinity soil large enough for most of the roots to function without ill effect. Because the soil water potential in the root zone remains high, water of higher salinity (> 3 dS m^{-1}) that could not be used for spray or furrow irrigation is often acceptable for trickle irrigation. However, salt accumulation towards the fringe of the wet zone necessitates periodic flushing of the whole soil if salinity problems for subsequent crops are to be avoided.

Quality of irrigation water

Kinds of salts and their concentration

The major ions in surface and underground waters used for irrigation are Ca^{2+}, Mg^{2+}, Na^+, Cl^-, SO_4^{2-} and HCO_3^- with low concentrations of K^+ and NO_3^-. The total concentration of dissolved salts (*TDS*), sometimes referred to as the '*salinity hazard*', is most conveniently

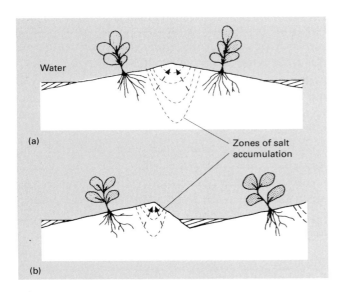

Fig. 13.5 Ridge and furrow designs to avoid salinity damage to plants. (a) Paired crop rows on broad ridges. (b) Single crop rows on asymmetric ridges.

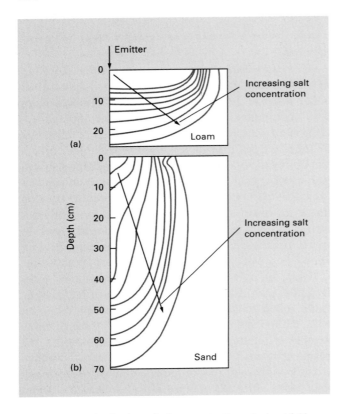

Fig. 13.6 The distribution of salt concentrations during trickle irrigation into soils of different texture (a) loam—input 4 L h^{-1}: (b) sand—input 4 L h^{-1} (after Bresler, 1975).

measured by the specific conductance, which is the electrical conductivity (*EC*) of the water, independent of the sample size. Analysis of many surface and well waters has revealed the approximate linear relationship

$$TDS\,(\text{mg L}^{-1}) \simeq 640\,EC\,(\text{dS m}^{-1}). \tag{13.2}$$

Note that *EC* is the reciprocal of the electrical resistance, hence the units of dS m^{-1} (SI units) which are numerically equivalent to the old units of mmho cm^{-1}. However, *EC* in Australia is commonly given in '*EC* units' which are μS cm^{-1} (dS m^{-1} × 1000). Equation 13.2 holds for waters of *EC* up to 10 dS m^{-1}. Note that *EC* increases with temperature and should be corrected to a standard temperature of 25 °C.

Experience in the western USA (Richards, 1954) suggests that *EC* class limits of < 0.25, 0.25–0.75, 0.75–2.25 and > 2.25 dS m^{-1} at 25 °C can be used to define waters of *low*, *medium*, *high* and *very high* salinity. The *EC* of most irrigation waters lies between 0.15 and 1.5 dS m^{-1} which is roughly equivalent to 1.5–15 mmol anion (−) or cation (+) charge per litre* (for a discussion of moles of charge and equivalents, see Box 2.3). In the soil, this water is concentrated 2–20-fold due to evapotranspiration, at which level many crops will experience a reduction in yield through osmotic stress. The osmotic potential ψ_s of the soil solution, which is a measure of the osmotic stress a plant might experience, can be calculated from the approximate formula

$$\psi_s(\text{bars}) = -0.36\,EC\,(\text{dS m}^{-1}). \tag{13.3}$$

Measurement of salinity in the soil solution and bulk soil is discussed in Box 13.2.

Sodium hazard

Apart from total salinity, the relative proportion of Na to Ca and Mg has an important effect on the quality of irrigation water. The Gapon equation (7.9), introduced in Section 7.2, shows that the proportion of exchangeable Na$^+$ to exchangeable Ca^{2+} and Mg^{2+} ions is determined by the Gapon coefficient $K_{\text{Na–Ca,Mg}}$ and the molar concentration ratio $(\text{Na}^+)/(\text{Ca}^{2+} + \text{Mg}^{2+})^{1/2}$ in the soil solution. In practice, when concentrations are expressed in meq L^{-1} we obtain the *sodium adsorption ratio* (*SAR*) as

$$SAR = \frac{[\text{Na}^+]}{\left[\dfrac{\text{Ca}^{2+} + \text{Mg}^{2+}}{2}\right]^{1/2}}. \tag{13.4}$$

In this form the *SAR* is approximately equal to the exchangeable Na percentage (*ESP*), in cmol charge (+)

---

* Because salt concentrations are usually expressed in milliequivalents per litre (meq L^{-1}) in the soil salinity literature, these units will be used throughout this chapter.

Box 13.2 Measurement of soil salinity.

Traditionally, soil salinity has been measured through the electrical conductivity (*EC*) of the soil solution. Sufficient distilled water is mixed with an air-dry soil sample to make a glistening paste and the saturation extract is then obtained by filtering under suction or centrifuging. The electrical conductivity of the *saturation extract* (*EC*$_e$) is measured with a conductivity meter. The *EC*$_e$ value is two to three times less than the true *EC* of the soil solution before dilution. In the USA and many other countries, the distinction between *saline* and *non-saline soils* is drawn at an *EC*$_e$ of 4 dS m^{-1} (Section 9.6). Very sensitive crops can be adversely affected at *EC*$_e$ values as low as 2 dS m^{-1} (see Fig. 13.9).

In Australia, *EC* is measured in a 1 : 5 soil/water suspension. After mixing the air-dry soil and distilled water, the suspension is allowed to settle before the electrodes of the conductivity meter are inserted into supernatant liquid. The *EC*$_{1:5}$ value will be much less than *EC*$_e$ for the same soil because of the dilution effect, although if solid salts are present (e.g. gypsum) these will dissolve and increase the *EC* value. Hence the method is of little use in such soils. An approximate value for the *TDS* (%) can be obtained by multiplying *EC*$_{1:5}$ (dS m^{-1} at 25°C) by 0.34.

Several field methods have been developed to measure the bulk soil *EC* (solid and solution phases combined) using a soil's response to pulses of electromagnetic radiation. The technique of *time domain reflectometry* (*TDR*) (Box 6.4) for measuring soil water content *in situ* can be adapted in saline soils (at constant water content) to measure bulk soil *EC*. The attenuation in the amplitude of an electromagnetic pulse transmitted down wave guides in the soil is related to the electrical conductivity of the surrounding soil. More popular, however, are portable *EM meters* which measure the soil *EC* to considerable depths from the attenuation of a pulse of electromagnetic radiation sent into the soil.

kg^{-1} soil or meq per 100 g soil, and is an index of the *sodium hazard* or *sodicity* of water. In the USA and many other countries, an *ESP* of 15 (or *SAR* $\simeq$ 15 meq$^{1/2}$ l$^{-1/2}$) has generally been accepted as the criterion of a *sodic soil*, which is a generic term for soils that exhibit poor physical properties due to the influence of Na (Box 13.5).

Prediction of a soil's *ESP* is vital to irrigation management because swelling pressures are enhanced as the *ESP* increases, especially if the total salt concentration of the soil solution is lowered (Section 13.3). To predict the likely *ESP* attained by a soil under irrigation, a

knowledge of the *SAR* of the irrigation water is essential. As the irrigation water is concentrated in the soil by evapotranspiration, assuming insoluble salts are not precipitated, the concentrations of all the ions increase in the same proportion, but the *SAR* increases disproportionately. For example, for a *twofold* increase in the ion concentrations, the *SAR* increases by $2/\sqrt{2} = 1.4$ (Equation 13.4). Exchange occurs between cations in solution and on the colloid surfaces until equilibrium is regained at a higher *ESP* value (Equation 7.9). Because the quantity of exchangeable cations in a given soil volume is usually much greater than the quantity of

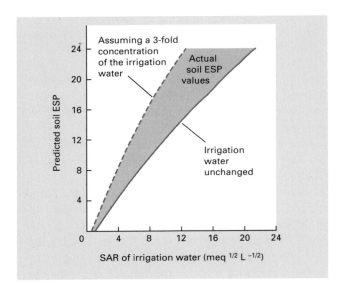

Fig. 13.7 Relationship between surface soil *ESP* and *SAR* of irrigation water (after Richards, 1954).

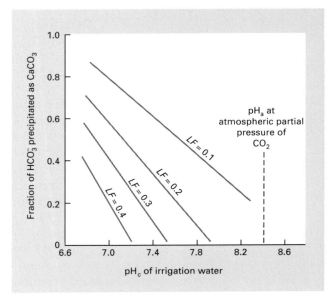

Fig. 13.8 Precipitation of insoluble carbonates in relation to pH_c of the irrigation water and *LF* (after Bower *et al.* 1968).

cations in solution, the changes in *ESP* are small when compared with the change in *SAR* of the irrigation water as it evaporates. Nevertheless, continued use of water of high *SAR* eventually leads to a high *ESP*. This difference in the cation buffering capacities of the solution and solid phases explains why, for a given *SAR*, waters of high salinity are considered to be of higher sodium hazard than waters of low salinity.

In Fig. 13.7, the solid line represents the relationship between predicted *ESP* and the *SAR* of a number of irrigation waters in western USA. The dashed line shows the predicted relation if the irrigation water were to be concentrated threefold in the soil. The actual *ESP* for a number of surface soils fell between the two lines, indicating reasonable agreement between prediction and measurement. However, *ESP* values deeper in the soil profile were less well predicted, probably because the simple *ESP–SAR* relation does not take account of:
1 precipitation of insoluble carbonates (Box 13.3), and
2 the differences in ion-pair formation and the effect of ionic strength on the activities on the Na^+, Ca^{2+} and

Mg^{2+} ions as the water becomes more concentrated through evapotranspiration.

Salt balance

Even good quality irrigation water contains some salts. Because crops take up little of this salt (about one-tenth) but transpire nearly all the water, salts inevitably accumulate in the soil under sustained irrigation. Good irrigation practice seeks to minimize this salt build-up.

A complete salt balance for an irrigated soil can be written as

$$S_p + S_{iw} + S_r + S_d + S_f = S_{dw} + S_c + S_{ppt}. \qquad (13.5)$$

In this equation, the cumulative contribution of S_p (atmospheric inputs), S_r (residual soil salts), S_d (salt released by weathering) and S_f (fertilizer salts) is usually small when compared with S_{iw} (salts in the irrigation water) and tends to be balanced by S_c (crop removal of salt). Usually, the effect of S_{ppt} (precipitation of carbonates and sulphates in the soil) has been discounted and

Box 13.3 Carbonate precipitation in irrigated soils.

Precipitation of $CaCO_3$ in soil can be predicted from the Langelier *saturation index* (*SI*) defined as

$$SI = pH_a - pH_c \tag{B13.3.1}$$

where pH_a ($= 8.4$) is the pH of water just saturated with respect to $CaCO_3$ at atmospheric pressure of CO_2 (Table 11.5), and pH_c is given by the equation

$$pH_c = (pK_2 - pK_{sp} + p\gamma_{HCO_3} + p\gamma_{Ca}) + p[Ca] + p[HCO_3]. \tag{B13.3.2}$$

The terms in round brackets are the negative logs of the second dissociation constant of H_2CO_3, the solubility product of $CaCO_3$ and the activity coefficients of HCO_3^- and Ca^{2+} ions, respectively. Because the sum of these terms is relatively constant, pH_c is influenced mainly by the concentration of calcium and bicarbonate in solution: as these increase (and $CaCO_3$ is more likely to precipitate), $p[Ca]$ and $p[Ca]$ will decrease and pH_c decreases. Thus, as pH_c falls below 8.4, $CaCO_3$ is likely to precipitate. This is illustrated in Fig. 13.8, which also shows that the smaller the fraction of applied water draining through the soil (the *leaching fraction*, *LF*), the greater is the precipitation of $CaCO_3$. The effects of *LF* and *SI* on the *SAR* of irrigation water, and hence on the predicted *ESP*, have been incorporated in two empirical equations

$$ESP \simeq SAR_u = SAR(1 + SI) \tag{B13.3.3}$$

for the upper (*u*) root zone, and

$$ESP \simeq SAR_l = kSAR \tag{B13.3.4}$$

for the lower (1) root zone. The coefficient k depends on the *LF* and the rate of soil mineral weathering (Rhoades, 1968). An alternative to this empirical approach is to adjust the *SAR* value for the effects of ionic strength and ion-pair formation (see point 2 on p. 290).

---

management practices have concentrated on matching S_{iw} to S_{dw} (salt removed in drainage water). This approach leads to the simplified equation

$$EC_{iw}D_{iw} = EC_{dw}D_{dw} \tag{13.6}$$

where the subscripts *iw* and *dw* refer to irrigation and drainage water, respectively, D is the volume of water applied per unit area (units of mm), and EC measures the salt concentration. Rearranging Equation 13.6 gives

$$\frac{EC_{iw}}{EC_{dw}} = \frac{D_{dw}}{D_{iw}} = leaching\ requirement\ (LR)\ . \tag{13.7}$$

The *LR* is therefore defined as the fractional depth of water, of given EC_{iw}, that must pass through the soil to maintain *EC* in the upper two-thirds of the root zone below a specified value. This critical value of *EC*, expressed as EC_e, is set by the crop's tolerance to salinity, which can be interpolated from diagrams of the kind

Box 13.4 Total salinity management.

Unnecessary leaching of salts from soil into 'return' flow creates salinity problems for water-users downstream. The Murray–Darling Basin of Australia provides an unfortunate example of steadily rising salinity proceeding downstream in the Murray River as successive irrigation areas discharge their return flows into the river and its tributaries. Proper irrigation management must therefore focus on the dual objective of:
- minimizing saline return flows, and
- avoiding a build-up of soluble salts in the soil and groundwater immediately below the soil.

In this context, the distinction between *LR* and *LF*, the *actual* fractional depth of the applied water that passes through the root zone, is important. Ideally *LF* should equal *LR*, but because of variability in the application and infiltration of water on a field scale, the average *LF* is generally > *LR*. Thus, crucial aspects of salinity management under irrigation are to:
- keep the *LR* as low as possible (< 0.05) by using water of low EC_{iw}, and
- optimize the uniformity of water application and infiltration into the soil so that *LF* is as low as possible ($\simeq LR$).

For irrigation waters high in Na, a low pH_c which promotes the precipitation of $CaCO_3$ (Fig. 13.8) may be a disadvantage; but for non-sodic waters, maximizing the precipitation of salts in the profile can be an advantage for off-site salinity control.

shown in Fig. 13.9. *LR* values usually lie between 0.1 and 0.3.

Equation 13.6 can be rewritten by replacing the term D_{dw} by D_{iw} minus D_{cw}, the total depth of irrigation water required to satisfy the crop's consumptive use, that is

$$EC_{iw}D_{iw} = EC_{dw}(D_{iw} - D_{cw}) \qquad (13.8)$$

from which the depth of irrigation water to be applied is calculated as

$$D_{iw} = \frac{D_{cw}}{1 - LR}. \qquad (13.9)$$

However, good salinity management must consider not only the soil salt balance, but also the off-site effects of salt (Box 13.4).

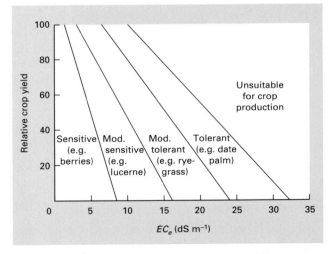

Fig. 13.9 Salt tolerance classes for crops based on the EC_e of the saturation extract (after Maas and Hoffman, 1977).

Specific ion effects

In addition to the osmotic effects of high salt concentrations on plant growth, specific toxicity symptoms arise when individual elements exceed certain concentrations. For example, fruit trees are likely to show marginal leaf burn and necrotic spots when Na and Cl concentrations exceed 0.2 and 0.5%, respectively, in the dry matter. Lithium can prove toxic at concentrations > 0.1 mg L^{-1} in the soil solution and irrigation water with > 0.3 mg B L^{-1} is suspect for use on sensitive crops like citrus.

13.3 Reclamation of salt-affected soils

Permeability changes on leaching

As discussed in Section 13.2, the evaporation of water from the soil–plant system leads to a rise in the *SAR* of the soil solution and eventually to an increase in *ESP*. The effect is exacerbated when high-salt irrigation water is used. A saline soil ($EC_e \geqslant 4$ dS m^{-1}) has been called *saline–sodic* when the *ESP* value is $\geqslant 15$. A more detailed description of saline–sodic and sodic soils is given in Box 13.5.

Box 13.5 Saline–sodic and sodic soils.

Australia and Africa have large areas of soils that are naturally *sodic*. The definition of sodicity is based on a soil's *ESP*, and ranges from $\geqslant 6\%$ in Australia to $\geqslant 15\%$ in the USA. The lower *ESP* threshold for Australian soils is atttributed to the generally low electrolyte concentration of the soil solutions and lack of weatherable minerals than can maintain the electrolyte concentration during leaching. Rengasamy and Olsson (1991) have proposed a classification of Australian sodic soils based on:
- sodium adsorption ratio (*SAR*),
- a threshold electrolyte concentration, *TEC* (measured as $EC_{1:5}$), and
- soil pH (1:5 soil/water suspension).

SAR is a surrogate for *ESP* and when measured in a 1:5 soil/water extract is approximately *half* the corresponding soil *ESP*. The *TEC* is defined as the *minimum* concentration of salts required to maintain the soil in a permeable condition. It is not a fixed value, as assumed in the general definition of a saline-sodic soil, but depends on the *SAR* (see Fig. 13.11). The classification is as follows

Saline–sodic ($SAR > 3$, $EC > TEC$)	Sodic ($SAR > 3$, $EC < TEC$)	
Alkaline sodic (pH > 8.0)	Neutral sodic (pH 6.0–8.0)	Acidic sodic (pH < 6.0)

Transformation from a saline-sodic soil to an acidic sodic soil can occur naturally (Fig. 9.16). In addition, change from the saline–sodic to sodic category can occur under irrigation, as discussed below. Sodic soils occupy some 60% of the area of Victoria, their distribution reflecting the influence of parent materials and effective rainfall. Few sodic soils occur in the wetter, mountainous country to the east and the alkaline-sodic soils occur mainly in the drier northwest of the State (Fig. 13.10).

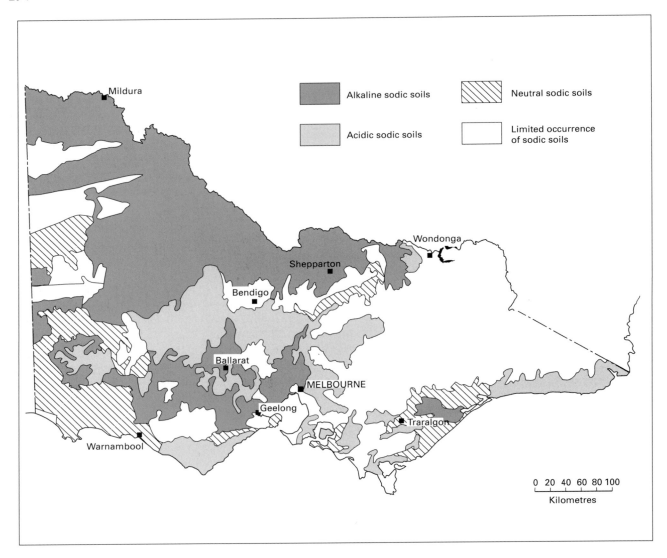

Fig. 13.10 Major sodic soil classes of Victoria.

Saline–sodic soils need very careful management because, although the structure remains stable as long as the soil solution $EC > TEC$, when soluble salts are removed by leaching, adverse changes in structure occur. Figure 13.11 illustrates the interactive effect between the SAR and the salt concentration of percolating water on a saline–sodic soil's permeability. The concentration below which the hydraulic conductivity falls by 10–15% is 10 meq L^{-1} at an SAR of 17 ($ESP \simeq 17$). The decrease in hydraulic conductivity is caused by the *swelling* of clay domains and quasi-crystals (Section 7.4), which decreases the porosity and may also lead to deflocculation of clay particles. These may then be eluviated and block the water-conducting pores. Although

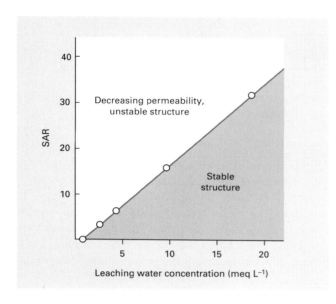

Fig. 13.11 Interactive effect of the *SAR* and concentration of leaching water on soil permeability (after Quirk, 1994).

swelling of flocculated clay is reversible (by increasing the salt concentration), deflocculation *and* clay translocation are not, and they may cause a permanent deterioration in soil structure (Section 9.6).

The relationship in Fig. 13.11 is described by the equation

$$SAR = 1.79\,C - 0.6 \qquad (13.10)$$

where *C* is the salt concentration of the percolating water (meq L^{-1}). Note that Equation 13.10 does not include any solid phase properties. However, it has been found that, for the same soil profile, the surface soil (high in organic matter) requires a higher salt concentration to remain stable at a given *SAR* than the subsoil (low in organic matter). Conversely, a high iron oxide content can lower the salt concentration required for stability at a given *SAR*.

Reclamation of saline–sodic soils

Reclamation of sodic soils requires leaching with water of *SAR* low enough to initiate Ca^{2+} exchange for Na$^+$,

but of sufficiently high total salt concentration to preserve the soil's permeability. As the *ESP* is gradually reduced, the salinity of the leaching water may also be reduced until reclamation is complete. The right conditions can be created by adding soluble calcium salts, usually as gypsum CaSO$_4$.2H$_2$O, which has a solubility of 30 meq L^{-1} in distilled water. Gypsum can be spread on the soil or dissolved in the irrigation water, its solubility increasing in more concentrated solutions due to the effect of ionic strength and ion-pair formation in reducing the activity of the Ca^{2+} and SO$_4^{2-}$ ions. Approximately 3.4 t is required to reduce by one unit the *ESP* of 1 ha of soil to a depth of 15 cm. Calcium carbonate, if present, is normally too insoluble to lower the *SAR* of the leaching water significantly; but in soils containing *Thiobacillus* bacteria the addition of elemental S enhances the solution of CaCO$_3$ due to the acidity generated when S is oxidized to H$_2$SO$_4$ (Equation 12.7).

Metal sulphides in marine sediments behave similarly when first exposed to air (Box 9.4), a reaction which benefits reclamation of some of the Dutch polders. The H$^+$ and SO$_4^{2-}$ ions produced react with indigenous CaCO$_3$ to produce gypsum sufficient to saturate the soil solution for several years as the reclaimed land is leached by rainwater.

Blending of low and high salinity waters

The theory of saline-soil reclamation was put to an interesting test with a saline–sodic soil from the Coachella Valley in the USA. The soil, which had an *ESP* of 39 and an EC_e of 4.4 dS m^{-1}, was leached with Salton seawater that was progressively diluted with good quality Colorado riverwater. Each water mixture, of which the range in composition is given in Table 13.1, was leached through the soil until the soil was in equilibrium with that mixture. Using four stepwise reductions in the leaching water concentration, the soil permeability did not fall below 0.12 m d^{-1} and the *ESP* was lowered to 5 in only 12 days. However, when the soil was leached with Colorado riverwater alone, the permeability fell to 0.005 m d^{-1} and reclamation took 120 days.

Table 13.1 Reclamation of a saline-sodic soil with water of stepped down salt concentration and *SAR*. After Reeve and Bower, 1960.

Dilution ratio (seawater + riverwater)	Water properties		Soil properties	
	*TDS** (meq L^{-1})	*SAR*[†] (meq$^{1/2}$ $L^{-1/2}$)	Hydraulic conductivity (m d^{-1})	*ESP*[#] (%)
Salton Sea	562	57	—	—
1:3	149	27	0.13	28
1:15	45.4	12	0.12	14
1:63	19.6	5	0.11	6
0:1	11.0	2	0.12	5
Colorado River	11.0	2	0.005	5

* total dissolved solids, [†] sodium adsorption ratio; [#] exchangeable sodium percentage.

Method of leaching

Intermittent leaching is more efficient in removing salts than continuous ponding. Water percolating through ponded soil tends to flow preferentially down the larger channels between aggregates (Section 6.3), bypassing much of the salt located in the small intraped pores.

Breaks in leaching process therefore allow time for salts to diffuse into the depleted outer regions of the aggregates, whence they are removed during the next leaching period. Thus, although 1 m of water may be required to leach 80% of the salt from 1 m of soil under ponding, as little as 30 cm of water applied intermittently can achieve the same result.

Removal of surplus water

It is a corollary of good irrigation management that surplus water, arising out of a leaching requirement or through seepage from a shallow groundwater table, should be disposed of efficiently. In permeable soils, the natural hydraulic properties may be adequate to cope with the excess water: where this is not a case, soil drainage must be improved. This can be achieved by:
• *surface drains* designed to remove as quickly as possible irrigation water and any surplus rainfall that has not infiltrated the soil, or

• *underdrainage* – pipes installed underground through which water is rapidly led away (see below).
Groundwater levels can be controlled by pumps, but such technology is outside the scope of this book.

13.4 Soil drainage

The purpose of drainage

Irrigation is practised predominantly in dry climates on soils which are reasonably permeable. Because large quantities of water are applied (up to 1000 mm y^{-1}), disposal of surplus water and salts and control of the regional groundwater table are the prime objectives. Achieving these objectives necessitates a drainage approach different from that for soils in humid temperate climates, where the natural permeability may be inadequate to prevent the soil becoming waterlogged for unduly long periods (> 1 day), when precipitation exceeds evaporation. Irrigation schemes are also very capital intensive, producing high returns per hectare, so that more expensive drainage systems (and groundwater pumps) may be justified on irrigated, but not on non-irrigated, land. These two broad categories of problem soils therefore warrant discussion individually.

Irrigated soils

The variables that interact to determine the watertable height at steady-state equilibrium in a soil with pipe drainage are:
• the saturated hydraulic conductivity K_s of the soil (Section 6.3),
• the depth of drains d and their height H above any impermeable in the soil; that is, a layer with a K_s value ⩽ 0.1 of the layer above,
• the distance L between drains, and
• the mean rate of water percolation J_w through the soil to the watertable.

The direction of flow to the drains is predominantly vertical throughout the unsaturated zone. Below the watertable, flow becomes two-dimensional (both vertical and horizontal) and then radial in the zone around

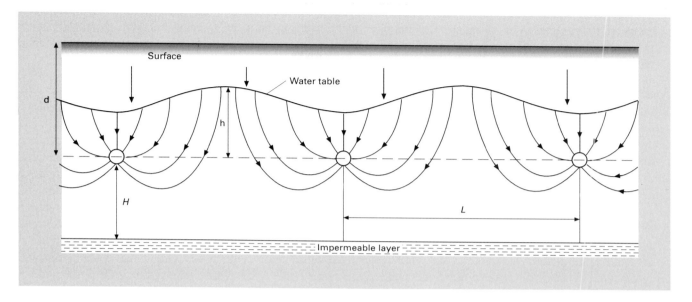

Fig. 13.12 Idealized flow lines of water moving to pipe drains after rain or irrigation.

the drain, as illustrated in Fig. 13.12. The relative extent of these flow zones depends on the magnitude of L, H and h, the last being the overall fall in hydraulic head from the watertable to drain level. When the rate of percolation J_w to the watertable is equal to the drain discharge rate, the system is in steady-state equilibrium and *Hooghoudt's equation* can be used to estimate the required value of L for given values of d, h, H and pipe diameter (Box 13.6).

Drainage of clay soils and duplex soils presents a different problem in that the impermeable layer usually lies at or just below drain depth ($H \rightarrow 0$), so that deep flow to the drains is inhibited. In this case, Equation B13.6.2 simplifies to

$$L = 2h\sqrt{K_s/J_w}. \tag{13.11}$$

Often the calculation of a theoretical drain spacing is vitiated by the great spatial variation in K_s so that predictions based on Hooghoudt's equation, or any other model of flow to drains, must be tempered by experience and practical judgement. Nevertheless, the

equation demonstrates that drain spacing can be increased if K_s increases or J_w decreases, and also if H increases (provided L is large) or h increases. To increase h, either the watertable at the mid-drain point must be allowed to rise or, more usually, drain depth must be increased. There are, however, practical limits to the depth at which drains can be placed, particularly when the permeability of the subsoil is low.

Soils of humid temperate regions

In reasonably permeable soils (including fen peats), drainage is sometimes achieved by a network of *surface ditches*, as illustrated in Fig. 13.14 This technique has also been used to remove surplus water in flood irrigation schemes, such as in much of the irrigation area of the Murray–Darling Basin of Australia. Design of such systems is essentially an engineering exercise to ensure adequate control of the regional watertable and flow rates in the ditches or channels. However, the drainage of wet, impermeable soils of humid regions generally requires a system of underdrainage, involving *pipe* (or *tile*) *drains* with or without supplementary *mole drains*. A widely accepted design criterion for draining soils in

Box 13.6 Calculation of optimal drain spacing and depth.

Hooghoudt's approach was to reduce the complex flow pattern around pipe drains to the equivalent of horizontal flow to vertical trenches, at the same spacing as the pipes, but filled with water to a lesser height H_e above an impermeable layer (Fig. B13.6.1). Because $H_e < H$, there is less cross-sectional area available for horizontal flow, so the head drop is greater than in the real case. The difference in head loss is just equal to the head loss in radial flow in Fig. 13.12, so that the total head loss in the simulated and real cases is the same. The average thickness through which water flows in the equivalent case is $H_e + h/2$ (Fig. B13.6.1) and the equation giving the discharge rate J_w is

$$J_w = 8K_s h(H_e + h/2)/L^2. \tag{B13.6.1}$$

Rearranging Equation B13.6.1 gives

$$L = \sqrt{8K_s h(H_e + h/2)/J_w}. \tag{B13.6.2}$$

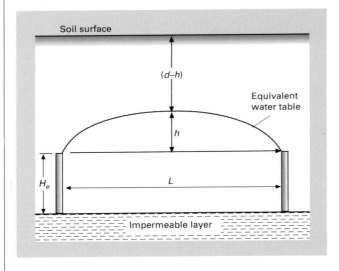

Fig B13.6.1 Diagram for Hooghoudt's 'equivalent flow' model to drains (after Smedema and Rycroft, 1983).

The effective height H_e above an impermeable layer is a function of H, the true height, L, and p, the wetted perimeter of the drain. Equation B13.6.2 must therefore be solved by successive approximations using trial values of L and H_e until the calculated value of L is very close to the trial value chosen. The procedure is:

Continued

Box 13.6 *Continued*

1 Choose the design criteria – J_w and the mid-drain watertable height $(d–h)$ (see Fig. B13.6.1). To prevent salt accumulation in an irrigated soil, J_w should be $\geqslant 0.001$ m d^{-1}. To prevent the capillary rise of saline groundwater bringing salts to the surface (Fig. 13.13), $(d–h)$ must be $\geqslant 2$ m.
2 Choose the drain depth d consistent with (1).
3 Establish the values of K_s and p.
4 Knowing H, the true depth to the impermeable layer, find from tables values of H_e for trial values of L.

these regions is to lower the watertable to 0.5 m below the surface at the mid-point between drains by 24 h after the end of rainfall. This will create a moisture suction of 5 kPa at the surface, corresponding to the FC of a well-drained soil. There are two major constraints on the effectiveness of such drainage.

• The *pore volume* drained at 5 kPa suction, because this determines the maximum air-filled porosity of the drained soil. Its value depends on the moisture characteristic curve of the soil (Section 6.4), but if it is < 10%, the improvement in aeration that drainage can achieve will be limited.

• The *hydraulic conductivity* K_s of the saturated soils, which determines the maximum rate of drainage when

the hydraulic head gradient is one. The conductivity of clay soils in Britain ranges from 10 to 100 m d^{-1} in the dry state, if deep fissures are prominent, to 0.1– 0.001 m d^{-1} when they are saturated and fully swollen. Given that the drainage system should be capable of removing 10 mm of surplus water per day, one can calculate that for an impermeable layer close to the drain depth, the theoretical drain spacing for a soil of K_s between 0.001 and 0.1 m d^{-1} is as small as 0.3–3 m (from Equation 13.11). Such close spacing of pipe drains is rarely economically feasible.

Underdrainage in practice

Drain depth

In order that the watertable in a soil does not come within 50 cm of the surface, pipe drains need to be placed more deeply, the actual depth depending on the position of any impermeable layer and the adequacy of

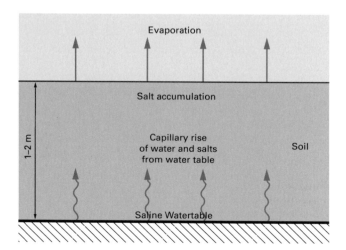

Fig. 13.13 Capillary rise of water and salts from saline groundwater.

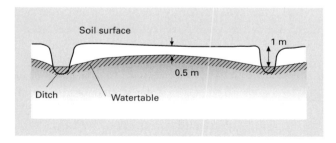

Fig. 13.14 Drainage ditches in a reasonably permeable soil with a high water table.

the outfall into the collection ditch. The drains are laid with a gentle gradient at a depth between 70 and 120 cm. The deeper the drains the drier the soil in the plough layer once hydraulic head equilibrium is established, but this condition is not always attained in clay soils in winter due to their low K_s values.

Tile drains are either short lengths of porous clay pipes, usually 75 mm in diameter, laid end-to-end, or continuous plastic pipe of the same diameter which is corrugated for strength and perforated to allow water entry. Because water enters the pipe from all sides, the drain's performance is much improved if it is imbedded in a permeable filling material (Fig. 13.15). Permeable fill, usually consisting of gravel or crushed stone of size range 5–50 mm, is of benefit for three reason:
1 It filters out very small soil particles which may otherwise block the drain,
2 it reduces the loss in head caused by the restricted

number of entry points for water into the pipe, and
3 it serves as a connection between the tile drains and any mole drains drawn at shallower depth.

If the hydraulic conductivity of the permeable fill is at least 10 times the K_s of the surrounding undisturbed soil, the head entry loss of the drain should be close to zero. Because gravel permeable fill is relatively expensive, well-structured soil from the A horizon of the trench dug for the tile drains can be used as 'permeable fill'.

Drain spacing

It is clear that laying tile drains at the spacing necessary to drain effectively soils with K_s values as low as 0.1–0.001 m d^{-1} is very expensive. The practical alternative is to choose an economic spacing between 20 and 80 m, commonly 20–40 m, and to try and increase water flow

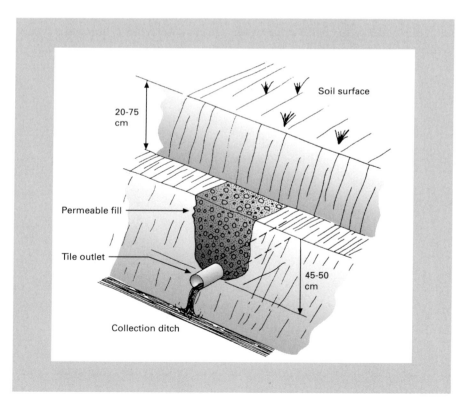

Fig. 13.15 Diagram of a tile drain imbedded in permeable fill.

to the drains by secondary treatments, of which the most important are:

- *moling.* This provides cheap drains at the spacing required for efficient drainage in impermeable soils. The depth of the mole drains determines the steady-state watertable position and the spacing governs its rate of rise and fall during and after rainfall. Closer spacings delay the rise and hasten the fall of the water-table.

- *subsoiling.* This has the aim of improving the hydraulic conductivity of the soil to the point where a drain spacing of 20–40 m provides efficient drainage.

As Equation 13.11 shows, however, a 10-fold increase in drain spacing requires a 100-fold increase in K_s for the same values of h and J_w. This is not easily achieved because the soil at depth is often too moist for effective subsoiling. Nevertheless, subsoiling can be effective where a plough pan markedly reduces the permeability of the soil above the drains.

Methods of moling and subsoiling

Moles are drawn by the implement illustrated in Fig. 13.16. The 'bullet' of 75 mm diameter is pulled through the soil at a depth of 40–50 cm and a spacing of 2–3 m. Behind the bullet, the slightly larger 'expander' consolidates the walls of the channel and seals the silt left by the vertical blade. Passage of the bullet and blade through the soil should fracture the structure to some extent and improve flow to the mole drain. Mole drains should lie at right angles to the tiles, provided their gradient can be kept between 2 and 5%, and pass through the top of the permeable fill (Fig. 13.16). Whereas tile drains should last for 50 years and more,

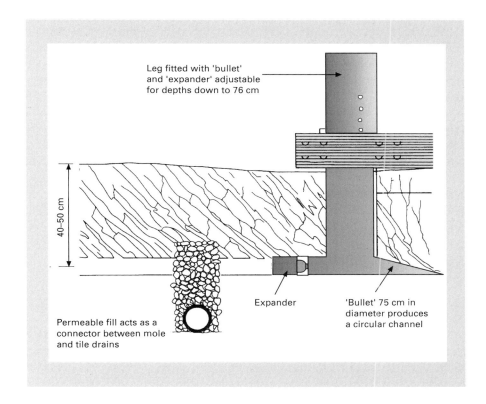

Fig. 13.16 Diagram of a mole drain plough in operation (after MAFF, 1980).

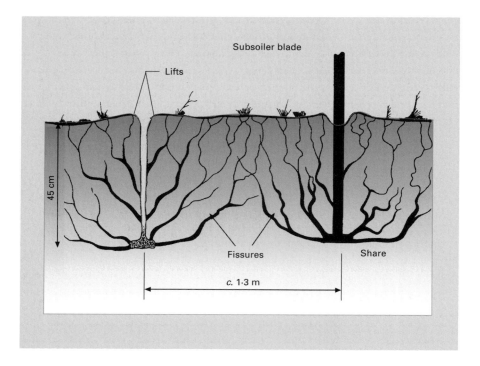

Fig. 13.17 Soil fissuring induced by effective subsoiling (after MAFF, 1981).

moles should be redrawn every 5–7 years for best results.

In subsoiling, a wedge-shaped share is pulled at right angles to the tile drains at 40–50 cm depth, if possible. Ideally, at this depth and a spacing of c. 1.3 m, the structure of all the upper soil profile should be disturbed by fracturing and fissuring (Fig. 13.17). The effect is more transient than moling. The conditions for effective moling and subsoiling are discussed in Box 13.7.

13.5 Summary

Excess soil water due to a consistently high $P:E$ ratio, an impermeable subsoil, or a high groundwater table, imposes severe limitations on land use.

In arid areas, groundwater is often saline and irrigation, if not properly managed, tends to raise the watertable. Capillary rise of salts and salinization then occur, with adverse effects on plant growth. Because most of the water added to soil is lost by evapotranspiration, but only about one-tenth of the salts are taken up by crops, irrigation itself increases soil salinity. The *salinity hazard* of irrigation water is determined by its total dissolved salts (*TDS*). *TDS* is conveniently measured by the electrical conductivity (*EC*) of the water, that is

$$TDS \,(\mathrm{mg\,l^{-1}}) \simeq 640 \; EC \,(\mathrm{dS\,m^{-1}}).$$

The concentrating effect of evaporation on the soil solution also causes the *sodium adsorption ratio* (*SAR*), defined by

$$SAR = \frac{[\mathrm{Na^+}]}{\left[\dfrac{\mathrm{Ca^{2+} + Mg^{2+}}}{2}\right]^{1/2}}$$

to increase. The equilibrium between exchangeable and solution cations then changes so that adsorbed $\mathrm{Ca^{2+}}$ and $\mathrm{Mg^{2+}}$ ions are replaced by $\mathrm{Na^+}$ and the exchangeable sodium percentage (*ESP*) of the soil increases.

Box 13.7 Effective moling and subsoiling.

The success of moling and subsoiling depends largely on the soil's *consistence* (Box 4.2) at the time of the operation. Consistence describes the change in the physical condition of the soil with moisture content, and reflects both the cohesive forces between particles (*ped shear strength*) and the resistance of the soil mass to deformation (*bulk shear strength*). The change in these two properties with change in moisture content θ for a clay soil is shown in Fig. B13.7.1. The ideal conditions for moling occur when:
• the soil at mole depth is near the *lower plastic limit* (maximum bulk shear strength and therefore stability of the mole channel when drawn), while the soil above is drier (in a *friable* state) and so tends to shatter as the mole is drawn, or for subsoiling when:
• the soil to 40–50 cm depth is friable, near to the *shrinkage limit*, when the bulk shear strength is lower and the soil should shatter as the subsoiler blade passes through.

These conditions are generally satisfied if the *SMD* in the whole soil to 1 m depth is at least 50 mm (for moling) and 100 mm (for subsoiling). The effect of subsoiling is usually more ephemeral than that of moling. Mole channels tend to have low stability in silty soils, and moling is quite unsuited to soils with sodic subsoils (Box 13.5), unless gypsum can be injected into the mole-channel walls to stabilize the clay.

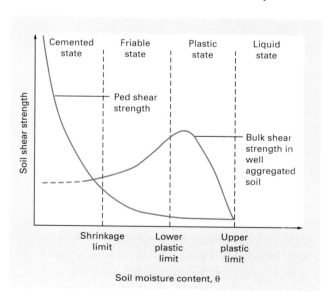

Fig. B13.7.1 Variation in soil shear strength and consistence with moisture content (after MAFF, 1975).

Thus, *SAR* can be used to asses the *sodicity* of irrigation waters. The rate at which the actual *ESP* increases with *SAR* depends on the total salinity of the irrigation water.

In the USA, a soil is classed as *saline* if the *EC* of its saturation extract (EC_e) $\geqslant$ 4 dS m^{-1}. If EC_e $\geqslant$ 4 dS m^{-1} and *ESP* $\geqslant$ 15, a soil is classed as *saline–sodic*. In Australia, the classification of *saline–sodic* and *sodic* soils is based on the *SAR*, a threshold electrolyte concentration, *TEC* (measured as *EC* in a 1 : 5 soil/water extract), and the soil pH. In this case, saline–sodic soils have an *SAR* > 3 and *EC* > *TEC*, and sodic soils have an *SAR* > 3 and *EC* < *TEC*. The *SAR* (1 : 5 soil/water ratio) in meq$^{1/2}$ L$^{-1/2}$ is approximately half the predicted *ESP*. Saline–sodic soils require careful management to avoid:

• swelling and loss of permeability, and possibly
• clay deflocculation and translocation,

when leached with good quality water. The first process is reversible; the second may lead to an irreversible deterioration of soil structure (solonization $\rightarrow$ sodic soils). Sodic soils are widespread in Australia and generally require *gypsum* to supply Ca^{2+} in solution and displace some of the exchangeable Na^+. In reclaiming saline or saline–sodic soils, leaching water of high EC and low *SAR* should be used and progressively diluted as salts and Na^+ ions are removed.

Drainage is often necessary to remove excess salts from irrigated soils. The ratio of drainage water to applied irrigation water that will maintain a salt balance in the soil is called the *leaching requirement* (*LR*). The ratio of *actual* drainage to irrigation water is the *leaching fraction* (*LF*). Ideally, *LF* = *LR*, but usually *LF* > *LR* because of unevenness in the application of water, particularly by *flood* or *furrow* methods. *Spray* irrigation is more efficient but the equipment is expensive. *Trickle* irrigation is the most efficient and allows water of a higher salinity than normal to be used. All irrigation should be scheduled according to water consumption by the crop (allowing for any *LR*).

In cool humid climates, problems of wetness, inadequate aeration, poor trafficability, invasion by flood-tolerant weeds, pests and disease, and low soil temperatures in spring can be mitigated by the installation of pipe or tile drains to create a suction of at least 5 kPa at the soil surface at the mid-drain point, 24 h after rain. The efficacy of this treatment will be limited if the drainable pore volume at 5 kPa suction is < 10%, or the saturated hydraulic conductivity K_s is < 0.01 m d^{-1}. The drain spacing necessary to achieve the necessary drainage rate can be calculated from Hooghoudt's equation, provided that certain soil characteristics (especially K_s) are known.

When K_s is so low that the pipe drains need to be placed at uneconomically close spacings (1–2 m), the practical solution is to use wider spacings (20–40 m) with *secondary treatments* such as *moling* and *subsoiling* to improve flow to the pipes. The *consistence* of the soil at moling or subsoiling is critical for the success of these operations. Mole drains are drawn above the pipes at intervals of 2–3 m, preferably when the soil moisture deficit (*SMD*) is at least 50 mm in the top metre (and the soil at mole depth is near the lower plastic limit). Imbedding the pipes in 20–30 cm of permeable fill promotes peripheral flow into the pipe and provides a good connection with the moles above. For best effect, subsoiling should be carried out when the *SMD* is at least 100 mm, and the soil to 40–50 cm depth is friable. Subsoiling is cheaper than moling but does not last as long.

References

Anon. (1995a) *Data Sheets on Natural Resource Issues.* Occasional Paper No. 6/95. Land & Water Resources Research & Development Corporation, Canberra.

Anon. (1995b) *Sustaining the Agricultural Resource Base.* Office of the Chief Scientist, Commonwealth of Australia, Canberra.

Bower C. A., Ogata G. & Tucker J. M. (1968) Sodium hazard of irrigation waters as influenced by leaching fraction and by precipitation or solution of calcium carbonate. *Soil Science* **106**, 24–34.

Bresler E. (1975) Two-dimensional transport of solutes during non-steady infiltration from a trickle source. *Soil Science Society of America Proceedings* **39**, 604–613.

Holmes J. W. (1971). Salinity and the hydrologic cycle, in *Salinity and Water Use* (Eds T. Talsma & J. R. Philip), Macmillan, London, pp. 25–40.

Maas E. V. & Hoffman G. J. (1977) Crop salt tolerance – current assessment. *Journal of Irrigation and Drainage Division. Proceedings of the American Society of Civic Engineers* **103**, 115–134.

Ministry of Agriculture, Fisheries and Food (1975) *A Design Philosophy for Heavy Soils.* Field Drainage Experimental Unit, Technical Bulletin 75/5.

Ministry of Agriculture, Fisheries and Food (1980) *Mole Drainage.* Field Drainage Leaflet No. 11. MAFF Publications, Middlesex.

Ministry of Agriculture, Fisheries and Food (1981) *Subsoiling as an Aid to Drainage.* Field Drainage Leaftlet No. 10. MAFF Publications, Northumberland.

Quirk J. P. (1994) Interparticle forces: a basis for the interpretation of soil physical behavior. *Advances in Agronomy* **53**, 121–183.

Reeve R. C. & Bower C. A. (1960) Use of high-salt waters as a flocculant and source of divalent cations for reclaiming sodic soils. *Soil Science* **90**, 139–144.

Rengasamy P. & Olsson K. A. (1991) Sodicity and soil structure. *Australian Journal of Soil Research* **29**, 935–952.

Rhoades J. D. (1968) Mineral-weathering correction for estimating the sodium hazard of irrigation waters. *Soil Science Society of America Proceedings* **32**, 648–652.

Richards L. A. (1954) (Ed.) *Diagnosis and Improvement of Saline and Alkali Soils.* United States Department of Agriculture, Handbook No. 60.

Smedema L. K. & Rycroft D. W. (1983) *Land Drainage.* Batsford Academic, London.

Further reading

Bresler E., McNeal, B. L. & Carter D. L. (1982) *Saline and Sodic Soils – Principles, Dynamics and Modelling.* Springer, Berlin.

Ritzema H. P. (1994) (Ed.) *Drainage Principles and Applications.* International Institute for Land Reclamation and Improvement Publication No. 16. Wageningen.

Smart P. & Herbertson J. G. (1992) (Eds) *Drainage Design.* Blackie, Glasgow.

Sumner M. E. (1993) Sodic soils: new perspectives. *Australian Journal of Soil Research* **31**, 683–750.

Thomasson A. J. (1975) (Ed.) *Soils and Field Drainage.* Soil Survey of England and Wales, Technical Monograph No. 7 Harpenden.

Chapter 14
Soil Information Systems

14.1 Collation, codification and communication

Soil variability

The point is made in Chapters 1 and 5 that the distinctive character of a soil is shaped by a complex interaction of many physical, chemical and biotic forces. The result is a remarkable range of soil types in the landscape (Box 1.1). The existence of such variability has two notable consequences, namely:

1 that soil scientist have been stimulated to create order out of disorder by collecting, collating and codifying soil information – the process of *classification*. The main purpose has been to facilitate the *communication* of existing knowledge and encourage the acquisition of new knowledge about soils and their behaviour.

2 that information about soils tend to be *site specific*. Many soil users – farmers, orchardists, landscape gardeners – acquire this site-specific information primarily from their own practical experience of 'working the land'. Other users – engineers, hydrologists, ecologists, planners – do so in a more abstract way. They require generic information either about the properties and behaviour of individual soil types (broadly defined), or the effect that a particular soil property (e.g. pH, water content, clay content and type of clay) has on the behaviour of a system they are studying (e.g. runoff in catchments, distribution of plant species, waste disposal sites and so on).

Hopefully, the activities referred to in (1) should benefit the corporate body of soil scientists, and satisfy the needs of the variety of non-specialist users identified in (2). But this has not occurred consistently and effectively for the two following main reasons.

a Soil scientists have communicated mainly with each other, or with scientists in closely related disciplines. Their language has not been readily understood by non-soil scientists and even less by lay people.

b In many cases, the end-users do not know enough about the role of the soil in their systems to ask the right questions of soil scientists.

With respect to (a), *classification* has been seen as a *prerequisite* of effective communication between soil scientists. But the traditional approach to classification (see below) has difficulties for the reasons outlined in Box 1.2 – mainly because *the soil is continuously variable in space and time*. In this chapter, the principles of soil classification are outlined and some of the main systems are discussed.

However, in this information technology age, many of the constraints which diverted soil scientists into traditional classification methods are no longer relevant. Soil information can be collected in digital form, or converted from analogue to digital form, and stored in huge databases. These databases can be accessed to provide information in response to specific requests – for soil properties or the distribution of soil types. No longer is it necessary to generalize soil information through soil classes – the primary data can be accurately referenced in space, stored, easily retrieved and rapidly disseminated. With respect to point (b), therefore, soil information can be made widely available in a form intelligible to non-specialist users. Some of the principles of *modern information systems* are outlined in this chapter.

14.2 Traditional classification

Faced with variability in natural populations, the traditional approach to classification usually involves three steps:

1 identification of the full range of variability in the population,

2 creation of groups or classes within the population according to the similarity between individuals, and

3 prediction of the likely behaviour of an unidentified individual from a knowledge of the characteristics of the class as a whole.

Step (2) – the *creation of classes* – is called classification. Soil classification is more difficult and contentious than the classification of other natural populations because a soil lacks the hereditary characteristics by which individuals within one generation are distinguished, and which are transmitted from one generation to the next. The soil population presents a continuum of variation so that *arbitrary* judgements are inevitable in the creation of classes.

Definition of a soil entity

Soil variability can be crudely partitioned according to the distance over which discernible changes in properties occur. Differences associated with soil faunal and microbial activity and the distribution of pores, for example, achieve maximum expression within 1 m (*short-range variation*). Since as much as half of the total variation in any property can occur as short-range variation, it is impossible to define rigorously a fundamental soil unit or entity. Attempts to do so have resulted in the introduction of esoteric terms such as the 'pedon' and 'pedounit'; but whatever the theoretical merits of these terms, the field worker *describes* a soil from a profile in a pit (usually 1 × 0.5 m in area, dug to the parent material or to 1.5 m, whichever is the shallower), or from a freshly exposed road cutting or quarry face. Because the excavation of pits is laborious and expensive, the assessment of soil variability is usually based on the examination of *soil samples* collected by means of augers or coring devices, inserted vertically into the soil (Fig. 14.1). Details of the observations made in soil survey are given by Hodgson (1978) and McDonald *et al.* (1990).

Kinds of classification

Another problem confronting the soil classifier is the choice of properties on which to separate classes. Many properties can be evaluated qualitatively in the field –

Fig. 14.1 A selection of augers and core samplers used in soil survey.

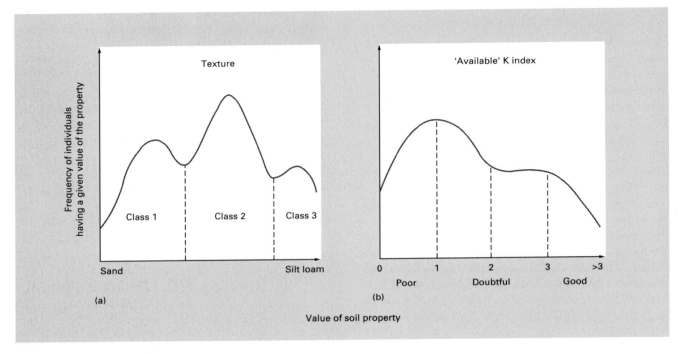

Fig. 14.2 (a) and (b) Subdivision of the variation in a soil property to define classes for a special-purpose classification.

soil colour, structure, type of humus and degree of gleying, for example. Other properties, such as pore-size distribution and the content or organic carbon, clay or calcium carbonate, must be measured in the laboratory. Practical constraints prevent all the soil properties being measured, and of those measured, not all are used. Some are selected as the *definitive* properties for class separation, the choice of which depends on the aim of the classification, which is discussed below.

Special-purpose classification

This kind of classification is made with a specific aim in mind and is based on one or a few soil properties. The texture classifications of Fig. 2.3 and the classification of soils for droughtiness and susceptibility to waterlogging (Fig. 11.6) are two cases in point. There are many other examples, especially relating to plant growth, such as the *ESP* and salinity (EC_e) status of irrigated soils, and the

assessment of a soil's 'available' P and K status by soil tests (Section 11.2). When classification is based on one or two properties, classes may be separated by subdividing the range of variation in the individual property at the minima in the frequency distribution (Fig. 14.2a), or according to the user's requirements (Fig. 14.2b).

General-purpose classification

Because this kind of classification is intended for many uses, foreseen and unforeseen, the classes are defined on as many properties as possible. Provided that the properties chosen are *relevant* to the use and management of the soil, the classification should be useful: if not, the classification may be academically elegant, but of little practical value. But where the important soil properties are those requiring tedious and expensive laboratory measurement, it is expedient to choose certain *diagnostic properties* for the separation of classes. These properties are not necessarily relevant to soil management, but are easily assessed in the field, and are assumed (or known) to be correlated with the important properties.

For example, the colour of the A horizon may be correlated with organic matter content, and mottling of the subsoil indicative of a low hydraulic conductivity. Where such a natural grouping or 'clustering' of soil properties exists, the classes created should be more meaningful to the general-purpose user than those defined by arbitrary criteria.

Quality of classifications

Implicit in the decision to classify soils is the assumption that the potential user will be able to make better generalizations about the properties of soils in a heterogeneous landscape than if there was no classification. In other words, better prediction of soil-dependent behaviour such as regional drainage, catchment sediment yield or crop productivity can be made using the class properties, rather than measurements made on the soil at a few individual sites. For this to be true, a substantial proportion of the variability in the properties of the whole population of soils must occur *between* classes rather than remain *within* classes. The accepted mathematical measure of the variability in a population is the variance σ^2. This parameter is a measure of the spread of individual values about the mean. This, and other statistical terms used in the quantitative analysis of the soil variability, are explained by Webster and Oliver (1990) to whom the reader should refer for a more detailed treatment. Assessment of the quality of soil classification using variances is discussed in Box 14.1.

14.3 Soil survey methods

What is soil survey

Soil survey primarily involves collecting information

Box 14.1 The reliability of soil classification.

It is found that, for a given categorical level of classification (Section 14.4), each class created has a similar range of variation so that a pooled *within-class* variance (σ_w^2) can be calculated by taking the weighted average of the class variances (the weighting factors are the number of degrees of freedom associated with each class variance). The effect of classification is then measured by the index

$$\tau = 1 - \sigma_w^2/\sigma_t^2 \qquad\qquad (B14.1.1)$$

where σ_t^2 is the total variance in the soil population. Clearly, when $\sigma_w^2 \simeq \sigma_t^2$ the spread of values within a class is comparable with the total spread of values, and the proportion of the total variance accounted for by the classification (τ) approaches 0 – the classification is therefore of little value. It is easiest to calculate τ for a special-purpose classification based on measurements of only one soil property. Typically, about the half the variance in a physical or mechanical property of the soil, but only about one-tenth of the variance in many chemical properties, can be attributed to differences between classes. The within-class variance can be decreased by increasing the number of observations from which the class mean and variance are calculated, but it is uncommon for σ_w^2 to be reduced much below 25% of the total variance unless a great deal of effort is expended. If the effort, and hence the cost of classification becomes too great, the alternative of measuring soil properties directly, when required, may be better than attempting prediction from a classification.

Devices used for remote sensing in soil and land survey detect electromagnetic radiation (EMR). The sensing may be:

- *active*, when the sensing device directs EMR at an object and detects the amount of that energy which is reflected back. Examples are radar and radio wave emissions (wavelength $> 10^3$ μm) and laser-imaging radar (UV, visible and near infrared – 10^{-3}–10 μm),
- *passive*, when the sensing device detects EMR originating from another source, principally the sun.

Remote sensing relies on detecting *differences* in the reflected or emitted radiation from different areas on the land surface over a range of wavelengths. It can be done from a variety of platforms, but mainly aircraft or spacecraft, and the data are recorded either photographically or in digital form. A complete remote sensing exercise usually involves integrating more than one 'level' of imagery with on-site observations, including ground-truthing.

Data must be in digital form for computer processing, so air photographs must be converted to a digital format before undergoing *image analysis* and sophisticated data processing. Most satellite and airborne scanner imagery is initially recorded in digital form. In digital form, remotely sensed data provide a natural input to GIS where they can be put into layers along with other spatial data (cadastral boundaries – roads, buildings, etc.), and linked to *attribute* data which describes the properties of a spatial feature in the GIS. For soil, attribute data could include pH, texture or drainage class, up to the predominant soil series in a defined area. Further details on remote sensing applications are given in Harrison and Jupp (1989).

Box 14.2 Remote sensing for soil and land survey.

about soils in the field. This information will usually be augmented by laboratory measurements on samples from the field. It is required so that informed decisions can be made on land use and management best suited to different land uses. The effort put into survey and the methods adopted are determined by the size of the area of interest in relation to the human and financial resources available, the intensity of the proposed land use, and the exclusiveness of any classes to be created. The *output* of soil survey may be a classification, and a *map* showing the distribution of the different soil classes at a scale commensurate with the density of sampling on the ground. Alternatively, the survey data

may be put into a spatially referenced database (a *geographic information system*, GIS – Box 14.2), which can be accessed to provide either:

- the physical distribution of soil classes according to the soil surveyor's best knowledge, or
- the distribution of soil 'classes' defined by the information needs of a specific user (Section 14.4).

Sampling to assess the full range of soil variability, and describing the soil at known locations within a target area, are necessary first stages of soil classification. Soil variability influences the density of sampling required to achieve a certain precision in the classification (measured by τ in Equation B14.1.1). Normally,

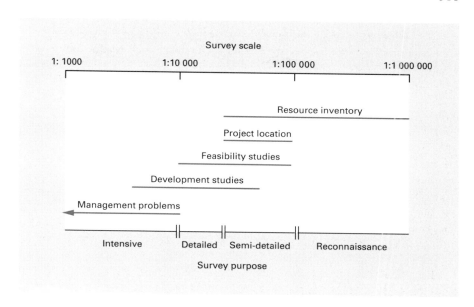

Fig. 14.3 Scale of soil survey in relation to purpose (after Young, 1973).

variability occurring within blocks $< 100 \text{ m}^2$ (i.e. < 0.01 ha) cannot be mapped at an acceptable scale,* so that this restriction sets an upper limit to the density of sampling.

Purpose of soil survey

The relation between the type and purpose of the survey and the map scale is summarized in Fig. 14.3. These are two main types of survey.

Reconnaissance and semi-detailed surveys. These range in scale from 1:2 000 000 to 1:25 000 and have the broad objective of providing a resource inventory for large areas (provinces, states, countries). The preliminary work relies heavily on the interpretation of remotely sensed data (Box 14.2) to delineate differences in vegetation, land form and management that may be the outward expression of soil differences. Once bound-

aries separating *notional* soil classes have been marked on the satellite images or air photographs, surveyors go into the field to check their significance, relocating them if necessary – the process of *ground-truthing*. They also describe the main features of the soils and environment within each unit. Areas suitable for various forms of land use can be identified and guidance given as to the most suitable type of survey to be performed at the next (more detailed) stage.

Detailed and intensive surveys. These range in scale from 1:25 000 to 1:1000 and are carried out for one of three reasons:
1 to assess the technical and economic feasibility of a proposed project in a given area,
2 to assess the preliminary work necessary for the development of a specific project, or
3 within a developed area, to provide data for the solution of particular management problems.
The last-mentioned surveys are usually special-purpose; for example, the mapping of salt-affected soils in an irrigation area, or soils subject to erosion in the catchment of a reservoir.

---

* The practical limit is a scale of *c.* 1:1000; that is, 1 cm on the map represents 10 m on the ground. This is a large scale, whereas 1:1 000 000 is a small scale.

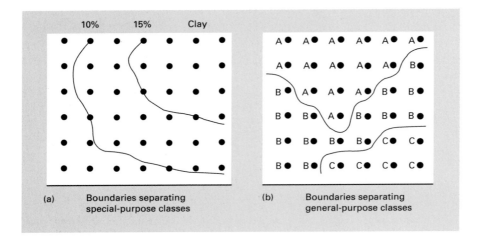

Fig. 14.4 (a) and (b) Soil boundaries drawn from grid surveys (after Beckett, 1976).

Survey methods

Any soil survey should begin with reconnaissance and general fact-finding in the area. The use of remotely sensed data has been mentioned. In addition, the surveyor needs to study the geology, climate, hydrology and vegetation of the area, as well as to consult local land users and agricultural historians. When the preliminaries are complete, the surveyor goes into the field to sample the soil and create a data set that will provide the basis for his mapping. He or she may intend to create special-purpose classification, or may work within the framework of an existing general-purpose classification. The sampling procedure is determined primarily by the scale of mapping and may be one of the following.

Grid survey

This procedure is preferred for special-purpose surveys that are mapped at larger scales (>1:25 000). The sample sites are located on a grid, and the density of sampling should be adjusted according to the area surveyed so that the number of observations per square centimetre on the final map is independent of the scale. Commonly, four to five observations per hectare are recommended, which at a scale 1:10 000 produces four to five observations per square centimetre of the map.

Boundaries are drawn to join points of equal value of a specific property, or where general-purpose classes are to be defined, boundaries are drawn between points at which dissimilar soils occur (Fig. 14.4). Grid survey is expensive, the cost increasing roughly in proportion to the square of the map scale.

One advantage of grid survey is that the observation points can be very accurately located, which facilitates the entry of the data into a GIS. Grid surveys (or regular sampling along a transect) with a large number of observation points (> 100 pairs) also permit the analysis of any spatially dependent variation in soil properties. Where this is identified, the techniques of 'spatial statistics' or *geostatistics* can be used to obtain best estimates of soil properties within localized areas (Box 14.3).

Free survey

In free survey, the surveyor chooses sampling points on the assumption that changes in surface features (soil colour, relief, vegetation or land use) are indicative of soil differences. The density of sampling can be varied as the surveyor concentrates on confirming the inferred boundaries and checking the uniformity of the soil within each boundary (Fig. 14.5a). At small scales, often the inferred boundaries must be accepted and the limited field effort directed to discovering which soilslie

Box 14.3 Geostatistics and its use in soil survey.

As stated in Section 14.1, soil is continuously variable in space. Thus, the value of a soil property at any position in the landscape is a function of that position. Intuitively, one might expect that the values of a property at close positions would be more similar than values at positions farther apart. This is the basis of *regionalized variable theory* which provides statistical tools for estimating with known precision the value of a soil property at unsampled sites. For intensive studies of specific soil properties in relatively small areas, it is a powerful alternative to traditional classification.

The estimation procedure relies on the spatial dependence in a soil property (z) being quantified through the relationship

$$E[\{z(x) - z(x + h)\}^2] = 2\gamma(h) \qquad \text{(B14.2.1)}$$

where $z(x)$ is the property value at point x and $E[\ ]$ is the expected value of the semi-variance $\gamma(h)$ for the $z(x)$ values and values of z at a distance h away, $z(x + h)$. Figure B14.3.1 is an example of a plot of $\gamma(h)$ against the lag h, showing some features of the *variogram*. Note that the semi-variance (or the variance per site when the sites are considered in pairs) increases as h increases up to a local maximum (the *sill variance*). The lag at this point is called the *range* and defines the limit of spatial dependence. The finite variance at $h = 0$ is called the *nugget variance*, and is due primarily to variation occurring over distances less than the shortest sampling interval (short-range variation as discussed in Section 14.2).

The variogram function is used to derive weighted averages of the property z within the local area – a technique called *kriging*. If most or all of the variance is nugget, kriging is not necessary – the variance is not spatially dependent and can be calculated from classical statistics as σ^2 (Box 14.1). Further details of the application of geostatistics to the spatial analysis of soil variability are given by Webster and Oliver (1990).

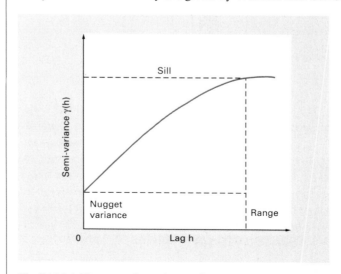

Fig. B14.3.1 Elements of a variogram for a soil property. (after Webster, 1985).

(a)

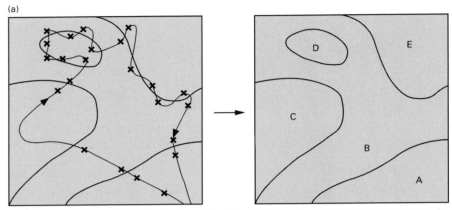

Soil surveyor's traverse (—▶) and observation sites (✖) giving the soil map on the right at a scale of 1:50,000. Boundaries of map units A and C were inferred from external features; those of units D and E were located by sampling

(b)

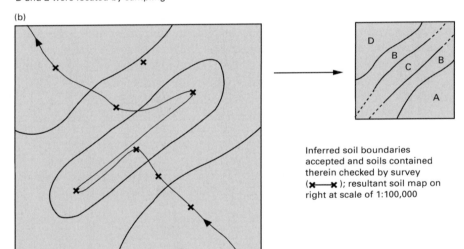

Inferred soil boundaries accepted and soils contained therein checked by survey (✖—✖); resultant soil map on right at scale of 1:100,000

Fig. 14.5 (a) and (b) Sampling pattern and soil boundaries drawn from free survey (after Beckett, 1976).

within the boundaries and the elucidation of any recurrent patterns, such as catenas (Fig. 14.5b).

Soil survey output

Classification

The surveyor's aim may be to place any soil in the area in an appropriate class. With special-purpose classifica-

tions, the class limits are rigidly defined according to the values of one or two properties. For example, one classification of soils for irrigation purposes sets the following class limits:

- saline soils $EC_e > 4$ dS m^{-1}, $ESP < 15$
- saline–sodic soils $EC_e > 4$ dS m^{-1}, $ESP > 15$
- sodic soils $EC_e < 4$ dS m^{-1}, $ESP > 15$.

For general purposes, however, each class is usually built around a *central concept*, covering a limited range in the

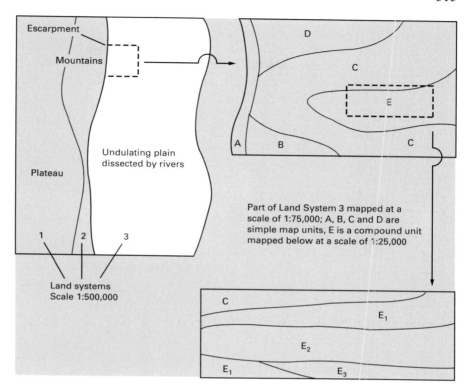

Fig. 14.6 'Nested' arrangement of soil mapping units.

value of several properties. A certain of overlap in the values of any one property is permissible at the class limits.

The list of classes is the classification. Data for individual profiles are subsumed into a definitive description of each class, and one or more properties may be chosen as *diagnostic* or *key* criteria that enable an unknown soil to be allocated to a class. Each class is given a locality name and referred to as a *soil series*. Soils of the same series, which differ only in the texture of the A horizon, are separated as *types*.

Soil mapping

All soil survey should produce a *map* to show the geographic distribution of the soils in an area. Because all the outcrops of any one soil series need not be contiguous, their distribution in the landscape is frequently complex. Depending on the map scale, this complexity

may necessitate some simplification for mapping. This is achieved by a *mapping unit*, consisting of one or more series, which is the smallest area of the map that can be delineated by a single boundary at the scale used. At large map scales, the objective is to use *simple* mapping units, each representing a narrowly defined class (i.e. a series), and to attain a purity > 80% within that unit. But as the map scale decreases (the threshold lying between 1:30 000 and 1:75 000), or even at larger scales where a complex mosaic of classes occurs in a small area, *compound* mapping units must be used. The *legend* to the map lists the classes in each unit, their proportions and if possible their pattern of occurrence. At the smallest scales (< 1:500 000) complex units appear as:

• *soil associations*, in which the classes are recognized and located first and then grouped for ease of presentation, or as

• *land systems*, which are first differentiated by their

Box 14.4 A numerical approach to soil classification.

Numerical classification uses objective, similarity-grouping techniques. The profiles are characterized in terms of those soil properties that are observable or measurable; that is, inferred genesis is not used. Some properties, such as soil water content, are not used because they are transient and influenced by external factors not germane to the objective of the classification. They may need to be specified, however, because of their effect on definitive properties, such as the effect of water content on soil colour and grade of structure. Generally, each property is given equal weighting when soil profiles are compared, although some taxonomists consider that the value of a property in a surface horizon should be given greater weight than that property's value in the subsoil. Some disadvantages of numerical classifications are:

• although the classes contain maximum information, they frequently do not correspond to any natural soil groupings in the field and so are difficult to map, and
• numerical classifications are unstable because the inclusion of newly described profiles in the similarity-grouping process may completely alter the classification.

Further details are given by Webster (1977).

surface expression (e.g. from air photos) and the included classes identified by subsequent sampling.

A 'nested' arrangement of mapping units, showing land systems and their included map units, is illustrated in Fig. 14.6.

In addition to the legend, most soil maps provide a *memoir*, which contains a detailed description of each soil series, with general comments about the geology and vegetation of the area and the management of soils grouped in mapping units.

Land evaluation

The output from soil survey can make a valuable contribution to *land evaluation*, which is the assessment of the suitability of land for particular purposes (also referred to as *land use capability*). This exercise takes account of the influence of climate, geology, soil and vegetation on the use of land for different purposes. The framework for such evaluation at a small scale (i.e. for large areas) has been established by FAO (1976) and is

useful for broad-brush planning purposes. However, for planning land use at a large scale (small areas), most government agencies have developed their own land evaluation methods which can be quite different from the generic FAO scheme, but still rely heavily on soil information.

14.4 Soil information systems

The role of classification

Divergent approaches

As an output from soil survey, classification provides a framework for making local generalizations about soil, based on the properties of the soil series identified. The upwards expansion of a classification to allow progressively broader generalizations at higher levels or *categories* has preoccupied many individual soil scientists and corporate bodies concerned with soil survey, resulting in national and even global classifications of a hierarchical structure.* Such classifications, which are *general purpose*, are normally developed from the bottom up by combining small, relatively homogeneous classes to form more inclusive groups; but identification of an unknown soil always proceeds stepwise from the highest category downwards, usually by means of a *taxonomic key*.

Opinions differ widely on the criteria to be used to establish the different categories and to separate classes within categories. Traditionally, the *definitive properties* are chosen on the basis of:

• the inferred *genesis* of a soil, on the assumption that the soil is in steady state with its environment and reflects the influence of prevailing soil-forming factors, and/or

• a soil's *morphology* which is the outward expression of the processes which shaped the soil.

The success of the general-purpose classification produced depends on the taxonomist's skill and experience in choosing definitive properties that are correlated with a large number of other soil properties. Less traditional taxonomists have employed *numerical methods* to sort soil profiles into classes (Box 14.4).

Soil classification in the USA, where most effort has been devoted to soil survey over a very large area (> 10 000 series defined), has followed a broadly 'traditionalist' approach which places primary emphasis on soil properties that can be observed or measured, and preferably those which are quantitative, such as percentage clay, rather than qualitative, such as colour or grade of structure. It takes no account of the imperfectly understood relationships between pedogenesis and the soil-forming factors in establishing classes. Nevertheless, one of its underlying premises is that 'the properties selected should be those that either affect soil genesis or result from soil genesis' (Soil Survey Staff, 1960). To make the classification as comprehensive as possible, the taxonomists consulted extensively with pedologists in other countries as the classification progressed through several 'approximations', culminating

in the publication of *Soil Taxonomy* (Soil Survey Staff, 1975). The outcome of this and other attempts at national or supra-national soil classification are summarized in Table 14.1.

Conceptual classes

No matter how divergent current classifications are in principle, language and format, two main levels of generalization stand out:

• a lower level encompassing the surveyor's series, which are narrow classes in harmony with the landscape, and

• a higher level of broad classes with somewhat overlapping boundaries, each of which embodies a central, usually pedogenetic, concept.

Table 14.1 Structures of a selection of national and supra-national soil classifications.

Classification	Categories (the number of actual or potential classes in each category is shown in parentheses)
USDA Soil Taxonomy (Soil Survey Staff, 1975)	Orders (10) → Suborders (47) → Great Groups (185) → Subgroups (970) → Families (4500) → Series (> 10 000)
World Soil Classification (FAO–Unesco, 1988)	World Classes (28) → Soil Units (153) → Phases
Soil Classification for England and Wales (Avery, 1980)	Major Groups (10) → Groups (41) → Subgroups (109) → Series
Russian Soil Classification (Rozov and Ivanova, 1967)	Genetic types (c. 110) → Subtypes → Species → Subspecies
Factual Key for Australian Soils (Northcote, 1979)	Divisions (3) → Subdivisions (11) → Sections (54) → Classes (271) → Principal Profile Forms (855)
Canadian System (Canada Department of Agriculture, 1970)	Orders (8) → Great Groups (22) → Subgroups (165) → Families (c. 900) → Series (c. 3,000)
Australian Soil Classification (Isbell, 1996)	Order (14) → Suborder → Great Group → Subgroup → Family

*A structure in which individuals are collected into small groups, the small groups belonging to larger groups, and so on.

Order	Brief description	Included Great Soil Groups
Entisols	Recently formed soils, no diagnostic horizons	Azonal soils and some Low Humic Gley soils
Vertisols	With swell-shrink clays, high base status	Grumusols
Inceptisols	Slightly developed soils without contrasting horizons	Acid brown soils, some Brown Forest soil. Low Humic Gley and Humic Gley soils
Aridosols	Soils of arid regions	Sierozem, Solonchak, some Brown and Reddish Brown soils and Solonetz
Mollisols	Soils with mull humus	Chestnut, Chernozem, Prairie, Rendzinas, some Brown Forest, Solonetz and Humic Gley soils
Spodosols	Soils with iron and humus B horizons	Podzols, Brown Podzolic soils and Groundwater Podzols
Alfisols	Soils with a clay B horizon and > 35% base saturation	Grey-brown Podzolic, Gray Wooded soils, Degraded Chernozem and associated Planosols
Ultisols	Soils with a clay B horizon and < 35% base saturation	Red-yellow Podzolic, Reddish-brown Lateritic soils and associated Planosols
Oxisols	Sesquioxide-rich, highly weathered soils	Lateritic soils, Latosols
Histosols	Organic hydromorphic soils	Bog soils (Peats)

Box 14.5 Soil taxonomy orders and some of the included Great Soil Groups. After Butler (1980).

Classes based on a *central concept* are valuable in teaching soil science because each conveys a measure of knowledge about the soil, distilled from the experience of scientists in many areas. For this reason, the USDA classification of Baldwin *et al.* (1938), revised by Thorp and Smith (1949), was presented in Table 5.4. The sub-order descriptions and Great Soil Group names denote the influence of one or more of the soil-forming factors on pedogenesis. The same can be said of Soil Taxonomy at the higher categorical levels (from *order* down), so this widely used classification may be used to illustrate the meaning of '*conceptual classes*'. The relationship between the 10 orders of Soil Taxonomy and the Great Soil Groups of the older Baldwin *et al.* classification is summarized in Box 14.5.

With the exception of the Entisols, which have no distinctive horizons, and Histosols (organic soils), each order has a broad climatic connotation. Below the order level the classes in Soil Taxonomy are defined progressively more narrowly according to measured soil properties and the microclimate in which a soil forms. For example, the *suborders* are separated on the basis of differentiae associated with hydromorphism (Section 9.5), vegetation (as it influences podzolization, lessivage – Section 9.2), parent material, time and intensity of weathering (resulting in ferrallitization – Section 9.4) and so on.

Diagnostic horizons

Separation of classes in the next inferior category – the *Great Group* – is intended to divide the soil continuum into more or less equally spaced groupings according to the variation in soil genesis. Much use is made of the concept of *diagnostic horizons*, which are connotative of an underlying pedogenic process (just as the symbol Ea in the soil classification for England and Wales connotes a bleached sandy horizon from which sesquioxides have been eluviated). However, a diagnostic horizon need not coincide with a pedological horizon, such as A1, B2 or Bt; it may include parts of more than one pedological horizon or soil layer. For example, a *mollic epipedon* – one of the six diagnostic horizons recognized at the surface – may include all of an A horizon and the upper part of a B horizon when organic matter has been uniformly incorporated by faunal activity to a considerable depth. More weight is given in Soil Taxonomy to subsurface horizons, of which there are 10, because they are less prone to disturbance through ploughing or truncation by erosion. Thus, the Great Groups are separated on the presence or absence of diagnostic horizons and their arrangement in the profile.

Other soil classifications

Soil Taxonomy has been widely deployed outside the USA and purports to be a supranational classification. It is not of great value for teaching general soil science because the terminology at any level below order lacks intuitive meaning and is intimidating for the non-taxonomist. Many countries have persevered with developing and refining their own national classifications, as for example in Australia where a new *Australian Soil Classification* (Isbell, 1996) has been adopted. This scheme is general purpose and multi-categoric (Table 14.1), and designed to incorporate as many as possible of the desirable features of schemes it would replace, such as the Great Soil Group classification (Stace *et al.*, 1968) and the Factual Key (Northcote, 1979), and schemes with which it would compete (Soil Taxonomy and the World Soil Classification (FAO–Unesco, 1988). Some of the general principles underlying the development of the Australian Soil Classification are that it:
• is based on diagnostic properties, horizons or materials,
• assumes concepts of pedogenesis in the development

of horizons in a vertical sequence in the soil profile,
• is based on Australian data (mainly from the field although supported by laboratory measurements), with the selected properties to be relevant as far as possible to land use and soil management, and
• has a simple but unambiguous nomenclature.

Table 14.2 Orders of the Australian Soil Classification and approximate included Great Soil Groups. After Isbell, 1996.

Order	Connotation	Included Great Soil Groups
Anthroposols	Human influence	None
Calcarosols	Calcareous throughout	Solonized brown soils, grey brown and red calcareous soils
Chromosols	Often brightly coloured	Non-calcic brown soils, some red-brown earths and podzolic soils
Dermosols	Often with clay skins on peds	Prairie soils, chocolate soils, some red and yellow podzolic soils
Ferrosols	High iron content	Krasnozems, euchrozems, chocolate soils
Hydrosols	Wet soils	Humic gleys, gleyed podzolic soils, solonchaks and some alluvial soils
Kandosols	–	Red, yellow and grey earths, calcareous red earths
Kurosols	Pertaining to clay increase	Many podzolic soils and soloths
Organosols	Dominantly organic materials	Neutral to alkaline and acid peats
Podosols	Podzols	Podzols, humus podzols, peaty podzols
Rudosols	Rudimentary soil development	Lithosols, alluvial soils, calcareous and siliceous sands, some solonchaks
Sodosols	Influenced by sodium	Solodized solonetz and solodic soils, some soloths and red-brown earths, desert loams
Tenosols	Weak soil development	Lithosols, siliceous sands, alphine humus soils and some alluvial soils
Vertosols	Shrink–swell clay	Black earths, grey, brown and red clays

The 14 orders of this classification, their connotation and included Great Soil Groups from Stace *et al.* (1968) are given in Table 14.2.

Other countries, particularly developing ones where soil survey data are sparse, have adopted the *World Soil Classification*. This classification arose as the list of classes in the legend to the Soil Map of the World (FAO–Unesco, 1974) which has subsequently been revised. It conveys useful generalizations about the genesis of soils in relation to the interactive effects of the main soil-forming factors. The classification is multicategorical, the critical level being the '*soil units*'. Like Soil Taxonomy, it makes class separations using diagnostic horizons which connote the underlying pedogenesis. Some of the descriptive terminology has been adopted from Soil Taxonomy in simplified form. But many of the traditional Great Soil Group names have been retained, as well as new names coined which do not suffer in translation nor have different meanings in different countries. The units are mapped at a scale of 1:5 000 000 as *soil associations* designated by the dominant soil unit, with three textural classes (coarse, medium and fine) and three slope classes superimposed (level to gently undulating, rolling to hilly, and steeply dissected to mountainous). The Soil Map of the World is a valuable inventory of global soil resources and the legend has been used for correlation of different national classifications.

For the sake of logical presentation and generalization at a higher level, the soil units have been grouped on generally accepted principles of soil formation to form 28 *World Classes* (Table 14.3). Although based on inferred pedogenic processes, such as gleying, salinization and lessivage which have a bearing on soil use and management, the World Classes are so broad and the total variation within each class is so large that their value in the prediction of land use is very limited. Their merit is that they constitute a gallery of conceptual classes which convey some general understanding to a worldwide clientele. At a lower level, *phases* are now recognized which are intergrades between first and second level units or have characteristics additional to those used in defining the soil units.

Table 14.3 Higher classes (level 1) and their connotations from the Soil Map of the World. After FAO–Unesco, 1988.

Acrisols	Strongly acid hence low base saturation
Alisols	Connotes high aluminium content
Andosols	Soil material rich in volcanic glass commonly having a dark surface horizon
Anthrosols	Connotes human activity
Arenosols	Suggesting weakly developed coarse-textured soils
Calcisols	Connotes the accumulation of calcium carbonate
Cambisols	Suggesting changes in colour, structure and consistence
Chernozems	Soils rich in organic matter and black in colour
Ferralsols	Soils high in sesquioxides
Fluvisols	Connotative of alluvial deposits
Gleysols	Mucky soil mass connotative of excess water
Greyzems	Connotative of uncoated silt and sand grains present in layers rich in organic matter
Gypsisols	Connotes the accumulation of calcium sulphate (gypsum)
Histosols	Soils rich in fresh or partly decomposed organic matter
Kastanozems	Soils rich in organic matter and brown or chestnut in colour
Leptosols	Connotes weakly developed shallow soils
Lixisols	Connotes accumulation of clay and strong weathering
Luvisols	Connotes clay accumulation through lessivage
Nitisols	Connotes shiny ped faces
Phaeozems	Soils rich in organic matter with a dark colour
Planosols	Connotes soils generally developed in level or depressed relief with seasonal surface waterlogging
Plinthosols	Connotes mottled clay materials which harden on exposure
Podzols	Soils with a strongly bleached horizons
Podzoluvisols	Intermediate between Luvisols and Podzols
Regosols	Implying a loose mantle overlying a hard rock core
Solonchaks	Connotative of salty areas
Solonetz	Strongly expressed salt effects
Vertisols	Connotes the turning over of surface soil

Modern soil information systems

In this age, no soil classification is of much value *unless* it is supported by a comprehensive electronic database (Section 14.1). Only in this case can major deficiencies in the communication of soil information to, and use by, non-specialist users be addressed. Several national and international agencies for soil and land resources information have foreseen this need and acted to satisfy it. For example, the National Cooperative Soil Survey (NCSS) of the USA has recognized that many of the concepts and tools still being used in soil survey and classification pre-date present advanced computer hardware and software technology. These tools and concepts were developed mainly because it was not possible for humans to handle in an orderly way the huge amounts of data derived from soil and land surveys. This is no longer a constraint so the NCSS has developed a new National Soil Information System (Ernstrom and Lytle, 1993) intended to provide new capabilities for users of the system, such as:

• the preservation of individual site observation data (field and laboratory),

• the relation of these individual data and the spatial variability of soil properties to individual, delineated areas on soil maps,

• the provision of two types of map units – *primary units* which are composed of the most detailed, delineated spatial entities representing the soil scientists' best knowledge, and *user-defined units*, which are any unique grouping of primary units defined by a user to meet specific needs, and

• the interpretation of individual components of map units as well as what were formerly included soil classes within the units.

In Australia, coordination in the collection, collation, interpretation and dissemination of soil information has been hindered by the existence of several State and Federal agencies and the incompatibility of the databases used. The only contiguous coverage of soil information is provided by the Atlas of Australian Soils (scale 1:2 000 000) which, although now available in digital form, does not have soil attribute values

(Box 14.2) linked to individual map units. This problem is being addressed through the Australian Collaborative Land Evaluation Program (ACLEP) established in 1992. ACLEP has agreed with State and Federal agencies the data exchange protocols for use in describing soil profiles and collating data (spatial and attribute) from various sources.

At the international level, the International Society of Soil Science and agencies such as FAO and UNEP (United Nations Environment Program) have cooperated to set up a World Soils and Terrain Database (SOTER). This has the broad aims of:

• being a computerized database containing information on topography, soils, climate, vegetation and land use, ideally complemented by compatible databases of socio-economic factors, and linked to a GIS in which each item of information can be accurately located and overlays of such spatial information portrayed graphically, and

• providing crop yield models moderated by soil and climatic conditions and management inputs, as well as environmental impact models for particular land units and land uses.

This very ambitious programme is being tested in phases in several parts of the world.

14.5 Summary

Soil information is collected by the *survey* of soils in the field and analysis of samples in the laboratory. Field information is collected in many ways including:

• remote sensing from spacecraft and aircraft,

• geological and vegetation maps,

• the experience of soil users, and

• visual observations and measurements made by a soil surveyor on the ground.

The exact combination of methods used depends on the *purpose* of the survey, which will determine the intensity of sampling and eventually the scale of any map produced. Depending on the map scale, which may vary from 1:2 000 000 for reconnaissance surveys to 1:1000 for intensive surveys of soil management problems, surveying proceeds on a *grid* (for large scales) or *free survey* pattern (for small scales). *Geostatistical techniques* are useful for intensive surveys on small areas (large map scale) where there is quantifiable spatial dependence in the variability of soil properties.

Traditionally, soil survey output has been used in *soil classification*, which has been a prerequisite for the orderly presentation of accumulated knowledge to soil scientists and other users. Classification involves *grouping* soils into classes according to the degree of similarity between individual soil profiles. However, soil classification is made difficult and its results contentious because there is no definable *soil entity*; the soil mantle varies continuously so that subdivision of the variation in one or more properties to create classes is usually arbitrary. Grouping has been attempted using *numerical methods*, but the results are often difficult to present in a soil map. For a classification to be useful for predictive purposes, the variation remaining within classes should be substantially less than the variation between classes for a range of soil properties. Classes are defined in terms of one or two properties relevant to a specific purpose (*special-purpose classification*), or on many properties considered relevant for various soil uses (*general-purpose classification*). *Diagnostic* or key properties, which are not necessarily relevant to use but are easily assessed in the field, are often chosen to define classes on the assumption they are correlated with more important properties.

The classes recognized by a surveyor (*soil series*) are usually aggregated to form progressively less homogeneous (more inclusive) classes, which facilitate national and international exchange of soil knowledge. Because of the wide choice of criteria on which to separate classes at any one level and to establish the different levels or *categories* of generalization, many classifications have been proposed. The USDA Soil Taxonomy is a classification purporting to use only those soil properties that can be observed or measured, preferably quantitatively, in establishing criteria for class separation. The advantage of this approach, particularly at the lower, more exclusive category levels, is that it avoids reliance on the imperfectly understood relationships between observed soil properties, pedogenesis and the soil-forming factors. However, non-taxonomists tend to seek broader generalizations about soil differences, which are embodied in the higher level *conceptual classes*. For this purpose Soil Taxonomy's orders and suborders may be used, or the World Soil Classification, which comprises 153 soil units grouped into 28 World Classes.

In this information technology age, however, to rely solely on classification as a means of ordering soil knowledge and communicating this knowledge is akin to being satisfied to keep records in a hand-written diary rather than the latest laptop computer. Thus, many national and international agencies charged with making soil and land resource inventories are developing modern *soil information systems* which ideally will provide for:

• the preservation of individual site observation data (field and laboratory);

• the relation of these data and the spatial variability of soil properties to individual, delineated areas on soil maps (at chosen scales);

• at least two types of map units – *primary units* composed of the most detailed, delineated spatial entities representing the soil scientists' best knowledge, and *user-defined units*, being any unique grouping of primary units defined by a user to meet specific needs; and

• the interpretation of individual components of map units as well as what were formerly included soil classes (e.g. soil series) within the units.

References

Avery B. W. (1980) *Soil Classification for England and Wales*. Soil Survey Technical Monograph No. 14. Harpenden.

Baldwin H., Kellogg C. W. & Thorp J. (1938) Soil classification, in *Soils and Man*. Yearbook of Agriculture, Washington DC.

Beckett P. H. T. (1976) Soil survey. *Agriculture Progress* **51**, 33–49.

Butler B. E. (1980) *Soil Classification for Soil Survey*. Clarendon Press, Oxford.

Canada Department of Agriculture (1970) *A System of Soil Classification for Canada*. Queen's Printer for Canada, Ottawa.

Ernstrom D. J. & Lytle D. (1993) Enhanced soils information system from advances in computer technology. *Geoderma* **60**, 327–341.

FAO (1976) *A Framework for Land Evaluation.* FAO Soils Bulletin 32. FAO, Rome.

FAO–Unesco (1974) *Soil map of the World* 1:5,000,000 Vol. 1. Legend, Paris.

FAO–Unesco (1988) *Soil map of the World, Revised Legend.* World Resources Report No. 60. FAO, Rome.

Harrison B. A. & Jupp D. L. B. (1989) *Introduction to Remotely Sensed Data.* CSIRO Publications, Melbourne.

Hodgson J. M. (1978) *Soil Sampling and Soil Description.* Clarendon Press, Oxford.

Isbell R. F. (1996) *The Australian Soil Classification.* Australian Soil and Land Survey Handbook. CSIRO Publications, Melbourne.

McDonald R. C., Isbell R. F., Speight J. G., Walker J. & Hopkins M. S. (1990) *Australian Soil and Land Survey Field Handbook,* 2nd edn. Inkata Press, Melbourne.

Northcote K. H. (1979) *A Factual Key for the Recognition of Australian Soils,* 4th edn. Rellim, Glenside, South Australia.

Rozov N. N. & Ivanova E. N. (1967) Classification of the soils of the USSR. *Soviet Soil Science* 2, 147–156.

Soil Survey Staff (1960) *Soil Classification: A Comprehensive System 7th Approximation.* Washington, DC.

Soil Survey Staff (1975) *Soil Taxonomy. A Basic System of Soil Classification for Making and Interpreting Soil Surveys.* United States Department of Agriculture Handbook No. 436, Washington DC.

Stace H. C. T., Hubble G. D., Brewer R., Northcote K. H., Sleeman J. R., Mulcahy M. J. & Hallsworth E. G. (1968) *A Handbook of Australian Soils.* Rellim, Glenside, South Australia.

Thorp J. & Smith G. D. (1949) Higher categories of soil classification. *Soil Science* 67, 117–126.

Webster R. & Oliver M. A. (1990) *Statistical Methods in Soil and Land Survey.* Oxford University Press, Oxford.

Webster R. (1977) *Quantitative and Numerical Methods in Soil Classification and Survey.* Oxford University Press, Oxford.

Webster R. (1985) Quantitative spatial analysis of soil in the field. *Advances in Soil Science* 3, 1–70.

Young A. (1973) Soil survey procedures in land development planning. *Geographical Journal* 139, 53–64.

Further reading

Beckett P. H. T. & Webster R. (1971) Soil variability – a review. *Soils and Fertilizers* 34, 1–15.

Petersen G. W., Bell J. C., McSweeney K., Nielsen G. A. & Robert P. C. (1995) Geographic Information Systems in agronomy. *Advances in Agronomy* 55, 67–111.

Chapter 15
Soil Quality and Sustainable Land Management

The fabric of human life is woven on earthen looms – it everywhere smells of the clay' (from Mitchell *et al.*, 1950 – Saskatoon Soil Survey Report No. 13)

15.1 What is soil quality?

Options for defining soil quality

In Chapter 11, the point is made that the productivity of land, usually measured by the yield of crop or animal product per hectare, is a function of a number of factors – soil, climate, pests, disease, genetic potential of the crop/animal, and management. *Fertility* is probably the most important single *soil* factor affecting productivity. However, in recent years groups in society outside the agricultural sector have demanded a more comprehensive descriptor of the 'condition' of a soil than is provided by soil fertility. This reflects an overall concern in society about the state of the environment in which we live, of which soil is increasingly recognized as a vital part, along with air and water. Whereas the desirable qualities of air and water for human and animal life can be adequately defined and monitored, the same is not true of soil for several reasons.

• Soil is continuously variable in space and, being a complex, dynamic biological system, is variable in time. The rate of change of individual soil properties is itself very variable.

• In most cases, soil properties and behaviour do not impinge directly on human and animal well-being. The effect of soil is experienced indirectly through the diversity and health of the natural ecosystems it supports, the quality and quantity of crops (and animals) grown,

the water (and sometimes sediment) which runs off and drains from a soil, and the gases absorbed or released by a soil.

• The quality of a soil depends on the use to which it is put. This is the premise underlying the assessment of *land-use capability* (called land evaluation – Section 14.3) in which land is classed according to its suitability for various uses. Agriculture at varying levels of intensity has traditionally been a key use, and soil quality is a major factor determining the land capability class (ranging from the best class – no limitations, to the worst class – severe limitations). Syers *et al.* (1995) give an example of how the land-use capability approach has been adapted to determine the suitability of soils for rubber plantations in Malaysia. Using a comprehensive set of biophysical criteria (climate, soil physical properties, soil acidity and organic matter content), a close correlation between rubber yield and a soil quality index was obtained. However, the index is specific to rubber plantations and is unlikely to be applicable to a wider suite of land uses: nor does it necessarily indicate the long-term sustainability of this particular land use (Section 15.2).

• The community at large has concerns about soil suitability not only for agriculture but also as a reserve of biological diversity, for nurturing natural ecosystems of recreational and conservation value, as a foundation for stable landscapes, and as a component of integrated waste management systems.

Thus, although ideally we aspire to defining *soil quality* in objective terms, in reality the concept is diffuse and must reflect a range of social, cultural and economic factors as well as biophysical properties. Nor in linking the concept of soil quality to 'fitness for use' is it always possible to identify a good relationship between soil attributes and the intended use of the soil. Notwithstanding the specific example quoted above for rubber plantations in Malaysia, there are many opinions amongst scientists as to *how* soil quality should be defined, even for agricultural production systems which have been much studied. For example, in seeking to define the 'primary' *indicators* of sustainable agricultural systems in Australia, a multidisciplinary group of scientists selected the 'resource-base' attributes shown in Table 15.1. Note that the preferred indicators differ according to the type of agricultural system – some are very broad (e.g. nutrient balance, biota, soil productivity, chemical residue level), some can be measured with varying degrees of difficulty (e.g. pH, earthworm numbers, irrigation water use), many are qualitative and all require trends to be assessed over time.

For these reasons, soil quality is often assessed in terms of *management practices* designed to improve, or at least maintain, the performance of a soil for specific purposes – for example minimizing erosion under cropping, storing good quality water underground or delivering it to surface storage, or minimizing the movement of heavy metals applied in industrial wastes. The soil quality is inferred from:

- the degree to which a management practice protects the soil from degradation (caused by erosion for example), and
- the extent of adoption of the 'best' practice, or other practices, within an area of interest (farm, catchment or land region).

Production and environmental impact

Until such questions as:

- is there a limited number of key soil attributes which can serve as appropriate soil quality indicators for a range of land uses, and
- is it feasible to measure these indicators over time (monitoring) with proper spatial referencing at a range of scales,

are resolved, the 'inverse' approach of inferring soil quality from land management practices is likely to be more effective than the approach illustrated in Table 15.1. There is, however, an important proviso. The research designed to develop and test 'best practices' must consider the particular system (e.g. high rainfall pastures) in the context of the whole environment. Too often such research has followed a 'reductionist' approach, focusing on a narrow production problem, without considering the ecological processes that sustain biological activity overall, and the flows of energy, nutrient and water between the production system and the wider environment. The widespread introduction of an exotic legume (subterranean clover) into pastures for beef cattle and sheep in southern Australia provides an example of this point. The pastures were fertilized with superphosphate to encourage the legume, which fixed atmospheric nitrogen, and improved the quality and quantity of the pasture. With higher stocking rates, native perennial grasses were grazed out and more nitrogen cycled through the grazing animals. Mineral N accumulated in the soils at the end of the dry summers and substantial nitrate leaching occurred during autumn and early winter, contributing to soil acidification. The already acidic A horizons of the duplex

Table 15.1 Primary indicators for the resource base (mainly soil) for six major agricultural systems in Australia. After Hamblin, 1992.

Agricultural system	Resource-based primary indicators
Rain-fed crops	Soil health; pH, nutrient balance, biota
High rainfall pastures	Soil bio-indicators; worms, termites – numbers and species
Low rainfall rangeland	Pasture and soil conditions, percent bare ground, pasture composition
Irrigated crops and pastures	Soil health; infiltration rate, subsoil compaction, biomass activity, chemical residue level
Intensive horticulture and viticulture	Soil permeability and watertables; irrigation water use, watertable depth, soil infiltration
High rainfall tropical systems	Soil productivity; pH and organic matter trends, soil structure

soils in the region slowly became more acid, a condition which adversely affected productivity and the range of species which could be grown (see 'Accelerated soil acidification' – Section 11.3). Watertables have also risen in the higher rainfall areas (> 600 mm y^{-1}) because of greater drainage below the root zone of the short-lived annual grasses compared with the perennial grasses and native trees.

Contamination and soil quality

The definition of soil quality becomes more specific when contaminated land is considered (Box 10.6). *Soil contamination* by organic and inorganic chemicals occurs in urban and rural areas from various causes:

• mining activities — exploration, extraction and processing,
• industrial activities, including defence, and the disposal of industrial waste,
• urban living, including motor transport, and the concentration of urban waste through storm water runoff, sewage effluent and sewage sludge (Box 11.1), and
• agriculture, forestry and horticulture, through the excessive use of fertilizers and pest and disease control chemicals.

Contaminants affect human health via several exposure pathways: (i) directly, through ingestion of soil, inhalation of gases or dust, and skin contact, and (ii) indirectly through food consumed and water which is drunk or used for washing. To determine whether a soil is contaminated (and its quality adversely affected), the concentration of potential contaminants must be compared with their *background* or '*normal*' concentrations in soil. For some anthropogenic compounds (e.g. DDT), the background is close to zero and any detectable concentration indicates contamination. For most inorganic elements, such as the heavy metals, assessing contamination is more difficult because of their variable concentration in soils naturally, due to differences in parent material and rates of weathering (Chapter 5). In this case, background concentrations must be calculated after excluding high concentrations due to geothermal anomalies, localized pollution due to waste disposal,

Table 15.2 Soil quality guidelines (mg kg^{-1} soil) for environmental purposes. After Barzi *et al.*, 1996.

Substance	Background concentration	Environmental investigation threshold
Heavy metals		
Cd	0.04–2	3
Cr	0.05–110	50
Ni	2–400	60
Pb	2–200	300
Monocyclic aromatic hydrocarbons (MAH)		
Benzene	0.05–1	1
Polycyclic aromatic hydrocarbons (PAH)	0.95–5	5
Chlorinated hydrocarbons		
Polychlorobiphenyls (PCB)	0.02–0.1	1
(DDT)	< 0.001–1	0.2

deposition along roadways and excessive use of agricultural chemicals and fertilizers.

In recent years, most countries have become very aware of soil contamination, particularly in urban areas, and have developed strategies for managing and, if necessary, ameliorating the problem through remediation. In Australia and New Zealand, for example, Guidelines for the Assessment and Management of Contaminated Sites have been drawn up (ANZECC/NHMRC, 1992). These guidelines give background concentrations for a range of inorganic and organic substances, and indicate concentrations (thresholds) above which further investigation of the site is required to determine the extent and degree of contamination. A selection of these soil quality criteria for contaminated soils is given in Table 15.2. Many of the values were originally derived from European and North American experience, but they are continually being revised as new information from Australasia becomes available.

15.2 Concepts of sustainability

Land use and land management

Although the terms land use and land management tend

to be used interchangeably in the literature on sustainability, it is important to recognize the distinction in their meaning.

The *use* to which land is put may be *urban, industrial* or *rural*. Within each major category, such as rural land use, there are subcategories such as agriculture, forestry (native woodlands and plantations), conservation and recreation. Some of these subcategories may also apply to urban and industrial land uses. Agriculture may be further subdivided into field crops, pasture, horticulture and so on, and in some cases, such as with sugar cane grown on specially assigned land in Queensland, the particular crop grown can be specified. *Land management*, on the other hand, refers to the *practices* adopted for a particular land use, such as the method of cultivation (Section 11.4), erosion control measures such as contour banks, strip cropping and stubble mulching (Section 11.5), fertilizer practice and pest control (Chapter 12). Management should encompass *farm planning* whereby the various uses of land – cropping, grazing, shelter belts and wood lots, and irrigation where feasible – are assigned to the most suitable soils and landscape units, and informed decisions are made about financial investment in the farm and its enterprises. Land management also includes such activities as surface and subsurface drainage, groundwater pumping (for irrigation and watertable control), and decisions on fence positions, paddock sizes (especially for livestock) and waste disposal.

Clearly, a land use that is totally inappropriate for a particular soil will not be viable, irrespective of whether the management is good or bad. Similarly, even a land use that is appropriate for a particular soil can become non-viable if the management practices are bad. These distinctions are important in defining the meaning of *sustainability*.

Ecological, economic and social sustainability

Sustainable agriculture

The concept of 'sustainability' became a widely accepted environmental and political credo during the 1980s. However, as the word gained currency, it began to lose precision in its meaning (Viederman, 1993). Implied in its use are concepts of *resource quality* and *availability, economic viability, environmental impact, social justice, intergenerational equity* and *biodiversity*. There are also spatial and temporal dimensions to sustainability, emphasizing the *dynamic* nature of any operational definition of sustainability (Smyth and Dumanski, 1993).

In discussing sustainability, some authors have focused solely on agriculture. For example, Reganold *et al.* (1990) consider that many of the formerly productive and labour-efficient farms in the USA are now experiencing declining productivity, deteriorating environmental quality, reduced profitability and possible threats to human and animal health. This change in fortune is attributed to a number of causes including:

• over-dependence on external inputs such as fossil fuels, fertilizers and concentrated feed for animals,
• heavy use of pesticides,
• depletion of groundwater reserves through irrigation,
• cropping and grazing management which induces soil structural decline and accelerates soil erosion, and
• loss of soil biological diversity, particularly beneficial organisms.

As an alternative to *conventional* agriculture which tends to be intensive, high-input (most of which is sourced from outside the farm) and environmentally unfriendly, Reganold *et al.* (1990) advocate *non-conventional* agriculture – sometimes called *organic, alternative, regenerative, ecological* or *low-input* – in which a farmer works in harmony with the biological and ecological processes of the soil–plant system, rather than relying on a high input of chemicals and expenditure of energy. Some of the basic principles of alternative or organic agriculture are described in Box 15.1.

A key question in the debate on the relative merits of non-conventional and conventional agriculture is – can the former be as profitable as the latter has proven to be? The answer is crucial to persuading large numbers of profit-minded landholders to change their farming practices, because they argue it is 'difficult to be green when you are in the red!'. A number of studies have

Box 15.1 Alternative or organic agriculture.

Organic agriculture is sometimes described as 'biological farming', as opposed to the 'chemical farming' of conventional agriculture. The basic principles of organic agriculture are as follows.

• *Crop rotation* – this involves a planned succession of crops on each field on a farm to achieve weed, disease and insect control, efficient nutrient cycling and good soil structure. By appropriate sequencing of crops and management of residues, the populations of the natural predators of pest organisms can increase, allowing the use of pesticides to be restricted. This is the basis of *integrated pest management*.

• *Use of organic manures and crop residues* – by maximizing the return of manures and residues to the soil, the cycling on nutrients is enhanced and soil organic matter is maintained or increased. The latter has beneficial effects on soil structure, increasing water infiltration and storage, and improves soil tilth (Section 2.2). Good tilth favours easy seed germination and seedling emergence. Organic matter encourages soil micro-organisms and soil fauna, especially earthworms.

• *Use of green manures* – the benefits of green manures are outlined in Section 11.1. Green manures are especially important in organic farming not only for the return of residues, but also for the incorporation of legumes which contribute N fixed from the air (Section 10.2). In addition to N inputs from legumes, P usually needs to be supplied from external sources, the preferred form being unprocessed reactively phosphate rocks (Section 12.3).

• *Diversity* – by encouraging a diverse range of crops and livestock, farm income can be buffered against market, climatic and biological risks. This is in contrast to crop monoculture (e.g. cotton) or single enterprise animal farming (e.g. sheep for wool) where income is particularly vulnerable to fluctuations in commodity prices, extremes of weather and the incidence of pests and disease.

been carried out to address this question, with mixed results. Authors such as Reganold *et al.* (1990) conclude that although yields from non-conventional farms are lower than from conventional farms, the difference is frequently offset by lower production costs which lead to equal or greater returns from the non-conventional farms.

Notwithstanding the conclusion that non-conventional agriculture can be equally or more profitable than conventional agriculture, the generally acknowledged *lower yields* (and often lower quality of harvested prod-

uct due to pest and disease damage) from the former beg the question – what would happen to world food security if China, India and the USA (the three largest food producing countries) were to switch to organic or low-input agriculture on a massive scale? The success of the Green Revolution in dramatically increasing food production, mainy in Asian countries, through high yielding varieties, fertilizers and good water management is described in Box 11.2. But the rising trend in world food production per capita plateaued during the 1980s (Fig. 11.2). This problem is more acute in some regions than others, especially in Africa where three-quarters of the nation states produced *less* food per capita at the end of the 1980s than at the beginning. With continued growth in the world population (projected to double to 12 billion by the mid-21st century, with most growth occurring in the poorer developing countries), efforts to devise more sustainable agriculture based on the widespread adoption of low-input, organic agriculture are likely to be subverted.

Others such as Rovira (1990) believe it would be a 'tragedy' for agriculture if the term 'sustainable' became synonymous in people's minds with organic farming only. They argue that conventional agriculture, based on mineral fertilizers and the judicious use of chemicals, is sustainable and in many cases improving soil fertility. This generalization, made with reference to Australian agriculture, is examined in more detail in Section 15.3.

Sustainable societies

Although agriculture is a very important land use (and soil is essential to agriculture), agricultural sustainability is only a component of the larger issue of sustainable land systems, which in turn are a subset of the *sustainable use of global resources*. Worldwide concern over the last-mentioned issue led the UN General Assembly in 1983 to ask the World Commission on Environment and Development (WCED) to 'propose long-term strategies to achieve *sustainable development*, combining global economic and social progress with respect for natural systems and environmental quality'.

In proposing strategies for sustainable development, the WCED report (1987) illustrates an inherent paradox in the concept of sustainability, namely:
• people in the many diverse countries of the world have basic *needs* and a legitimate expectancy that those needs will be met, but
• technology and society impose *limitations* on the ability of the environment to meet those needs.
Thus, while the world population continues to grow and expectations of improved living standards also rise, the demands placed on the Earth's natural resources will

continue to increase. Technological change, coupled with economic and social adjustments, can help to meet these demands, but this inevitably places further strains on the Earth's resources and quality of the environment. The response to this paradox illustrates how the fundamentals of a *sustainable society* can be defined in different ways. See the following examples.
• The goal of sustainability must be to revive growth, but to make economic growth less energy intensive. The essential needs of an expanding population in the developing world must be met, but the population must reach a sustainable and stabilized level. The resource base must be conserved and enhanced, technology must be reorientated, and environmental and economic concerns must be merged in decision-making (Lebel and Kane, 1990).
• A sustainable society must keep population within the capacity of the resources needed to produce an acceptable quality of life; pollution must not exceed the capacity of the ecosystem to absorb its impacts; and consumption per capita should be minimized by maximizing the recycling of nutrients, minerals and other materials (Roberts, 1992).
• A sustainable society, with agriculture at its core, must address the human, cultural and ecological dimensions, and include wise management (conservation) of natural resources and restoration of ecological systems (Miller and Wali, 1995).

15.3 Sustainable land management

Sustainable production and land degradation

The foregoing discussion shows that the condition of the physical resource for food and fibre production, primarily soil and water, has a major influence on the sustainability of an ecosystem or society. There are many data at local, national and global levels to show the adverse impact of human activity on physical resources. For example, agriculture, deforestation and mining have moderately to severely degraded *c.* 1.2 billion hectares of land, representing nearly 11% of the Earth's vegetated surface. The main causes are soil erosion, salinization, weed infestation, severe nutrient decline and accelerated soil acidification. The impact of 'slash-and-burn' agriculture in tropical rainforests (c. 500 million hectares) on soil erosion is illustrated in Table 11.1, and an example of the effect of various tillage practices on soil erosion under crops in Illinois, USA is given in Table 11.7. In Australia, a recent report (Anon., 1995b) concluded that maintenance of agricultural productivity and competitiveness has been achieved at considerable cost to the condition of the land, and may have disguised the extent of damage taking place: 'the seriousness of the situation is reinforced by the fact that the effects of past decisions may still take many decades to be fully felt'. Approximate estimates of the extent of soil degradation through several causes are given in Table 15.3. Estimates of cost to the nation of such degradation are also provided, with the overall 'guesstimate' of production losses being more

Table 15.3 Soil degradation problems associated with current and past land-use and management practices in Australia. After White, 1997.

Problem	Area at risk*	Direct cost (million $)	Off-site costs (million $)
Soil acidity (pH < 4.8 in CaCl$_2$	24 million ha	134–300	Unknown
Soil erosion	2/3 of crop land 1/3 of non-arid land 1/2 arid rangelands	80	Comparable to direct costs
Decline in soil structure	Unknown	200	Unknown
Dryland salinity	5 million ha	200	Comparable to direct costs
Salinity under irrigation	0.5 million ha	10	44–65
Waterlogging	Unknown	180	Unknown
Decline in soil fertility – N deficiency	2/3 of crop land	Unknown	Not applicable

*Total land area 768 million ha; area of arid rangelands 370 million ha, area of crop land and improved pasture 48 million ha.

than A$1 billion, and a comparable level of costs attributed to off-site impacts (Anon., 1995a). Examples of the effects of rising saline groundwater and of wind erosion on land productivity are shown in Fig. 15.1a and b.

Compounding the problem of degradation of existing physical resources is the loss of land for agriculture through competition from other uses, for example urban and industrial. This is particularly true along the densely populated eastern seaboard of China, for example, where rapid industrialization and urban development are occurring on prime land previously used for wheat, maize and rice production. Globally, the reserves of good quality land that can be converted to agriculture are relatively small and FAO estimates suggest that for every new hectare made arable, one hectare of 'old' land is lost through some form of degradation (Anon., 1993). Certainly, the area harvested per person for grain production in the world decreased from c. 0.23 to 0.13 ha from the mid-1950s to early 1990s, reflecting not only increased yields per hectare but also loss of land by degradation and a shift in land use to other crops.

The squeeze between the demands of increasingly affluent and growing populations, and the ability of natural resources to produce with an acceptably low environmental impact, means that *sustainable land management* has become not only a challenge for science and technology but also a political issue. In Australia, landholders, scientists and governments (representing the interests of the whole community) have taken steps to meet the challenge through a range of actions, now coordinated through the *National Strategy for Ecologically Sustainable Development* (ESD). Both State and Federal Governments support improved management of soil, water and vegetation through the National Landcare Program, salinity, nutrient and total catchment management strategies in the Murray–Darling Basin and elsewhere, pest and weed control programmes, a national drought policy, the reform of legislation governing water use, and programmes to encourage tree planting and the conservation of natural biodiversity. The Landcare 'movement' has been especially successful, with 86% of the community being

(a)

(b)

Fig. 15.1 (a) Landscape affected by the rise of saline groundwater in northern Victoria. Note the contrast between bare soil and sparse salt-tolerant plants in the foreground and productive agricultural land in the distance. (b) Soil erosion by wind on sandy soils of the Mallee, northwest Victoria. Wind-blown soil accumulates along fences and tree shelter belts.

aware of Landcare in 1994, and a significant proportion of farmers in most settled regions of Australia being members of Landcare groups (Fig. 15.2).

A prerequisite of any national or international programme of ESD is a reliable system of monitoring progress towards achieving sustainable land management (SLM) over a realistic time period. To this end, scientists have developed a *Framework for Sustainable Land Management* (FESLM) (Dumanski *et al.*, 1991). FESLM provides a logical pathway to assess the *probability* of sustainability by connecting the form of land use with a range of environmental, economic and social factors that together determine whether that particular land use is sustainable. Key aspects of FESLM which relate to the soil are:

• assessment of the soil resource – its properties and their spatial distribution, and
• monitoring the change in the soil resource and/or the effect of management practices on the condition of that resource.

These two aspects are discussed below.

Evaluating sustainable land management

Resource assessment

The purpose and methods of collecting soil information by *soil survey* are discussed in Section 14.3. Remote sensing of soil and land properties and their input to a spatially referenced database – a geographic information system (GIS) – are briefly described in Box 14.2. These procedures form the core of *land resource assessment* (LRA) which traditionally has interpreted soil, geology and climatic information in terms of the suitability of land for generalized uses (land use capability). There is still scope for this type of LRA for broad-scale planning purposes, but increasingly more quantitative functions are needed to relate land resource attributes to the yield of specific crops, yield stability, options for alternative land use, and environmental impacts, especially on soil and water quality. For example, specific, quantitative functions of this kind are a prerequisite for 'precision farming' or site-specific soil management

(Box 11.4). Thus, in the quest for sustainable land management, increasing emphasis is being placed on LRA which marries estimates of productivity for specific enterprises with probabilities of their environmental impact. Advanced techniques are now being used to collect data for LRA and in modelling processes at a landscape scale. An example is given in Box 15.2.

Sustainability indicators

Ecosystems are naturally complex, and the complexity of their function is generally made greater by human intervention. In determining whether a particular agro-ecosystem or land use is sustainable or not, we need *indicators* of the behaviour of the system which are diagnostic of its underlying condition.

Ideally, an indicator should be measurable objectively and directly, although surrogate variables may be used to represent the true indicator. To monitor a system effectively, it is necessary to measure an indicator with respect to time. Spatial variability can affect the reliability of the measurements, and there is the need to set realistic *threshold* and *benchmark* values for an indicator. These terms are defined as follows:

• a threshold value is one below (or sometimes above which) the system changes significantly and is no longer sustainable;
• a benchmark value is one which identifies an optimum condition for that particular system.

Unless the unsustainability of a system is very obvious, a single indicator is unlikely to provide a reliable assessment of the sustainability or unsustainability of the system.

These concepts are illustrated in Fig. 15.3 (p. 335), which is based on studies of grazed pastures on acid soils in the Kyemba Valley of southeastern Australia under an average annual rainfall *c.* 600 mm. The land was originally woodland with native perennial grasses but very few native legumes, and grazed mainly by kangaroos. Following tree clearance and development of improved pastures with superphosphate and subterranean clover (to fix N), sheep and cattle were introduced. Stocking rates were gradually increased, but the native perennials disappeared and annual grasses became dominant. For the reasons given in Section 15.1, the A horizons of these soils (mainly red and yellow podzolics or Kurosols (A)) have become more acid, and recharge to the regional groundwater by deep drainage has increased. The pH decrease is approximately 1 unit (from *c.* 5 to 4 in 0.01M $CaCl_2$) over 30–50 years, depending on the level of N inputs. As the groundwater rises, the area of land affected by dryland salinity has increased. The profitability of these grazing systems is low and farmers can not afford the capital cost of applying lime (*c.* $200 per hectare at current prices) to counteract the accelerated soil acidification. The system is tending towards unsustainability, but information is needed as to which biophysical indicators can be used to determine whether this is true, how fast it is degrading, and most importantly whether alternative land management can arrest the degradation.

Figure 15.3 indicates a 'stack' of potential indicators, each one of which has a measurable value and is monitored through time (annually). The thresholds must be determined by field experimentation. Although these are well known for an indicator such as soil pH, they are not well known for recharge to groundwater which is measured as drainage below the root zone (> 1.5 m for grasses). In terms of a surrogate variable for acid inputs to the soil, the proportion of clover in the pasture through spring and early summer is a candidate because this may indicate how much mineral N (especially NO_3^-) accumulates in the soil profile by the end of summer, and is vulnerable to leaching during the autumn and early winter. As Fig. 15.3 shows, there is large variability from year to year, particularly in indicators such as drainage or per cent clover, due to the variability of the climate. In establishing thresholds, benchmarks and trends, measurements need to be made over several years (ideally 10 or more) to minimize the confusing effects of large annual variations. Such knowledge can then be applied on a farm-by-farm basis, although a composite index of sustainability might need to be developed which takes account of all the individual indicators, and weights them according to their influence on the system's condition.

But what of the larger scale and the sustainability of this land use in a catchment or a whole region? In this case, the distribution of the main soil types and man-

Box 15.2 Topographic modelling for LRA.

Topographic data are basic to most land resource databases and can be used to create a *digital elevation model* (DEM). The source data for DEMs are published topographic maps, stereo pairs of aerial photos, or stereo pairs of satellite images. The DEM provides a three-dimensional image of a landscape from which other physical information can be deduced, for example slope, aspect, catchment boundaries, the direction of water flow and accumulations of flow (wetness zones). Other layers of information – meteorological variables, vegetation type and soil properties – can be 'draped' over the DEM, and the combined information used to develop a model of catchment behaviour, such as TOPOG (Dawes and Hatton, 1993). TOPOG is used to predict water fluxes in a catchment – surface runoff (and hence the likelihood of soil erosion and gully formation), subsurface flow (and hence the location of likely soil saturation), and water yield over time. DEMs may also be combined with gamma-ray images obtained from airborne sensors, which reflect the surface geochemistry and can be analysed to give information on the relative rates of soil formation and weathering in a landscape (Wilford, 1995).

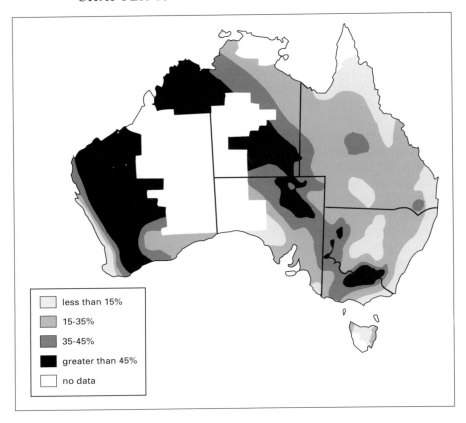

Fig. 15.2 Proportion of Australian farms with a 'landcare' member (from Mues *et al.*, 1994).

Legend:
- less than 15%
- 15-35%
- 35-45%
- greater than 45%
- no data

agement practices in the catchment must be known and basic information on thresholds and trends in the indicators obtained for each of these soils and managements. The indicators, or a composite index, can then be weighted according to the areal distributions of soils and management. Where land uses other than grazing on permanent pastures occur (e.g. cereal cropping), appropriate indicators and their trends must be identified. The weighted index should then take account of soils, land uses and management practices.

Alternatively, as happens at present, *surrogate variables* can be used as sustainability indicators (Table 15.4). While this approach may be acceptable, to be reliable the relationships between these higher level surrogates and the more fundamental biophysical indicators referred to in Fig. 15.3 need to be identified. Whereas much research has been carried out on these cause and effect relationships in agriculture, the knowledge gained has not been consistently incorporated into decision support systems that farmers and their advisers can use with confidence. One of the reasons for a lack of coupling between what is *known* about maintaining sustainability and the *implementation* of this knowledge is that less attention has been paid to the social and economic factors, which are often decisive in determining whether changes in land use and/or improved management practices are adopted. This is the current and future challenge for all those concerned with promoting SLM.

15.4 Summary

The essential components of *soil quality* are more difficult to define than for air and water quality because of

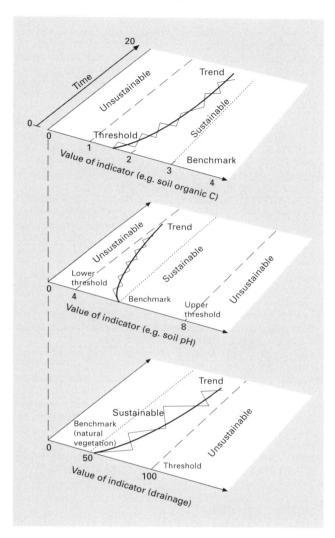

Fig. 15.3 A 'stack' of soil sustainability indicators.

Table 15.4 Sources of information and types of indicators for assessing sustainability at different scales. After Syers *et al.*, 1995.

Scale	Sources of information	Type of indicator
National overview	Agricultural census statistics, rainfall records	Yield per hectare, yield per capita, range of crops, mean net farm income
Regional	Terrain analysis, land use inventories, crop and animal surveys, river records	Proportion of land < 15% slope that is cropped, proportion of irrigated land without drainage, incidence of effluent controls on farms
Local (farm)	Farmer's records, fertilizer sales, advisers' reports	Gross margins : debt equity ratios

Sustainability is an imprecise concept and implies aspects of resource quality and availability, economic viability, environmental impact, social justice and bio-diversity. Its spatial and temporal dimensions must be recognized. *Conventional agriculture* which relies heavily on external inputs may be less sustainable than *non-conventional agriculture* (organic or alternative) which relies more on recycling nutrients within a farm, biodiversity and integrated pest management. However, on a global scale, the reduction in yields associated with low-input, alternative agriculture poses a serious problem for satisfying the legitimate needs of an expanding world population, within the environmental limitations imposed by society and technology.

There is much evidence that human activities have degraded natural resources on a national and global scale. Governments and international agencies have responded with strategies to promote *ecologically sustainable development* (ESD). A prerequisite of any ESD programme is a reliable system of monitoring progress towards achieving *sustainable land management* (SLM). Key aspects of SLM are assessment of the soil (land) resource – the relevant properties and their spatial expression, and monitoring changes in the resource and/or the effect of management on the condition of the resource.

Land resource assessment (LRA) depends heavily on

the spatial variability of soil and the fact that it is used for a variety of purposes. Except in the case of contamination from human activity, soil quality is inferred from the degree and extent to which soil management practices protect it from degradation. Establishing best practice in management requires that the soil–plant system be considered as a component of the wider environment, rather than focusing narrowly on production.

soil survey and advanced techniques of modelling land-scape form and behaviour, as in *digital elevation modelling* (DEM). *Indicators* are measurable variables that are diagnostic of the underlying condition of the resource. *Threshold values* showing when a system becomes un-sustainable, and *benchmarks* for good condition, need to be identified and the system's performance monitored against these values over time. *Surrogate* variables are often used as sustainability indicators, especially at the national or regional level, but the relationship between these surrogates and the more fundamental biophysical indicators should be well defined. Incorporating such relationships into *decision support systems* for SLM requires that social and economic factors, which often determine whether or not individual landholders adopt improved practices, be taken into account.

References

Anon. (1993) FAO sounds soil-loss siren. *Science* **261**, 423.

Anon. (1995a) *Data Sheets on Natural Resource Issues.* Occasional Paper No. 6/95. Land & Water Resources Research & Development Corporation, Canberra.

Anon. (1995b) *Sustaining the Agricultural Resource Base.* Office of the Chief Scientist, Commonwealth of Australia, Canberra.

ANZECC/NHMRC (1992) *Australian and New Zealand Guide-lines for the Assessment and Management of Contaminated Sites.* Australian and New Zealand Environment and Conservation Council and National Health and Medical Research Council, Canberra.

Barzi F., Naidu R. & McLaughlin M. J. (1996) Contaminants and the Australian soil environment. In *Contaminants and the Soil Environment in the Australasia–Pacific Region* (Eds R. Naidu *et al.*). Kluwer, Dordrecht, pp. 451–484.

Dawes W. R. & Hatton T. J. (1993) Topog. IRM: Model description. CSIRO Division of Water Resources Technical Memorandum 93/5, Canberra.

Dumanski J., Eswaran H. & Latham M. (1991) A proposal for an international framework for evaluating sustainable land management. In *Evaluation for Sustainable Land Management in the Developing World*, Vol. 2 (Eds J. Dumanski *et al.*). Technical Papers, International Board for Soil Research and Management, Bangkok, pp. 24–45.

Hamblin A. (1992) *Environmental Indicators for Sustainable Agri-culture.* Report of a National Workshop, Bureau of Rural Resources, Canberra.

Lebel G. G. & Kane H. (1990) *Sustainable Development. A Guide to our Common Future.* Oxford University Press, Oxford.

Miller F. P. & Wali M. K. (1995) Soils, land use and sustainable agriculture: a review. *Canadian Journal of Soil Science* **75**, 413–422.

Mues C., Roper H. & Ockerby J. (1994) *Survey of Landcare and Land Management Practices 1992–93.* Australian Bureau of Agricultural & Resource Economics Research Report 94.6, Canberra.

Reganold J. P., Papendick R. I. & Parr J. F. (1990) Sustainable agriculture. *Scientific American* **262** (6), 112–120.

Roberts B. (1992) An international perspective on sustainable land management. In *Sustainable Land Management* (Ed. P Henriques). Hawke's Bay Regional Council, Napier, New Zealand, pp. 10–17.

Rovira A. (1990) Biological – organic – biodynamic or conventional – which is it? *Soil News* **82**, 1–2.

Smyth A. J. & Dumanski J. (1993) *FESLM: an International Framework for Evaluating Sustainable Land Management.* World Soil Resources Report 73. FAO, Rome.

Syers J. K., Hamblin A. & Pushparajah E. (1995) Indicators and thresholds for the evaluation of sustainable land management. *Canadian Journal of Soil Science* **75**, 423–428.

Viederman S. (1993) A dream of sustainability. *Renewable Resources Journal* **11**(2), 14–15.

White R. E. (1997) Soil science – raising the profile. *Australian Journal of Soil Research* **35** 961–977.

Wilford J. R. (1995) Airborne gamma-ray spectrometry as a tool for assessing relative landscape activity and weathering developing of regolith, including soils. *Australian Geological Survey Organization Research Newsletter* No. 22.

World Commission on Environment and Development (1987) *Our Common Future.* Oxford University Press, Oxford.

Further reading

Barr N. & Cary J. (1992) *Greening a Brown Land. The Search for Sustainable Land Use.* Macmillan, Melbourne.

Lawrence G., Vanclay F. & Furze B. (1992) (Eds) *Agriculture, Environment and Society.* Macmillan, Melbourne.

Roberts B. (1995) *The Quest for Sustainable Agriculture and Land Use.* University of New South Wales Press, Sydney.

Index

Principles and Practice
of Soil Science

The Soil as a Natural Resource